T0254903

Elementary Flight Dynamics with an Introduction to Bifurcation and Continuation Methods

Elementary Flight Dynamics with an Introduction to Bifurcation and Continuation Methods

Second Edition

Nandan K. Sinha
N. Ananthkrishnan

CRC Press is an imprint of the
Taylor & Francis Group, an **informa** business

Second edition published 2022
by CRC Press
6000 Broken Sound Parkway NW, Suite 300, Boca Raton, FL 33487-2742

and by CRC Press
2 Park Square, Milton Park, Abingdon, Oxon, OX14 4RN

First edition published by CRC Press 2014

CRC Press is an imprint of Taylor & Francis Group, LLC

ISBN: 978-0-367-56207-6 (hbk)
ISBN: 978-0-367-56211-3 (pbk)
ISBN: 978-1-003-09680-1 (ebk)

DOI: 10.1201/9781003096801

Typeset in Times
by KnowledgeWorks Global Ltd.

Contents

Preface

This book is intended for a first course in flight dynamics or airplane stability and control, either at the undergraduate or the graduate level. However, with the corrected aerodynamic model and the improved literal approximations, this book should also serve as a useful reference to the experienced practitioner and the engineer in the field. The original intention of the book has not changed with the second edition. From the overwhelming responses from readers all over the world, we feel gratified that the purpose of writing this book has been well served. An excerpt from one feedback is reproduced below verbatim.

> First of all, thank you for great introductory book to subject of flight dynamics. Book is very easy to read and follow with many real world examples which make it even more interesting. Hopefully nicely written books like yours will make students more interested in the subject.

This textbook is the outcome of a logical progression of events, to provide a written account of our simpler, cleaner and correct approach to airplane flight dynamics. This is the first textbook to present an updated version of the aerodynamic model with the corrected definition of the rate (dynamic) derivatives. This is also the first time that the corrected literal approximations to the various dynamic modes are being placed before the student. These core aspects from the first edition have been retained untouched in this version.

Several oversights and typos were reported by keen readers which we have tried our best to correct in this edition; otherwise, Chapters 1–4 have no change. A more intuitive derivation of Lanchester's formulation for Phugoid mode frequency based on energy exchange has been introduced in place of the presentation in the first edition in Chapter 5. In this edition, some more exercise problems have been added to Chapter 6. The issue of dynamic modes and timescales with scaling in airplane size has been introduced as additional reading in an appendix to Chapter 7. In Chapter 8, a brief discussion on aircraft motion in the presence of wind has been added. The major change in this second edition is the addition of an appendix in the form of Chapter 9, which presents two case studies. The first case study presents aerodynamic estimation and stability calculations for a General Aviation airplane. The second case study involves writing down the 6DOF equations of motion for a rigid airship and the study of airship dynamic modes in flight.

This second edition was prepared in the midst of the Covid-19 pandemic making it a somewhat challenging exercise in many respects. We cannot thank enough Senior Editor Gagandeep Singh from CRC Press for his guiding hand throughout the preparation of this edition. Lakshay Gaba and Nishant Bhagat, project coordinators, deserve special mention for providing editorial assistance. Faculty colleagues, staffs, and students at IIT Madras have been immensely helpful throughout during the preparation of this manuscript, a big thanks to them. We hope that the mistakes are few and we would certainly appreciate being informed of them. Getting permissions for aircraft images taken from the internet has been a daunting task; we would like to thank original photographers from the bottom of our heart for readily sharing the images for use in the book. They have been duly acknowledged in the figure captions. Last but not the least, we want to thank several readers who directly wrote to us their appreciation for the first edition and to the reviewers of the first edition—J. Whidborne, R.J. Poole, J. Salga and others—for their kind words.

Nandan K. Sinha and N. Ananthkrishnan
IIT Madras (www.iitm.ac.in), Chennai and IIT Powai (www.iitb.ac.in), Mumbai

Authors

Nandan K. Sinha is a professor of aerospace engineering at the Indian Institute of Technology (IIT) Madras, India. He has been teaching and guiding research in the areas of aircraft flight dynamics and control, aircraft performance and design, and nonlinear dynamics and bifurcations methods applied to flight dynamics for over a decade and half. He has previously taught at the Department of Mechanical Engineering (formerly known as Institut für Mechanik), Technical University of Darmstadt, Germany and carried out research in the areas of vibrations of continuous and discrete systems. His research work in aircraft flight dynamics, stability and control is well recognized internationally. He, along with N. Ananthkrishnan, has authored the textbook *Advanced Flight Dynamics with Elements of Flight Control* (CRC Press). He has been a nominee and participant of the prestigious IVLP (International Visitors Leadership Program) organized by Cultural Vistas USA on the theme of US-India Space Cooperation. He is a senior member of the American Institute of Aeronautics and Astronautics and serves as subject expert for many reputed journals as reviewer and for other organizations in various activities.

N. Ananthkrishnan has taught and researched in the areas of aircraft flight dynamics and control, air-breathing propulsion systems, and aircraft design at IIT Bombay (India), Caltech (USA) and KAIST (S. Korea). In 2000, he received the Excellence in Teaching award at IIT Bombay. He is credited with having discovered and fixed a 100-year-old fundamental error in the formulation of the small-perturbation theory of aircraft flight. His work on inventing a Panic Button algorithm for recovering aircraft after loss of control and that on devising a criterion for aircraft spin susceptibility have received wide acclaim. He, along with Nandan K. Sinha, is also responsible for making possible computations of bifurcations in multi-parameter systems which has proved to be a useful tool for flight dynamic analysis. Along with Nandan K. Sinha, he has authored another textbook, *Advanced Flight Dynamics with Elements of Flight Control* (CRC Press). He is an Associate Fellow of the American Institute of Aeronautics and Astronautics and serves as consultant and expert reviewer for several organizations.

1 Introduction

1.1 WHAT, WHY AND HOW?

'Aircraft' is a generic term for airplanes and other flying vehicles such as rockets, missiles, launch vehicles and helicopters, and could even cover objects such as balloons, boomerangs, parachutes and airships. Flight dynamics is the science of the motion of any of these flying 'vehicles' in air.

We can divide these 'vehicles' into two categories: those that generate lift *aerodynamically* and those whose lifting mechanism is *aerostatic*. A brief recap of aerostatic versus aerodynamic lift is given in Box 1.1. As seen, there is a fundamental difference between the motions of these two kinds of vehicles. In this book, we focus primarily on flying vehicles that depend on *aerodynamic* forces for their lift while touching upon airship dynamics as a special case study in an appendix.

The aerodynamic lift generation mechanism can be further subdivided into two types: fixed-wing, such as conventional airplanes, and rotary-wing, such as helicopters. This book is limited to the fixed-wing type of aerodynamically lifting vehicles. Thus, for the purposes of this book, 'aircraft' means any fixed-wing, aerodynamically lifting, flying vehicle. This includes, besides conventional airplanes, gliders, almost all kinds of rockets, missiles, launch vehicles, parafoils (also called ram-air parachutes) and even boomerangs.

How an aircraft behaves in flight is a matter of great interest and often concern to many people. For instance, anyone who has been on a commercial airplane flying through a thunderstorm and seen the coffee spill out of its cup will wonder if the airplane has actually been designed to shudder and sway in that manner. Or, someone watching the news about yet another space vehicle launch that did not go as planned will wonder why this is called 'rocket science'. But, to a flight dynamicist, when an unmanned vehicle, for example lands on target, exactly as predicted, it is a matter of great satisfaction. Above all, it is a fascinating subject, a wonderful interplay of math and physics.

To an aerospace engineer, flight dynamics is the central point about which the other disciplines of aerodynamics, structures, propulsion and control may be integrated. Most of the requirements for aircraft design are naturally stated in terms of flight dynamic quantities. Hence, a sound knowledge of flight dynamics is absolutely essential to make a well-rounded aerospace engineer.

In the following sections, we take you through the theory and practice of aircraft flight dynamics with several applications and examples, showing you how flight dynamicists think and work.

DOI: 10.1201/9781003096801-1

BOX 1.1 AEROSTATIC VERSUS AERODYNAMIC LIFT

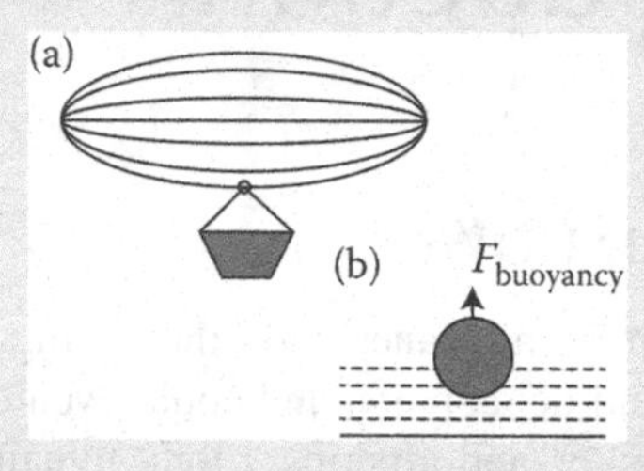

The aerostatic lift used by lighter-than-air (LTA) vehicles is based on the Archimedes' principle of buoyancy and is independent of the speed of the aircraft. An expression for aerostatic lift is given by:

$$L_{buoyancy} = \text{weight of the displaced air}$$

Such vehicles, collectively known as aerostats, do not require relative motion of air; hence thrust-producing devices are not required to remain afloat, but only for moving forward.

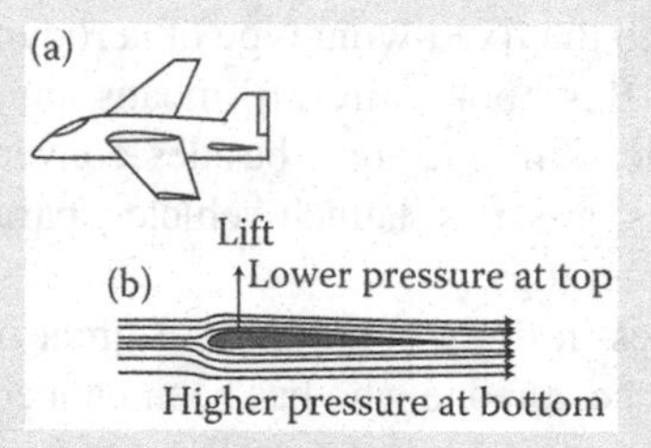

The aerodynamic lift used by heavier-than-air (HTA) vehicles, also called aerodynes, is based on the use of Bernoulli's principle of change in pressure. Properly designed airfoils constituting lifting surfaces (e.g., a wing) are used on vehicles to create an appropriate difference in pressure over the top and bottom surfaces of the airfoil. A statement for the aerodynamic lift of a lifting surface is given by:

$$L_{aerodynamic} = \text{integral of pressure difference over the top and bottom of the lifting surface in a direction perpendicular to the free-stream flow}$$

1.2 AIRCRAFT AS A RIGID BODY

The simplest and the most useful way to visualize an aircraft in flight is to assume it to be a rigid body, but with control surfaces that can be deflected. These control surfaces are usually the elevators, rudder and ailerons, but there could be other surfaces such as flaps, slats, spoilers or foldable fins. Figure 1.1 shows a three-dimensional view of an airplane with the aerodynamic control and other surfaces marked.

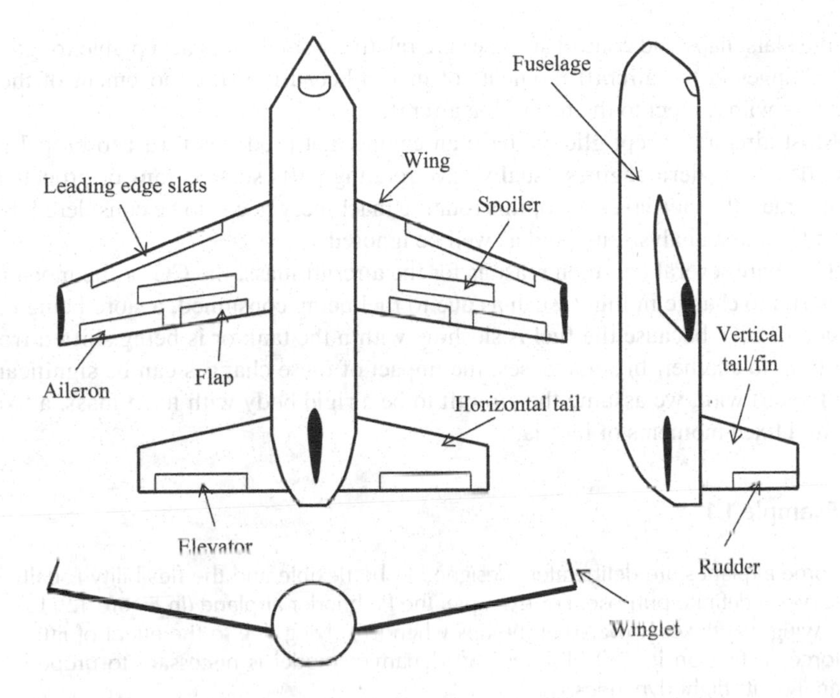

FIGURE 1.1 Components of an airplane.

The aerodynamic surfaces are made up of airfoils that are designed to produce lift when immersed in a relative flow as mentioned in Box 1.1. However, the same pressure distribution that creates lift also causes a drag force opposing the motion of the airplane. This is in addition to the drag due to viscosity (also called 'skin friction drag'). When these lift and drag forces act at a point away from the airplane's centre of gravity (CG), they also cause moments about the CG.

Aerodynamically, different components on an airplane may be characterized in terms of the nature of the force or moment they generate and the purpose for which that force or moment is employed.

- *Wing*—to produce most of the lift that an airplane needs to remain aloft while in motion.
- *Leading edge slats*—to enhance the lift on special occasions.
- *Ailerons*—to provide roll control.
- *Spoilers*—to create drag during landing and also to provide roll control.
- *Winglets*—to decrease the induced drag created due to the wing tip vortex.
- *Flaps*—to change the lift and drag over the wing, typically used during take-off and landing.
- *Horizontal and vertical tails*—also called *fins*, are used for trim and to provide stability to an airplane in pitch and yaw, respectively. They are therefore also known as *stabilators* or *stabilizers*.
- *Elevator and rudder*—Small deflectable flaps on the horizontal and vertical tails used to provide the airplane with control in pitch and yaw, respectively.

As the slats, flaps and control surfaces are relatively small, it is acceptable to ignore any changes in the aircraft moments of inertia because of the movement of these surfaces with respect to the rest of the aircraft.

Most aircraft, except gliders, have an engine that produces thrust to propel the aircraft and modern engines usually have rotating parts such as fans or propellers. Sometimes, the inertial effect of this rotating machinery needs to be considered; but, often for a first analysis, it could as well be ignored.

There are several common reasons for the aircraft mass, the CG or the moments of inertia to change in flight, such as due to fuel being consumed, a store being discarded or even because the fuel is sloshing within the tank or is being shifted from one tank to another. In some cases, the impact of these changes can be significant. But to start with, we assume the aircraft to be a rigid body with fixed mass, a fixed CG and fixed moments of inertia.

Example 1.1

Some airplanes are deliberately designed to be flexible and the flexibility usually serves a definite purpose. For example, the Pathfinder airplane (in Figure 1.2) has a wing that flexes upwards at the tips when it is flying due to the effect of lifting forces acting on it. A flexible aircraft dynamics model is necessary to properly analyse its flight dynamics.

Homework Exercise: The wing tips of a Boeing 747 also flex upwards in flight by as much as 6 feet, but a rigid-body dynamics model is usually adequate to study its flight dynamics. Can you guess why?

Example 1.2

Wonder why spoilers are used on the Boeing 737 to augment the roll control moment obtained from the ailerons (the primary roll control device) at low speeds? See Figure 1.3. Note that the spoilers provide roll control by 'dumping' lift and yaw control by differential drag, that is, when deployed on one wing and not on the

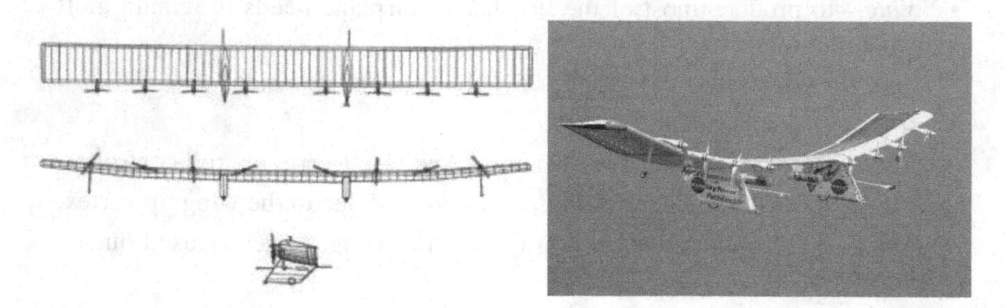

FIGURE 1.2 (a) Three-dimensional view of the Pathfinder airplane showing the wing in a non-lifting condition on the ground (http://www.nasa.gov/centers/dryden/images/content/107948main_pathfinder_drawing2.jpg) and (b) flexed wing shape in flight (https://www.nasa.gov/sites/default/files/images/330281main_ED02-0161-2_full.jpg).

FIGURE 1.3 The outboard spoilers on the Boeing 737 (http://www.b737.org.uk/images/spoilers.jpg).

other, it results in a loss of lift and an increase in drag on that wing. As a result, the airplane banks in the direction of that wing and yaws in the same sense too. Unlike the aileron roll moment that is a function of dynamic pressure (airplane speed and density at the flight altitude) and hence less effective at low speeds, the spoiler roll moment though disruptive (it loses wing lift) is still effective at low speeds.

Example 1.3

Airplanes designed to fly at supersonic regimes of speed (e.g., the Concorde) have to transition from subsonic to transonic to supersonic speeds and finally back to subsonic flight in a typical take-off, cruise and landing sequence. An outcome of increasing speed through three different regimes is that the aerodynamic centre, X_{AC} (a point on the airplane about which the pitching moment does not change when airplane orientation changes with respect to relative airflow) moves backward as shown in Figure 1.4 (top figure). As X_{AC} moves axially in flight while transiting through Mach 1.0, the CG also needs to be shifted in a similar manner. One can also observe from Figure 1.4 that the shift in X_{AC} is accompanied by a shift in the centre of gravity (X_{CG})—this is obtained by fuel shift, that is, pumping fuel between fore and aft tanks. Can you guess why the shift in X_{CG} is needed on the Concorde? The difference ($X_{AC} - X_{CG}$), as we will see later, is directly related to the airplane stability and control in pitch, and must be held within a certain margin.

Homework Exercise: One of the effects of the shift in X_{CG} is to change the moments of inertia, especially about the pitch and yaw axes. In this case, it is necessary to model the change in moments of inertia while simulating the flight of the Concorde. Also, the Concorde nose was swivelled down during landing approach and when taxiing, as seen in Figure 1.4 (bottom figure). How about the moment of inertia change at this time? Does it need to be modelled?

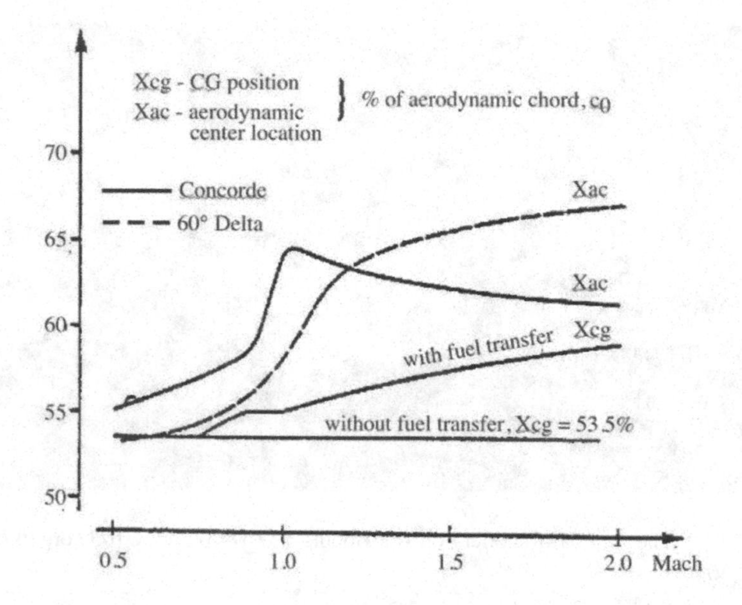

FIGURE 1.4 A plot showing the variation of X_{CG} (axial location of centre of gravity) and X_{AC} (axial location of aerodynamic centre) for the supersonic transport aircraft Concorde and also for a 60° delta wing (a wing with a triangular planform and semi-apex angle of 60°) for different values of Mach number (Jean Rech and Clive S. Leyman. *A Case Study by Aerospatiale and British Aerospace on the Concorde*, AIAA Professional Study Series. http://www.dept.aoe.vt.edu/~mason/Mason_f/ConfigAeroSupersonicNotes.pdf; https://www.airliners.net/contact-photographer/form/10908/2660715). (Photo courtesy: Anthony Noble.)

FIGURE 1.5 Image showing various sweep angle settings on the F-111—from high-sweep for supersonic flight (leader) to mid-sweep for high subsonic flight (middle) to low-sweep for low subsonic flight (trailer). (By Jason B from Australia—3 Configs, CC BY 2.0, https://commons.wikimedia.org/w/index.php?curid=11668023)

Example 1.4

A change in the centre of gravity and a consequent change in moment of inertia can also be caused by movement of airplane components in flight. For example, fighter aircraft such as F-111 and MiG-23/27 fitted with swing wings (variable-sweep) move their wings backward (sweep back) when they transition from low subsonic to high subsonic to supersonic regimes of speed. Changing wing sweep is another way of counteracting the effect of shift in aerodynamic centre with the Mach number, though the primary purpose for changing wing sweep is aerodynamic (see Figure 1.5). Can you remember from aerodynamic theory why this needs to be done? How about the change in moments of inertia at this time? Does it need to be modelled?

1.3 SIX DEGREES OF FREEDOM

A rigid body in free flight has six degrees of freedom. To describe these, let us pin down the CG of the aircraft and attach to it a set of orthogonal axes. These are body-fixed axes as they are stuck to the body and go along with the aircraft. So let us label them as $X^BY^BZ^B$, where the superscript 'B' indicates that it is a body-fixed axis. Figure 1.6 shows how these axes appear for two example cases—(a) an airplane in approximately horizontal flight and (b) a space launch vehicle in nearly vertical flight.

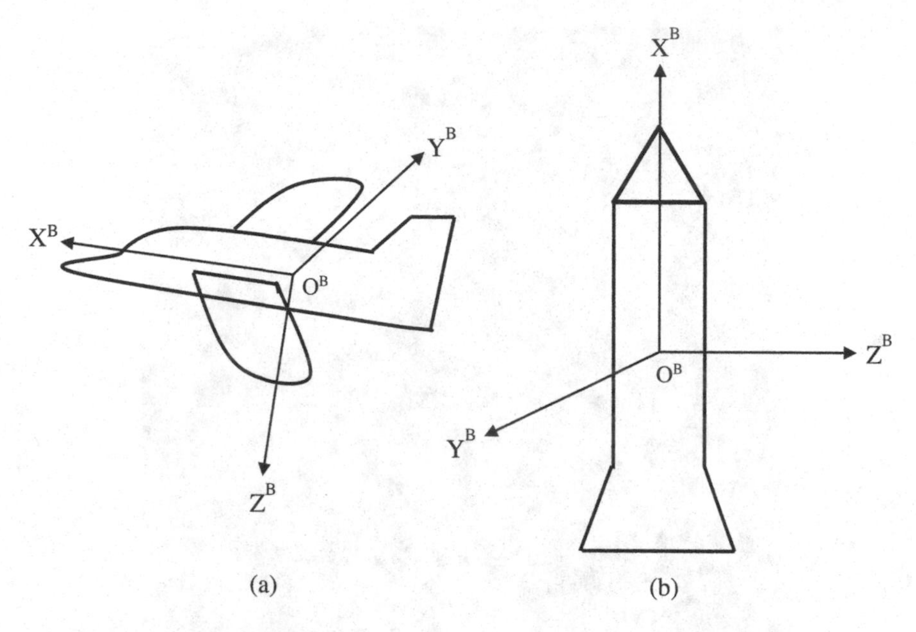

FIGURE 1.6 Body-fixed axes system attached to the centre of gravity (O^B) for: (a) a conventional airplane and (b) a launch vehicle.

In either case, the origin of the body-fixed coordinate system $X^BY^BZ^B$ is at the centre of mass of the body. The body-fixed axes system translates and rotates with the vehicle.

Following the usual practice for airplanes, the axis X^B is taken to point towards the nose of the airplane, but there are various choices for how it is located. For example, it can be along the fuselage reference line (FRL), a straight line in the plane of symmetry X^B–Z^B passing through the centre of fuselage sections at various locations, or it can be selected such that $X^BY^BZ^B$ form the principal axes for the airplane, that is, the cross products of inertia, are all zero. The Z^B axis is orthogonal to X^B such that the plane X^B–Z^B defines the plane of symmetry for the airplane. Finally, the Y^B axis is taken along the starboard and completes the coordinate system according to the right-hand thumb rule. The X^B–Z^B plane is also known as the longitudinal plane of the airplane and the X^B axis as the longitudinal axis.

The launch vehicle in Figure 1.6 has two planes of symmetry, X^B–Y^B and X^B–Z^B.

Example 1.5

Most airplanes have symmetry about the X^B–Z^B plane, but not oblique wing airplanes. An example of an oblique wing airplane is the NASA AD-1 (see Figure 1.7) whose wing is pivoted about a central point on the fuselage. As the wing tip moves forward on the starboard (right) side, the other tip moves backward on the port (left) side, making the arrangement asymmetric about the X^B–Z^B plane. The oblique wing concept was introduced by Richard Vogt of Germany during World War II primarily to avoid undesired effects due to changing aerodynamics during the transition from subsonic to supersonic flight.

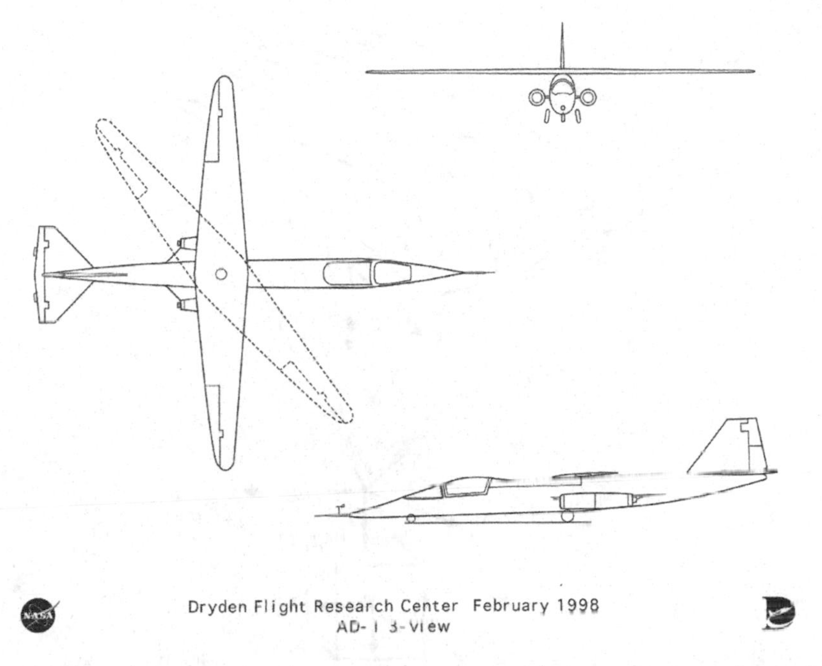

Dryden Flight Research Center February 1998
AD-1 3-View

FIGURE 1.7 A 3-view view of the NASA AD-1 oblique wing aircraft whose wing could be swivelled in flight about a central hinge (https://www.dfrc.nasa.gov/Gallery/Graphics/AD-1/Medium/EG-0002-01.gif).

An aircraft can have three kinds of *translational* motion along the three axes, and three kinds of *rotational* motion about the three axes. That makes six kinds of motion in all. So, an aircraft in free flight has six degrees of freedom.

Example 1.6

Consider a parafoil with an underslung payload, as pictured in Figure 1.8. The connecting wire can twist about its length. As one can see, the motion of the parafoil–payload system consists of motion of the parafoil and the payload under physical constraints (the wire) that force them to stay together. Let us try to determine the number of degrees of freedom of this system.

This question can be approached in two ways: firstly, the parafoil being a rigid body has six degrees of freedom. The payload likewise has six degrees of freedom. When linked together, their translational motions are no longer independent; this introduces three constraints. So, the net number of degrees of freedom works out to 6 (parafoil) + 6 (payload) – 3 (constraints) = 9. In the second approach; the connection point *C* has three translational degrees of freedom. The parafoil and payload then each have three rotational degrees of freedom about the connection point. So, that makes 3 (translation of *C*) + 3 (rotation of parafoil axes) + 3 (rotation of payload axes) = 9.

Homework Exercise: What happens if the connecting link between the two bodies is rigid?

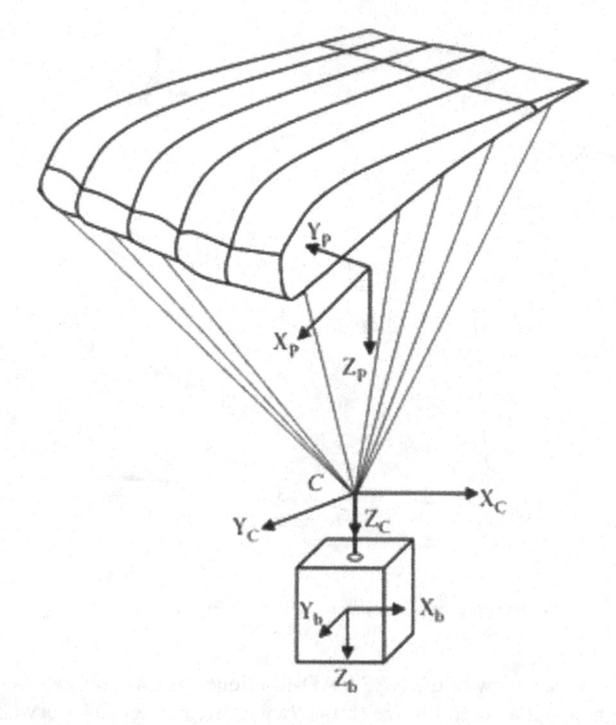

FIGURE 1.8 Schematic diagram of a parafoil–payload system showing the body-fixed axes $X_PY_PZ_P$ attached to the parafoil and the body-fixed axes $X_bY_bZ_b$ attached to the payload. A third set of axes $X_CY_CZ_C$ is placed at the connecting point *C* where the two bodies are linked.

1.4 POSITION, VELOCITY AND ANGLES

Of course, the translational and rotational motions have to be reckoned with respect to an inertial reference frame. For most work in aircraft flight dynamics, an axis system fixed with respect to Earth serves as an inertial reference frame. Let us call these axes as $X^EY^EZ^E$. Then, as shown in Figure 1.9, the aircraft's position in space can be denoted by the position vector *R* from the origin of the Earth-fixed axes $X^EY^EZ^E$, O^E, to the origin of the body-fixed axes $X^BY^BZ^B$, O^B, which is selected to be at its CG.

Homework Exercise: As we all know, Earth is rotating, so it is truly a non-inertial reference frame. Under what conditions can we take it to be an inertial reference? Does it have anything to do with atmospheric versus space flight, the speed of flight (low speed vs. hypersonic), the length of time or distance covered by the flight?

The aircraft's inertial velocity is given by the velocity vector $\underline{V}$ of the aircraft CG with respect to $X^EY^EZ^E$. Likewise, the angular velocity of the aircraft about its CG, again with respect to $X^EY^EZ^E$, is denoted by $\underline{\omega}$. The angular position (orientation)

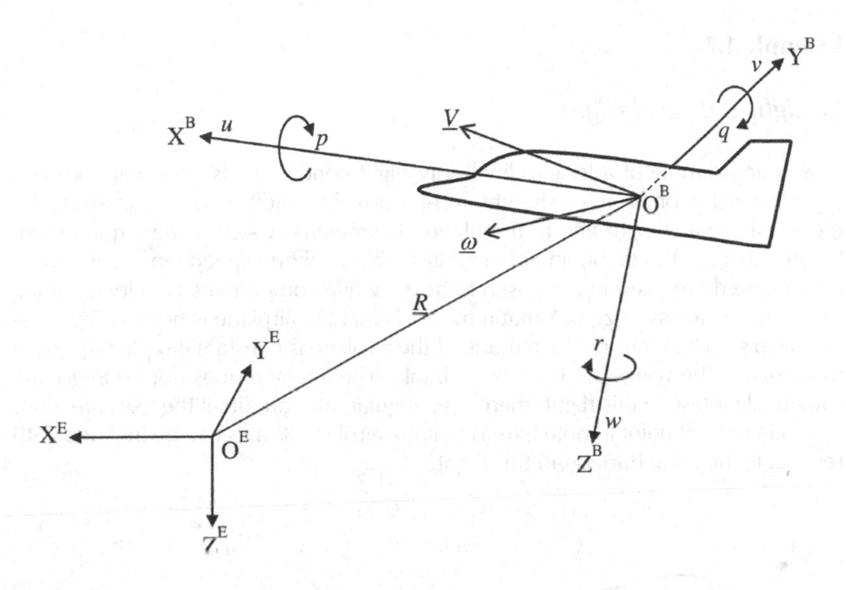

FIGURE 1.9 Sketch of an aircraft in space with $X^BY^BZ^B$ axes fixed to its centre of gravity (CG) O^B, velocity vector $\underline{V}$ at CG and angular velocity vector $\underline{\omega}$ about the CG, axes $X^EY^EZ^E$ fixed to Earth, position vector $\underline{R}$ from origin (O^E) of $X^EY^EZ^E$ to aircraft CG (O^B).

of the aircraft with respect to the Earth-fixed axes is a little harder to describe right now. We can let that follow in a future chapter.

So, the motion of the aircraft can be described by just two vector quantities: $\underline{V}$ and $\underline{\omega}$. Remember, $\underline{V}$ is the velocity of the aircraft (CG) with respect to the Earth-fixed inertial axes and $\underline{\omega}$ is the angular velocity of the aircraft about its CG, again with respect to the Earth-fixed inertial axes. There are three components each of $\underline{V}$ and $\underline{\omega}$—that is, six variables in all. And, integrating the velocities gives the aircraft position and orientation with respect to the Earth-fixed axes—that is, another three plus three, so, six variables. Thus, a total of 12 variables are needed to completely describe the motion of an aircraft in flight. Remember, from Newton's law, that a six-degree-of-freedom system can be mathematically represented by a set of 6 second-order differential equations in 6 velocity variables or equivalently a set of 12 first-order ordinary differential equations in 6 velocity and 6 position/orientation variables. So, it all adds up just right!

The three translational degrees of freedom are characterized by the three components of the velocity vector $\underline{V}$ (u along X^B, v along Y^B and w along Z^B axes). The three rotational degrees of freedom—called *roll, pitch* and *yaw*—are characterized by the three components of the angular velocity vector $\underline{\omega}$ (p about X^B, q about Y^B and r about Z^B axes) as shown in Figure 1.9.

Airplanes doing aerobatic manoeuvres can exhibit complicated motions involving rotations *about* and translations *along* several of the three axes simultaneously. However, some of the more sedate aircraft flight motions are described below.

Example 1.7

Straight and Level Flight

This is an example of a longitudinal-only flight condition, also popularly known as *cruise* flight condition. A straight-and-level flight condition is characterized by motion along a straight line in the plane of symmetry X^B–Z^B of the airplane (see Figure 1.10a). In this flight, an airplane flies with a uniform speed (no acceleration) along a fixed direction here chosen to be X^E, while components of velocity along the Y^E and Z^E axes are zero. What it means is that the airplane is not moving sideways (no sideslip motion), the altitude of the airplane is constant (no plunge/heave motion) and the wings are level (zero bank). The airplane does not undergo any rotational motion in this flight, therefore, angular rates about all the axes are zero.

An important point to note here is that the airplane X^B axis can be inclined with respect to the local horizon in this flight.

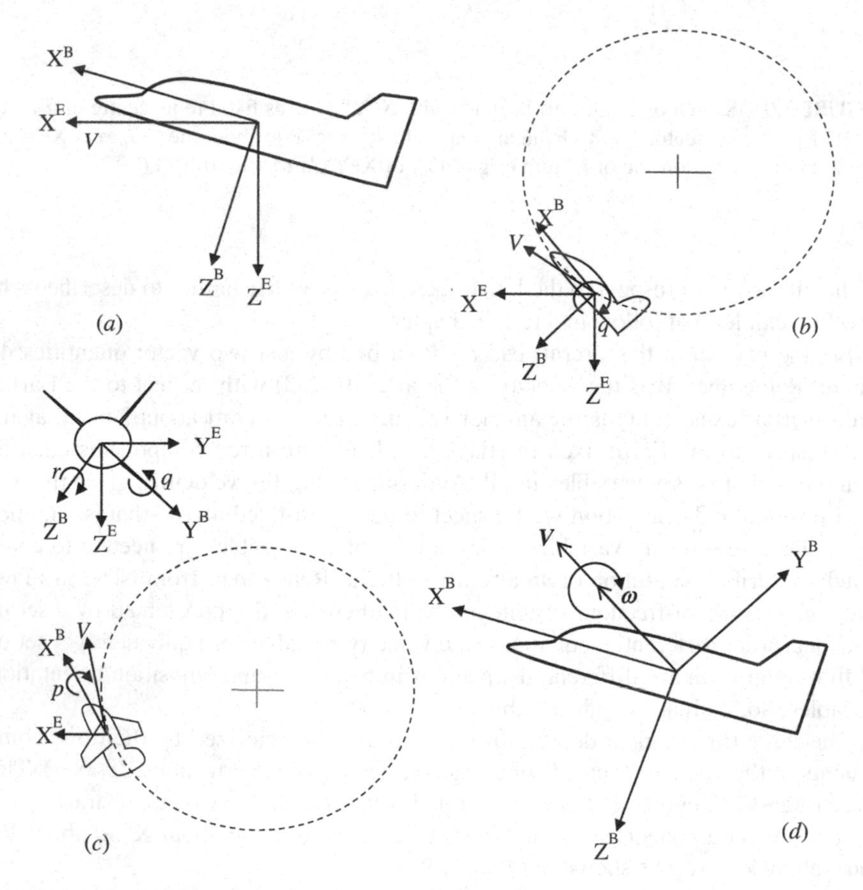

FIGURE 1.10 Sketch of airplane axes, velocity vector and trajectories for some standard airplane motions: (a) straight and level flight, (b) vertical pull-up, (c) horizontal level turn and (d) roll about the velocity vector.

Example 1.8

Vertical Pull-up

Vertical pull-up for an aerobatic airplane can be associated with executing a complete vertical circular loop, or for a commercial airliner, it could be a matter of clearing the boundary obstacles on a runway. This is another example of a flight purely in the longitudinal plane X^B–Z^B, but an accelerated one in which the airplane has to develop a force normal to the flight path acting towards the centre of the vertical circular loop, as shown in Figure 1.10b. For a perfectly circular loop, the airplane speed changes along the trajectory to maintain a constant radius in the loop. The airplane is in continuous pitch motion in this flight such that the velocity vector is tangential to the circular loop at all times.

Example 1.9

Horizontal, Level Turn

This is approximately the flight manoeuvre executed by commercial airplanes, for example, while loitering over a busy airport or by airplanes in combat and aerobatic categories. In this flight condition involving turning at constant altitude, the airplane trajectory is constrained in the X^E–Y^E plane (local horizon). In a horizontal level turn, an airplane is banked (as shown in Figure 1.10c) to develop a centripetal acceleration towards the centre of the loop and is required to follow a horizontal circular loop in a plane parallel to the ground. The airplane velocity is tangential to the circular loop at all times and the airplane has *non-zero* angular rates about all the axes.

Do note from Figure 1.10c that the angular velocity vector $\underline{\omega}$ is oriented along the local vertical, that is, along the Earth Z^E axis. Because the airplane is banked, there is a clear component of $\underline{\omega}$ along the Y^E axis; in other words, a pitch rate. But, the airplane nose does not bob up and down with respect to the local horizon. This is a useful example to understand the difference between the pitch rate q and the rate of change of the airplane attitude angle with respect to the local horizon $\dot{\theta}$. (We will introduce the Euler angle θ soon.)

Example 1.10

Roll about Velocity Vector

Also known as the *velocity vector roll*, this is a complex manoeuvre where the airplane's angular velocity vector $\underline{\omega}$ is constrained to be aligned with the translational velocity vector $\underline{V}$. In this manoeuvre, generally all three components of $\underline{V}$ and $\underline{\omega}$ are non-zero. As shown in Figure 1.10d, the airplane nose appears to make a cone with the vertex at its centre of gravity while moving forward. This kind of manoeuvre may sometimes be used at high angles of attack for tactical advantages by fighter aircraft.

Homework Exercise: Think of two other airplane manoeuvres, sketch the airplane trajectory in those manoeuvres and figure out which velocity and position/orientation variables are zero and which are non-zero (and whether they are constant or how they vary with the manoeuvre).

1.5 AIRCRAFT MOTION IN WIND

Figure 1.11 shows an airplane flying with velocity V in the presence of: (a) head wind and (b) tail wind conditions, denoted by wind velocity V_W. V is the velocity of the airplane with respect to the ground (inertial axes $X^EY^EZ^E$ fixed on Earth), and therefore is also known as ground velocity or inertial velocity. When there is no wind, V is also the relative velocity of the airplane with respect to wind or, in other words, one can say that the airplane sees air coming onto itself at a velocity V_∞, which is equal to its own velocity V. In the no-wind condition, ground velocity is also the aerodynamic velocity. Aerodynamic velocity is defined as the relative wind velocity and is used in the calculation of aerodynamic forces and moments acting on the airplane.

Now, consider an airplane with inertial velocity V in the presence of wind (having velocity V_W with respect to the inertial frame fixed on Earth). The aerodynamic velocity of the airplane now changes due to the presence of the wind. The aerodynamic velocity V_∞ would now be $V+V_W$ in a head wind condition and $V-V_W$ in a tail wind condition, as shown in Figure 1.11. The ground velocity, of course, remains V, as before.

While writing the equations of aircraft motion using Newton's law, it is the inertial velocity and the inertial angular velocity (i.e., velocities with respect to an inertial [Earth-fixed] axis system), which describe the motion of the airplane with

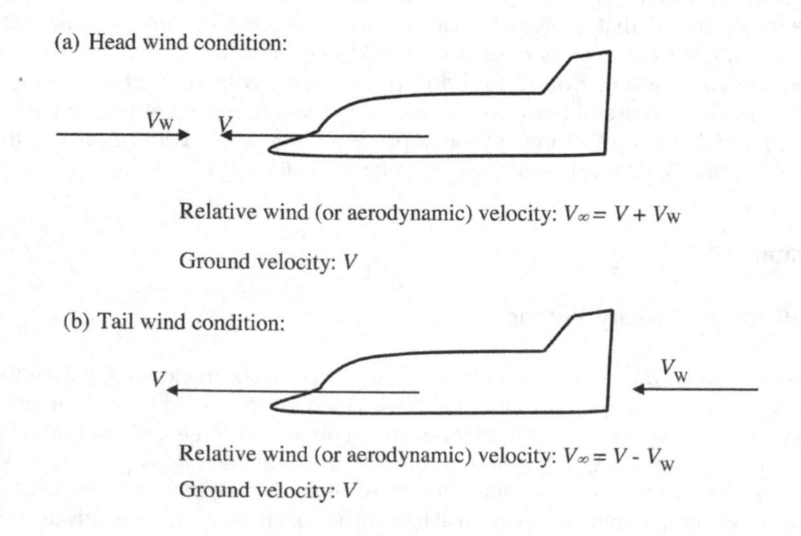

FIGURE 1.11 Airplane flight in: (a) head wind and in (b) tail wind conditions.

respect to the inertial frame even when flying in wind condition. Since the aerodynamic forces and moments of an airplane depend on the aerodynamic velocity, the forces and moments acting on the aircraft would be different in wind condition; the equations of motion per se would remain unaltered, whether it is a head wind, a tail wind or no wind.

When there is a head wind, the aerodynamic velocity V_∞ of an airplane is more $(V + V_W)$, which results in an increase in aerodynamic lift and drag (proportional to the square of aerodynamic velocity) and, in the tail wind condition, aerodynamic velocity is less $(V - V_W)$, which results in a decrease in aerodynamic lift and drag acting on the airplane. Thus, the sudden introduction of wind can cause the airplane to speed up or slow down (depending on the thrust and the new value of drag) and climb or descend in its trajectory (depending on its weight and the new value of lift). However, more important is the effect of sustained head/tail wind on the flight of commercial airliners, which we will discuss below.

Note: Aerodynamically, the Mach number is defined as the ratio of the aerodynamic velocity V_∞ to the local speed of sound (at that altitude), $Ma_\infty = (V_\infty / a)$. So, the Mach number would change depending on whether there is a head/tail wind or no wind at all. The Mach number is a key aerodynamic parameter (see Box 1.2).

BOX 1.2 NON-DIMENSIONAL PARAMETERS OF INTEREST TO AIRCRAFT FLIGHT DYNAMICS

There are several non-dimensional parameters used often in aircraft flight dynamics. Some of them are listed below:

Mach number is defined as the ratio of the relative air velocity V_∞ and the speed of sound a; $Ma_\infty = (V_\infty / a)$. A flow is subsonic when $Ma_\infty < 1$, sonic when $Ma_\infty = 1$ and supersonic when $Ma_\infty > 1$.

Reynolds number is defined as the ratio of inertia and viscous forces, $Re = (\rho l V_\infty / \mu)$, where ρ is the density of air, l is a characteristic length of the object and μ is the coefficient of viscosity. A low Reynolds number refers to low inertia force (smaller length scale and/or speed) and/or large viscous force (medium such as water). A high Reynolds number refers to large inertia force (longer objects and/or speeds) and/or low viscous force (medium such as air). Obviously, the latter is the case for most regular airplane flights though micro air vehicles may fall in the low Re category.

Angle of attack, α, is defined as the angle between the component of relative wind velocity vector in the longitudinal plane and a reference axis fixed to the airplane. This reference axis could be the longitudinal axis X^B of the airplane, the fuselage centerline or a line joining the leading and trailing edges on the wing.

Example 1.11

Wind Effects on Commercial Airplane Operation

Figure 1.12 is a typical plot showing the variation of the airplane drag coefficient (see Box 1.3) with the freestream Mach number. Critical Mach number, Ma_{cr}, is the freestream Mach number at which *for the first time* sonic speed is observed on the wing surface, usually at a point where the pressure coefficient is the lowest. Below the critical Mach number, the flow is subsonic everywhere on the airplane surface and the drag is mainly attributed to the viscous forces generated in the boundary layer. The drag coefficient below Ma_{cr} is fairly constant as shown in Figure 1.12. Beyond the critical Mach number, a shock begins to form over the upper surface of the wing and the drag increases rapidly (due to the shock wave and its interaction with the boundary layer). At Ma_{DD}, an exponential growth in drag occurs, which finds a maximum around $Ma = 1$. Ma_{DD} is close to Ma_{cr}, but always greater that Ma_{cr}. A first-hand estimate often used by aircraft designers follows the relation $Ma_{DD} = Ma_{cr} + 0.02$. Ma_{cr} and Ma_{DD} thus set an upper limit on the cruise Mach number of an airplane expected to fly at high subsonic speeds. (Cruise Mach number for long-range airliners is usually high subsonic and in the range 0.75–0.85.)

With the limit cruise Mach number fixed, in a head wind, the pilot must reduce the inertial velocity to maintain the limit cruise Mach number, so the airplane flies slower with respect to Earth, and thus takes longer to cover a certain distance. Exactly the reverse happens in a tail wind condition, that is, pilots can increase the inertial velocity to maintain the limit cruise Mach number and the airplane flies faster and takes lesser time. So, a tail wind is good for long-haul commercial flights.

This can be a useful feature in planning flight routes (see Figure 1.13). Airlines flying westwards often use a 'Great Circle Route' along the latitude, which happens to be geometrically the shortest route along a spherical surface. However, when flying east, they may take advantage of the Jet Stream, which is a sustained wind from the west to the east. Even though following the Jet Stream may cause a longer ground path, it actually saves time (and also fuel) due to the tail wind as we saw above. The next time you fly long-distance back and forth between two sets of airports, watch out for the flight times each way. May be you rode a tail wind one way!

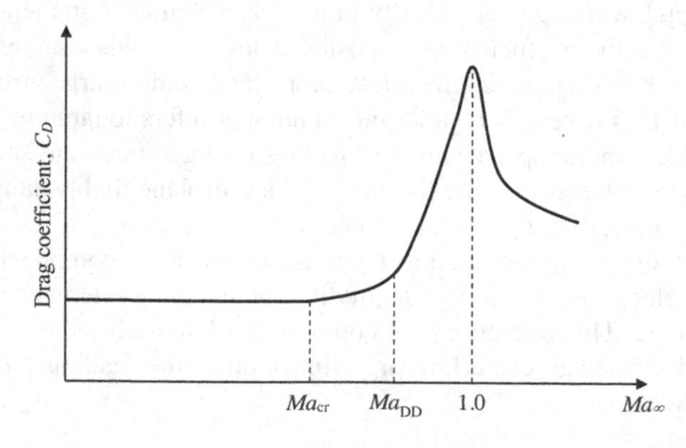

FIGURE 1.12 Variation of drag coefficient of an airplane as a function of Mach number.

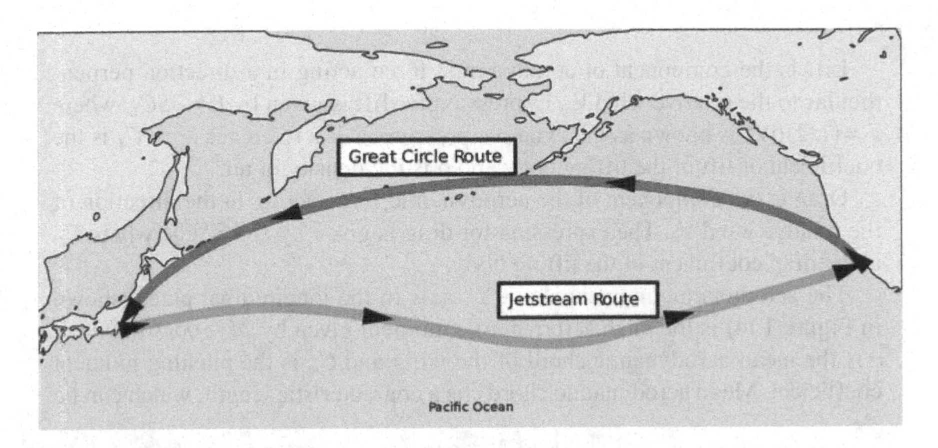

FIGURE 1.13 Airlines flying westward may choose the 'Great Circle Route', which is geometrically the shortest, but when flying east they may fly along the Jet Stream, which provides a sustained tail wind, gaining time and saving fuel. (http://upload.wikimedia.org/wikipedia/commons/7/79/Greatcircle_Jetstream_routes.svg). (Photo courtesy: Chaos Nil, Public domain, via Wikimedia Commons.)

BOX 1.3 AERODYNAMIC FORCES AND MOMENTS

The imbalance of pressures on the top and bottom surfaces of an airplane wing in relative motion with air results in a net aerodynamic force acting on the airplane (see Box 1.1). Lift and drag are components of the resultant aerodynamic force in the longitudinal plane. The resultant aerodynamic load acts on the airplane at a point known as the *centre of pressure*, which is at an offset distance from the centre of gravity of the airplane, thus causing a net moment about the centre of gravity. Moments caused by the aerodynamic forces about the centre of gravity are known as aerodynamic moments (see Figure 1.14).

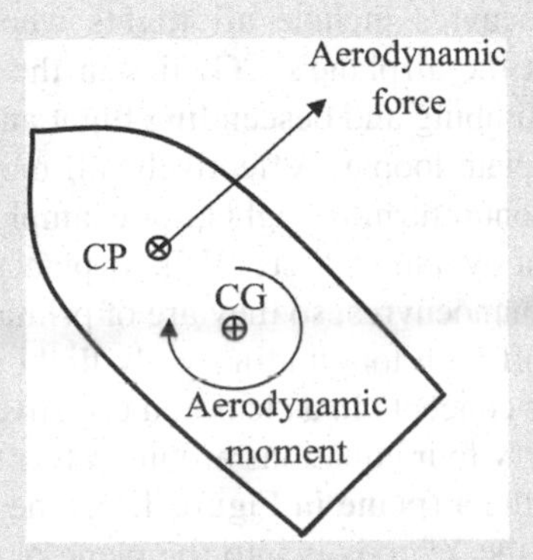

FIGURE 1.14 Aerodynamic moment caused by aerodynamic force acting at the centre of pressure.

Lift is the component of aerodynamic force acting in a direction perpendicular to the relative wind V_∞. Expression for lift is given by $L = \overline{q}SC_L$, where $\overline{q} = (1/2)\rho V_\infty^2$ is known as the dynamic pressure, S is a reference area, C_L is the coefficient of lift of the lifting body and ρ is the density of air.

Drag is the component of the aerodynamic force acting in the direction of the relative wind V_∞. The expression for drag is given by $D = \overline{q}SC_D$, where C_D is the drag coefficient of the lifting body.

The aerodynamic moment about Y^B axis in the longitudinal plane (shown in Figure 1.14) is the pitch aerodynamic moment given by $M = \overline{q}ScC_m$, where c is the mean aerodynamic chord of the wing and C_m is the pitching moment coefficient. Mean aerodynamic chord c is a characteristic length, which can be found using the relation $c = \left(\frac{1}{S}\right)\int_{-\frac{b}{2}}^{\frac{b}{2}} c^2(y)dy$, where $c(y)$ is the airfoil chord length at a distance y from the wing root along the span of the wing and b is the wing span.

While landing, there is a lower limit on the speed of an airplane, which is about 1.2–1.3 times the stall velocity. In a head wind, pilots can fly at a lower inertial velocity and still maintain the lower Mach limit, so touchdown can occur at a lower velocity, and this can result in a shorter landing ground run. The same applies for take-off. So, a head wind is good for landing and take-off. The reverse happens in a tail wind condition.

Homework Exercise: Sometimes airports have to suspend operations on some runway because of too strong a tail wind. Can you guess why?

1.6 LONGITUDINAL FLIGHT DYNAMICS

Longitudinal flight manoeuvres include all flights where the velocity vector V (remember it is placed at the airplane's CG) lies in the body-fixed X^B–Z^B plane. This covers level flight, climbing and descending flight and vertical loops (pull-ups, push-downs or even complete loops)—effectively, all motions that are confined to a vertical plane. So, horizontal turning flights, for example, are excluded for now.

Why longitudinal flight dynamics first? Well, airplanes spend a lot of their time flying longitudinal flight manoeuvres, so they are of primary interest. In addition, in our experience, starting off with longitudinal-only flight dynamics helps introduce almost all flight dynamics concepts in a clear and effective manner with less clutter. It is certainly easier to carry four equations around rather than eight or nine.

Refer to the sketch of the airplane in Figure 1.15. The axes X^B and Z^B are fixed to the airplane at its CG, the Y^B axis is into the plane of the paper. Remember, we assume the airplane to be a rigid body and the X^B–Z^B plane to be a plane of symmetry. These assumptions are good for a vast majority of airplanes in operation, so the contents of this section are widely applicable.

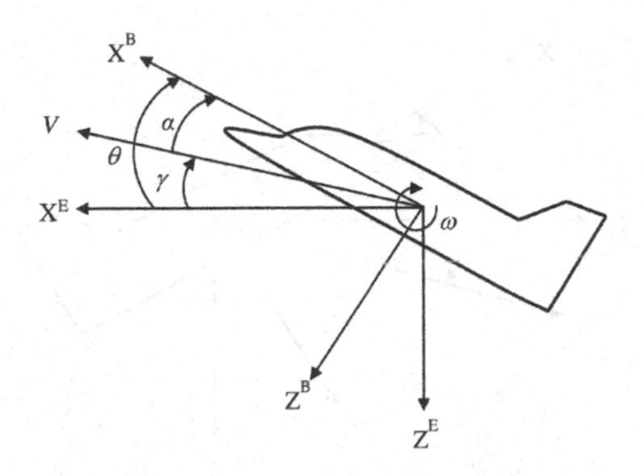

FIGURE 1.15 Airplane in longitudinal climbing flight showing Earth and body axes, V and ω vectors, angles α, γ and θ.

Figure 1.15 also shows the airplane velocity V and angular velocity ω at its CG, as we agreed earlier, although now the velocity vector $\underline{V}$ has only two components in the X^B–Z^B plane, whereas the angular velocity is actually just a scalar—the component about the Y^B axis. This scalar angular velocity is called the pitch rate, denoted by q. So, the airplane motion in longitudinal flight can be described by two components of $\underline{V}$ and q—three components in all.

We would also like to reckon the airplane's motion with respect to the ground (or Earth). So, let us transplant the Earth-fixed axes X^EZ^E from its origin on Earth to the airplane centre of gravity, the origin of the body-fixed axes. This helps us to mark out the angles and indicate the components more easily. The X^E axis marks the local horizontal (horizon) and the Z^E axis the local vertical. The angle the velocity vector $\underline{V}$ makes with the X^E axis is called the *flight path angle*, γ. This indicates whether the airplane is ascending or descending with respect to the ground (Earth). As shown in Figure 1.15, a positive γ is one when the airplane is ascending with respect to the horizon. The angle between the Earth- and body-fixed axes, marked as θ, is called the pitch angle. It indicates how the airplane's nose is oriented with respect to the horizon. With the nose above the horizon, as sketched in Figure 1.15, θ is taken to be positive.

Knowing the velocity V and the angular (pitch) velocity q, the airplane motion with respect to the ground is given by the following *kinematic equations*, which are fairly obvious:

$$\dot{x}_E = V\cos\gamma \tag{1.1a}$$

$$\dot{z}_E = -V\sin\gamma \tag{1.1b}$$

$$\dot{\theta} = q \tag{1.1c}$$

Here, x_E is the range, the distance covered along the X^E axis, and z_E is the vertical distance covered along the Z^E axis. We could replace z_E with the altitude $h = -z_E$ as:

$$\dot{h} = V\sin\gamma \tag{1.2}$$

And θ is the orientation of the airplane nose as described above.

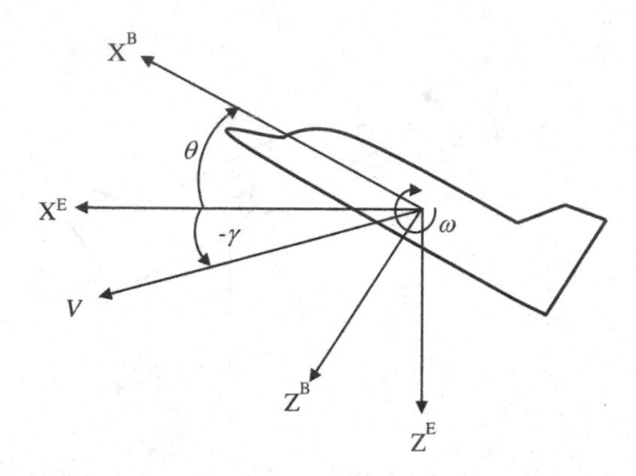

FIGURE 1.16 Sketch of an aircraft in landing approach showing the axes and various angles.

That was easy! If we know V, γ and q, that is all we need at each instant to integrate Equations 1.1 to obtain the airplane position and orientation with respect to the ground at any time. But, to know V, γ and q, we need to write down and solve the airplane dynamic equations. That is our next step.

A note of caution: Make sure not to get confused between the angles θ and γ. θ tells you where the airplane nose is pointing and γ tells you where the airplane is heading (direction of its velocity vector).

Example 1.12

γ and θ during Landing Approach

Figure 1.16 shows an aircraft in the landing approach. One can notice that, in this approach, the aircraft nose is pitched up (positive θ), but the velocity vector is below the horizon now, and therefore, the flight path angle γ is negative.

1.7 LONGITUDINAL DYNAMICS EQUATIONS

Figure 1.17 shows an airplane in flight with all the forces and moments acting on it. Engine thrust, T, acts at an angle δ from the body-fixed X^B axis. Note the manner in which δ is defined in Figure 1.17 with the thrust line below the X^B axis. L and D are aerodynamic lift and drag forces acting perpendicular and in line with the velocity vector, respectively. M is the aerodynamic moment about the Y^B axis. An introduction to the aerodynamic forces and moments is given in Box 1.3. The airplane weight acts downwards along the local vertical (axis Z^E). There are three different types of forces acting on an airplane, namely, propulsive, aerodynamic and gravitational. The airplane equations of longitudinal motion are derived using Newton's laws as follows.

Let us take the components of all the forces in the direction of the velocity vector $\underline{V}$ and in the direction normal to it. The net force along (and opposite to) the velocity vector $\underline{V}$, equated to the airplane mass times the acceleration dV/dt, yields the first equation, Equation 1.3a. The net force in the direction normal to the velocity vector $\underline{V}$

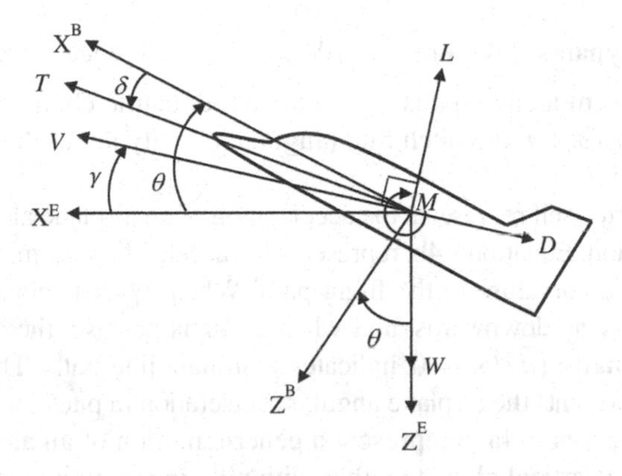

FIGURE 1.17 Free body diagram of an airplane showing all the forces and moments acting on it.

is equated to the mass of the airplane times the centripetal acceleration, giving the second equation, Equation 1.3b. The net moment M about the Y^B axis is equal to the angular acceleration in pitch times the pitch moment of inertia. This is the third equation, Equation 1.3c.

$$m\frac{dV}{dt} = T\cos(\theta - \gamma - \delta) - D - W\sin\gamma \tag{1.3a}$$

$$mV\frac{d\gamma}{dt} = T\sin(\theta - \gamma - \delta) + L - W\cos\gamma \tag{1.3b}$$

$$I_{yy} = \frac{dq}{dt} = M \tag{1.3c}$$

In Equation 1.3, m is the mass of the aircraft, $W = mg$ is the aircraft weight, V is the magnitude of the velocity vector and I_{yy} is the moment of inertia of the aircraft about the Y^B axis. Thus, obtaining the equations of motion of a rigid airplane in longitudinal flight is quite easy. This way, we can quickly get to the heart of the subject without labouring over the derivation of the complete six-degree-of-freedom equations of motion, which we introduce later.

Equations 1.3 can be rewritten with expressions for the aerodynamic forces and moments from Box 1.3 as follows:

$$m\frac{dV}{dt} = T\cos(\theta - \gamma - \delta) - \bar{q}SC_D - W\sin\gamma \tag{1.4a}$$

$$mV\frac{d\gamma}{dt} = T\sin(\theta - \gamma - \delta) + \bar{q}SC_L - W\cos\gamma \tag{1.4b}$$

$$I_{yy} = \frac{dq}{dt} = \bar{q}ScC_m \tag{1.4c}$$

where $\bar{q}$ is the dynamic pressure $(=\frac{1}{2}\rho V^2)$, S is a reference area, usually the airplane wing planform area and c is the mean aerodynamic chord. Please make sure to distinguish between q, the pitch rate (angular velocity about the Y^B axis) and the dynamic pressure $\bar{q}$.

Equation 1.4a by itself represents the acceleration of an airplane along the flight path. The second equation, Equation 1.4b, represents the acceleration normal to the flight path, which introduces a curvature to the flight path. When $d\gamma/dt$ is negative, the airplane follows a path curving downwards, and when $d\gamma/dt$ is positive, the airplane follows a path curving upwards. $(d\gamma/dt)=0$ indicates a straight-line path. The third equation, Equation 1.4c, represents the airplane angular acceleration in pitch motion about its CG.

Together, Equations 1.4a–c represent a generic motion of an airplane flying on a curved path in the vertical plane together with pitching motion along the flight path.

1.8 A QUESTION OF TIMESCALES

We derived the equations for longitudinal flight dynamics in two lots: the first set being the kinematic equations for the variables x_E, z_E and θ (Equation 1.1) and the second set being the dynamics equations (Equation 1.4) for V, γ and q. There is another way of grouping these six equations, which is more useful for our purposes and which is based on the natural timescale over which each variable changes.

Let us rewrite the V and γ equations first as follows:

$$\frac{\dot{V}}{V}=\left(\frac{g}{V}\right)\left[\left(\frac{T}{W}\right)-\left(\frac{\bar{q}S}{W}\right)C_D-\sin\gamma\right] \tag{1.5a}$$

$$\dot{\gamma}=\left(\frac{g}{V}\right)\left[\left(\frac{\bar{q}S}{W}\right)C_L-\cos\gamma\right] \tag{1.5b}$$

assuming that the thrust is acting along the velocity vector V. Then, $(\theta-\gamma-\delta)$ is zero—this is a reasonable assumption to make for most airplanes under common manoeuvres, but not really necessary; one may carry the angle $(\theta-\gamma-\delta)$ as well without much loss.

Next, let us merge Equations 1.1c and 1.4c for θ and q into a single equation as:

$$\ddot{\theta}=\left(\frac{\bar{q}Sc}{I_{yy}}\right)C_m \tag{1.6}$$

Finally, write the equations for x_E and z_E from Equation 1.1 as:

$$\frac{\dot{x}_E}{H}=\left(\frac{V}{H}\right)\cos\gamma \tag{1.7a}$$

$$\frac{\dot{z}_E}{H}=-\left(\frac{V}{H}\right)\sin\gamma \tag{1.7b}$$

where H is a length scale of the order of the airplane ceiling (altitude).

This arrangement naturally throws out three different timescales. The fastest of these from Equation 1.6 is:

$$T_1 = \sqrt{\frac{I_{yy}}{\bar{q}Sc}} \tag{1.8}$$

called the *pitch* timescale, which for most conventional airplanes is of the order of 1 second. This corresponds to the pitching motion, that is, nose bobbing up and down, and represents the rate of change of θ.

The next slower timescale from Equation 1.5 is:

$$T_2 = \left(\frac{V}{g}\right) \tag{1.9}$$

For most airplanes, this is of the order of 10 seconds and corresponds to the heaving motion, that is, the airplane alternately gains and loses altitude (or climbs and descends). The variables V and γ naturally change at this rate.

Note: By using the word 'naturally' we mean that, left to itself, this is how the airplane would respond. When forced by the pilot, the airplane could be made to respond differently, but that would not be 'natural'.

Finally, the slowest timescale from Equation 1.7:

$$T_3 = \left(\frac{H}{V}\right) \tag{1.10}$$

is the order of time taken by a typical airplane to cover a certain range or climb to a certain altitude. Typically, this is of the order of 100 seconds or several minutes. This is usually not a *dynamics* timescale and relates more to the airplane *performance*.

It can be seen that each slower timescale is one order (that is, around ten times) slower than its preceding faster timescale. A key physical rule is that phenomena that occur at clearly distinct timescales can be studied separately. It is this very rule that allows us to investigate airplane range and climbing performance (T_3 timescale) without concerning ourselves with its dynamic behaviour (T_1 and T_2 timescales). Further, one can study the pitch dynamics (T_1 timescale) independently of the heave dynamics (T_2 timescale) because they are so well separated. This is an important principle widely used in aircraft flight dynamics, but the actual timescales are rarely mentioned.

Example 1.13

In Figure 1.18, a time history is shown of an airplane undergoing heave motion having a time period of 10 seconds. The pitching motion superimposed over it has a period of 1 second. The figure clearly shows an order of magnitude difference in timescales over which these motions take place. Notice that the heave variable changes little over the short time for which the pitch dynamics is active. In other words, it is safe to assume a constant, average value of the heave variable and analyse the pitch dynamics equation alone over this short interval. In the same way, while studying several cycles of the heaving motion, the pitch dynamics appears as a short, little blip and can be ignored on the larger (slower) timescale.

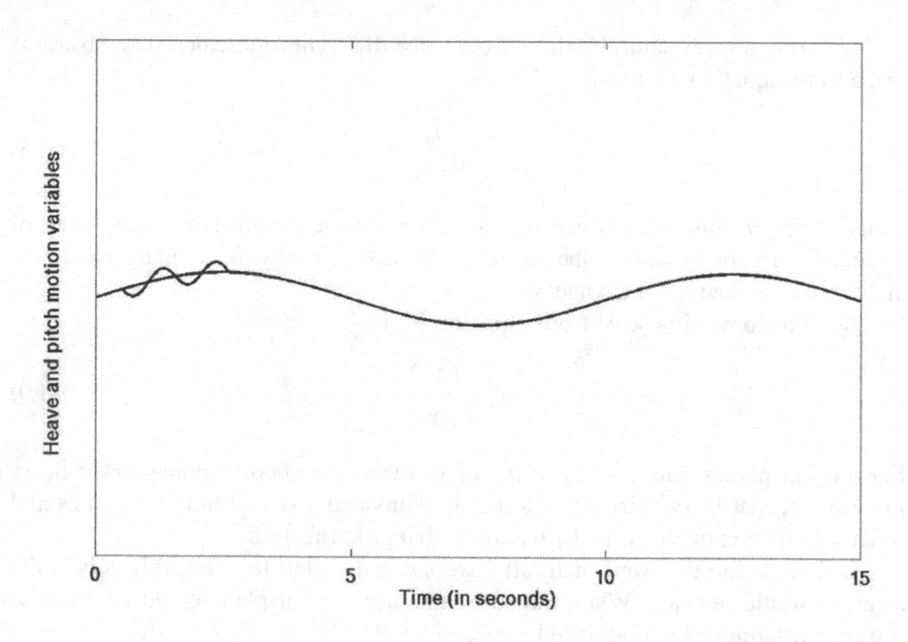

FIGURE 1.18 Suggestive time history of airplane motion with pitching motion (at quicker timescale) superimposed over heaving motion (slower timescale).

1.9 LONGITUDINAL TRIM

Equation 1.7 for $\dot{x}_E$ and $\dot{z}_E$ is concerned with the range and climb performance of the airplane, which forms the subject of books on *aircraft performance*, so it is not discussed here. We shall only examine Equations 1.5 and 1.6, which deal with the dynamics of the airplane. They are collected together and renumbered below:

$$\frac{\dot{V}}{V} = \left(\frac{g}{V}\right)\left[\left(\frac{T}{W}\right) - \left(\frac{\bar{q}S}{W}\right)C_D - \sin\gamma\right] \tag{1.11a}$$

$$\dot{\gamma} = \left(\frac{g}{V}\right)\left[\left(\frac{\bar{q}S}{W}\right)C_L - \cos\gamma\right] \tag{1.11b}$$

$$\ddot{\theta} = \left(\frac{\bar{q}Sc}{I_{yy}}\right)C_m \tag{1.11c}$$

The first step is to solve these equations for an *equilibrium* or *trim* state. This is a state where the rate of change of the variables is zero, the variables here being V, γ and θ. In other words, the airplane flies with:

$$V = constant, \quad \gamma = constant, \quad \theta = constant$$

These are typically straight-line path flights, level ($\gamma = 0$), ascending (γ positive) or descending (γ negative), with the nose held fixed with respect to the horizon (θ is constant).

What does it take to get the airplane into such a trim state? That is easily obtained by setting the left-hand sides of Equation 1.11a–c to zero, and solving the right-hand sides as below:

$$C_D^* = \frac{W}{\bar{q}^* S}\left[\frac{T^*}{W} - \sin\gamma^*\right] \tag{1.12a}$$

$$C_L^* = \cos\gamma^* \frac{W}{\bar{q}^* S} \tag{1.12b}$$

$$C_m^* = 0 \tag{1.12c}$$

where we use the (*) superscript to indicate that it is a trim value. The first and second of these equations are of course the same as what is used to study the *point performance* of the airplane. Its various forms must be familiar from a previous course on *aircraft performance*.

For instance, in level flight ($\gamma^* = 0$), the trim equations simply read as:

$$C_D^* = \frac{T^*}{\bar{q}^* S}\left(\text{or } D^* = T^*\right), \quad C_L^* = \frac{W}{\bar{q}^* S}\left(\text{or } L^* = W\right) \text{ and } C_m^* = 0 \tag{1.13}$$

For climbing flight, Equation 1.12a gives the climb angle:

$$\sin\gamma^* = \frac{\left(T^* - D^*\right)}{W} \tag{1.13a}$$

or the climb rate as:

$$\left(V\sin\gamma\right)^* = \frac{\left(T^* - D^*\right)V^*}{W} \tag{1.14}$$

Of course, when studying the long-time flight *trajectory* of the airplane, one would integrate the dynamic Equations 1.3 and kinematic Equations 1.1, all together. In the case of an ascending or descending trajectory, integrating the z_E equation would yield the new altitude at every instant. The atmospheric density, being a function of altitude, would change (see Box 1.4) and so would the dynamic pressure $\bar{q}$. Note that $\bar{q}$ affects the aerodynamic forces and moments in the dynamic equations. That couples the kinematic and dynamic equations both ways. However, when analysing (longitudinal) flight dynamics, as long as the equilibrium trajectory is a level flight or a shallow climb/descent (small absolute value of γ), since the timescale of the

BOX 1.4 STANDARD ATMOSPHERE

By standard atmosphere, we mean *standardized* properties of air, viz., pressure, density and temperature at different altitudes. These properties at sea-level conditions are:

Pressure $p_0 = 1.01 \times 10^5 \, N/m^2$ or 760 $mm\ Hg$, density $\rho_0 = 1.225 \, kg/m^3$, temperature $T_0 = 288.16 \, K$.

The standard atmosphere (refer Figure 1.19) accepted by International Civil Aviation Organization (ICAO) assumes a temperature profile that is linearly decreasing up to an altitude of 11 km above sea level and remains constant (at 216.65 K) thereafter up to 25 km (isothermal region). Most commercial and general aviation airplanes fly below 15 km altitude and therefore this altitude range is primarily of interest in a course on atmospheric flight mechanics.

In the linear-temperature-profile altitude regions, density and pressure variations with altitude can be calculated using the formulas:

$$\frac{\rho}{\rho_1} = \left(\frac{T}{T_1}\right)^{-\left[1+\left(\frac{g_0}{R\lambda}\right)\right]} \quad \text{and} \quad \frac{P}{P_1} = \left(\frac{T}{T_1}\right)^{-\left(\frac{g_0}{R\lambda}\right)},$$

where g_0 is the value of gravitational acceleration at sea level, $R = 287 \, m^2/\left(K \cdot s^2\right)$ is the gas constant and $\lambda = -0.0065 \, K/m$ is the rate of change of temperature with altitude, also known as lapse rate. Quantities in the above formulas with subscript '1' are values at the start of the linear region.

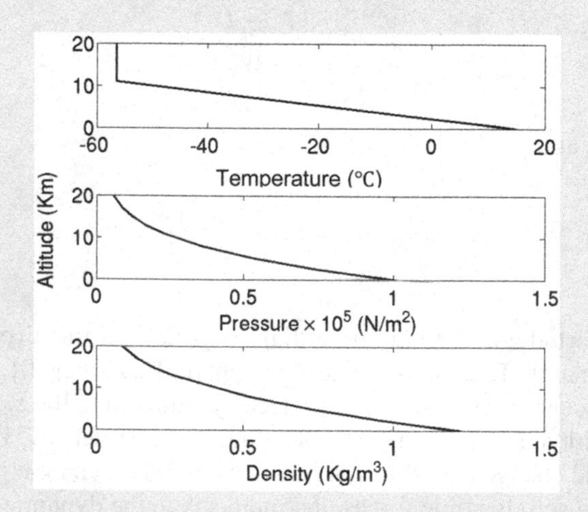

FIGURE 1.19 Standard atmospheric properties in normal atmospheric flight altitude range.

In isothermal regions, the corresponding formulas are:

$$\frac{P}{P_1} = e^{-\frac{g_0(h-h_1)}{(RT_1)}}; \quad \frac{\rho}{\rho_1} = e^{-\frac{g_0(h-h_1)}{(RT_1)}}$$

where subscript '1' now represents the values at the beginning of the isothermal region.

dynamic motions (T_1 or T_2) is small as against the timescale (T_3) of significant change in altitude, the change in density is usually small enough to be ignored. Then we can study the dynamic equations independently of the kinematic ones.

Homework Exercise: Consider the case of an airplane flying vertically up, that is, $\gamma = 90°$. The force balance gives $T = D + W$ and $L = 0$, at each instant. Can this 'trim' be studied without including the z_E equation in the analysis?

Note: The density of the surrounding air is a function of the altitude obtained by integrating the z_E equation, which must be accounted for to satisfy the 'trim' condition at each instant along the trajectory.

1.10 AERODYNAMIC COEFFICIENTS C_D, C_L, C_m

Before we proceed further, we need to understand better the static (longitudinal) aerodynamic coefficients—C_D, C_L and C_m. As seen earlier (in Box 1.3), and as you probably know from an earlier course in *aerodynamics*, they are *non-dimensional* coefficients—what is left over in the expressions for aerodynamic drag, lift and pitching moment after the dimensional terms have been isolated following a dimensional analysis (a review is presented in Box 1.5). Each non-dimensional coefficient can then only be a function of other non-dimensional quantities. Of these, the most important for our purposes are: (i) the angle of attack (α), and (ii) the Mach number (Ma) (see Box 1.2), and we shall discuss these below.

Another non-dimensional number that could be of interest is the Reynolds number, Re. The typical flight Reynolds number for general aviation airplanes varies from the order of 10^6 (for light airplanes) to 10^9 (for commercial jets). The flight Reynolds number for a commercial large transport airplane such as the Boeing 747 is approximately 2×10^9. For atmospheric flight at these Reynolds numbers, the aerodynamic coefficients C_L and C_m do not depend much on Re. As explained in the note below, neither does C_D vary much with Re at these conditions. Hence, there is usually no need to consider the aerodynamic coefficients as functions of Re. But, this may not be true for very-small-scale airplanes in low-speed flight where the Reynolds number may be of the order of 10^4 or lower.

Note: As you know, the lift generated over a (lifting) body such as a wing is mainly due to the pressure difference between its lower and upper surfaces, and therefore the lift coefficient C_L may be assumed largely independent of viscous effects represented by the Reynolds number, Re. The drag coefficient C_D has two parts, one due to a fore

BOX 1.5 DIMENSIONAL ANALYSIS

Aerodynamic forces and moments in general depend upon six parameters related to geometric shape of an airplane and the flow field. These parameters are air density (ρ), air viscosity (μ), speed of sound in air (a), airplane orientation with respect to the wind (α), airplane reference area (S) and relative airspeed (V_∞). The six dimensional parameters are usually reduced to non-dimensional *similarity* parameters, freestream Mach number Ma_∞ and Reynolds number Re and the angle of attack (α) as defined in Box 1.2. This helps in transferring results of wind tunnel measurements of forces and moments on a scaled-down model to the complete airplane in the form of aerodynamic coefficients. This is physically achieved by creating a *dynamically similar* flow in the wind tunnel over the scaled-down model, which is representative of the flow over the actual airplane in flight. The non-dimensional aerodynamic coefficients thus found by using the relations:

$$C_L = \frac{L}{\bar{q}S}, \quad C_D = \frac{D}{\bar{q}S}, \quad C_m = \frac{M}{\bar{q}Sc}$$

from the wind tunnel measurements of forces and moments are the same as those for the actual airplane, and are functions of the similarity parameters, Ma_∞, Re and α. What it means is, an airplane at a given angle of attack in flight will have the same values of the aerodynamic coefficients as measured on a scaled-down prototype at the same angle of attack in the wind tunnel if Ma_∞ and Re are the same, that is, irrespective of the values of the six dimensional parameters mentioned above. The coefficients determined in this manner are based on some reference area S (e.g., wing planform area or body cross-sectional area), which one needs to know before using the coefficients in a simulation.

and aft pressure difference (such as arising in case of flow separation at a point on the body), and the other due to skin friction (or viscosity). The 'pressure' part of the drag coefficient C_D may be taken to be largely independent of Re. The skin friction part of the drag coefficient, C_f, however depends strongly upon Re. A representative plot of C_f versus the Reynolds number for a flat plate is as shown in Figure 1.20. Flow transition from laminar to turbulent over airplanes typically occurs at $Re \sim 10^6$, which means that most airplanes operate at flight Reynolds numbers where the flow is expected to be turbulent. Note from Figure 1.20 that in the turbulent region of the Reynolds number, C_f is lower and fairly constant with Re.

Homework Exercise: Near about stall, the Reynolds number can have some effect on the airplane aerodynamic coefficients. Look up any book on *aerodynamics* to get information about this.

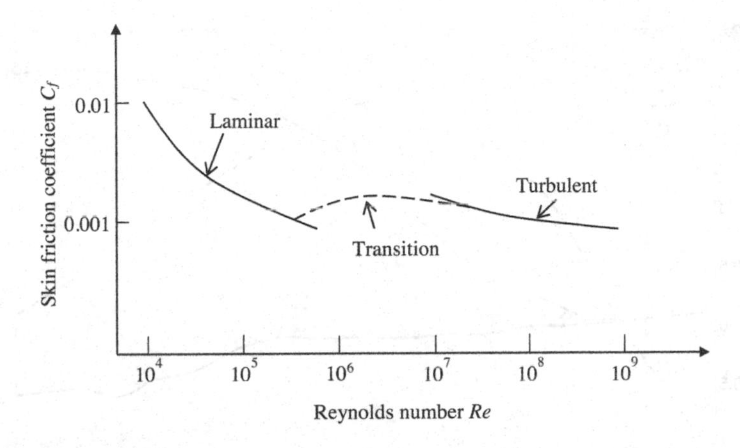

FIGURE 1.20 Variation of C_f as a function of the Reynolds number for a flat plate. (Adapted from *Fundamentals of Aerodynamics* by John D. Anderson, Jr., Fourth Edition, McGraw Hill Publication, 2007, pp. 77.)

1.10.1 Aerodynamic Coefficients with Angle of Attack (α)

Refer again to Figure 1.15 of an airplane in longitudinal flight, with the angle of attack marked by the symbol, α. The angle of attack is the angle between the velocity vector V (also called the relative wind velocity, denoted by the symbol V_∞) and a reference line on the airplane, here chosen to be the body axis, X^B.

It is important to distinguish between the three angles: α, γ and θ.

- γ is the angle the velocity vector V makes with the horizon—it tells you whether the airplane is flying level, climbing or descending. So, it is a *navigation* angle.
- θ is the angle the airplane X^B axis makes with respect to the inertial X^E axis (usually aligned with the horizontal)—it tells you which way the airplane nose is pointed with regards to the horizon. So, it is an *orientation* angle. Remember, as we saw in the landing approach Figure 1.16, the airplane may be descending ($\gamma < 0$) while the nose is pointed above the horizon ($\theta > 0$).
- α is distinct from θ and γ and has nothing to do with either navigation or orientation of the airplane *with respect to the ground*. Rather, α gives the *orientation* of the airplane with respect to the relative wind and so is of the utmost aerodynamic significance. In the simple case of longitudinal flight, as seen from Figure 1.15, $\alpha = \theta - \gamma$.

As you know from aerodynamic theory, the (static) aerodynamic force and moment coefficients in longitudinal flight depend on the angle of attack, α. Take a look at the typical variation of C_D, C_L and C_m for a combat airplane with α in Figure 1.21—these data are for a research model of F/A-18, known as high-angle-of-attack research vehicle (HARV). Such data are typically obtained from wind tunnel tests (see Box 1.6).

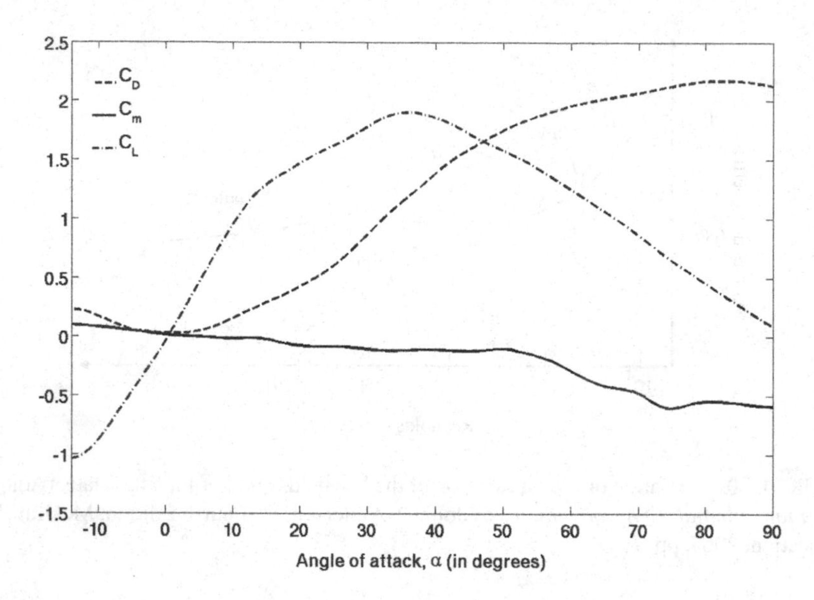

FIGURE 1.21 Plot of drag, lift and pitching moment coefficients as a function of angle of attack for the F-18/HARV airplane.

BOX 1.6 AERODYNAMIC COEFFICIENTS CAPTURED FROM WIND TUNNEL TESTS

Aerodynamic coefficients are measured in the wind tunnel by measuring the forces along the airplane fixed X axis (F_X) and the Z axis (F_Z). From the forces, force coefficients can be obtained by using the relations: $C_X = F_X / \left(\left(\frac{1}{2}\right)\rho V^2 S\right)$ and $C_Z = F_Z / \left(\left(\frac{1}{2}\right)\rho V^2 S\right)$. When the airplane nose is directly into the flow direction, that is, airplane X axis (body fixed) coincides with the X axis of the wind tunnel (inertial), these forces are the same as drag (D) and lift (L) at $\alpha = 0°$. A change in orientation of airplane X axis with respect to the flow direction or the wind tunnel axis (pitch angle) also changes the angle of attack of the airplane. Thus, the force coefficients, C_X and C_Z, are measured in the wind tunnel as functions of angle of attack.

From these coefficients, lift and drag coefficients, C_L and C_D, respectively, can be obtained by using the relations: $C_L = C_X sin\alpha - C_Z cos\alpha$ and $C_D = -C_X cos\alpha - C_Z sin\alpha$. One can notice from here that one needs to correctly take components of C_X and C_Z to arrive at C_D and C_L, before using the data for stability analysis and simulation work. Further, pitching moment (M) due to these forces about the centre of gravity can be measured and pitching moment coefficient C_m is determined using the relation: $C_m = M / \left(\left(\frac{1}{2}\right)\rho V^2 Sc\right)$, which is also a function of the angle of attack.

It is worthwhile pointing out some interesting facts from this graph:

- C_L is fairly linear over the (–10, +10)° α range with a positive slope—this is the typical arena of linear aerodynamic theory. Then, the slope is reduced, but C_L continues to increase until about $\alpha = 35°$ at which it peaks. This peak point is called the 'stall'. Airfoils and high-aspect-ratio wings usually stall at much lower angles of attack, around 15–18°. But, it is important for you to know that this is not an absolute fact, as this example shows. Also, $C_L \approx -0.036$ at $\alpha = 0°$.

  Homework Exercise: Can you guess why the intercept for C_L is slightly negative?

- The C_D curve looks approximately quadratic between (–5, +25)°, as 'linear' aerodynamic theory would suggest, popularly modelled by the 'drag polar': $C_D = C_{D0} + KC_L^2$, where K is taken as $(1/\pi ARe)$. But, in practice, often a good fit to the C_D curve is of the form: $C_D = C_{D0} + k_1\alpha + k_2\alpha^2$. It is good to recognize that the linear term '$+k_1\alpha$' is often present. And C_D keeps increasing almost all the way with α.
- C_m is approximately linear with a negative slope over most of the range, the significance of which we shall soon see in chapter 2 and further.

Example 1.14

Many commercial and general transport airplanes fly in what is called the 'linear regime', where C_L and C_m appear to be linear functions of α. C_D is of course quadratic because of the nature of dependence of induced drag on α. For instance, consider the C_L and C_m variation with α for an unknown airplane (let us call it airplane X) as they appear plotted in Figure 1.22. These follow the linear relations (α in degree):

$$\begin{aligned} C_L &= -0.027 + 0.0865\alpha \\ C_m &= 0.06 - 0.0133\alpha \end{aligned} \tag{1.15}$$

Notice once again that C_L at zero α is negative.

Homework Exercise: From Figure 1.22, you can reckon that C_m is zero at an angle of attack, $\alpha = 4.5°$. At this α, $C_L \sim 0.36$. What is the significance of this α?

1.10.2 Aerodynamic Coefficients with Mach Number (Ma)

The variation of the coefficient of pressure with the Mach number is given by the Prandtl–Glauert rule from aerodynamic theory:

$$C_p = \frac{C_{p,\,incomp}}{\sqrt{1 - Ma^2}} \tag{1.16}$$

in subsonic flow, where $C_{p,\,incomp}$ is the incompressible value of the pressure coefficient. Pressure-dominant coefficients, such as C_L and C_m, do vary with the Mach

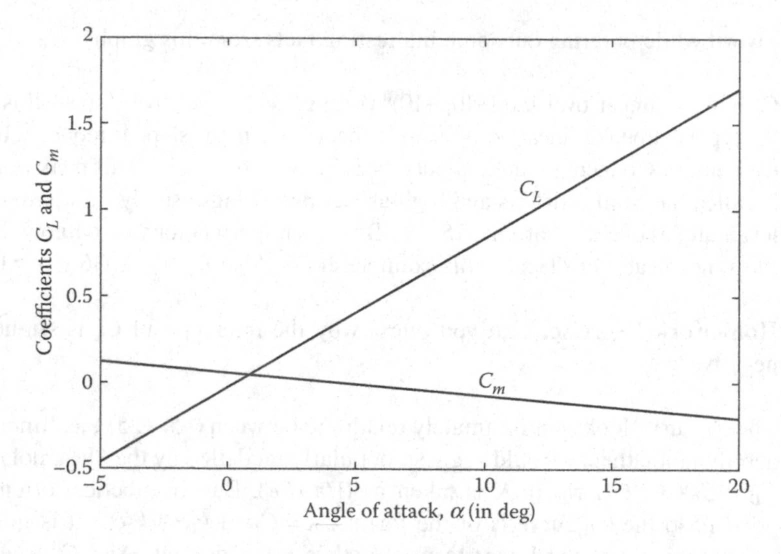

FIGURE 1.22 Plot of variation of C_L, C_m with angle of attack α for airplane X.

number in a fashion close to that predicted by the Prandtl–Glauert relationship, whereas the skin-friction-dominant coefficient C_D does not quite do so. In supersonic flow, C_L, C_m and C_D (due to wave drag) vary with the Mach number closely as per the ratio $\frac{1}{\sqrt{Ma^2-1}}$, also sometimes called the supersonic Prandtl–Glauert rule.

A typical variation of C_L for an airplane with Mach number is shown in Figure 1.23. The C_L profile follows the Prandtl–Glauert rule both in the subsonic and in the supersonic speed regimes, but the Prandtl–Glauert rule does not apply at transonic Mach numbers. As the major contribution to C_m arises from C_L, the variation of C_m with Mach number is quite similar to that of C_L. The drag coefficient C_D (shown in Figure 1.12) follows the Prandtl–Glauert rule at supersonic Mach numbers where the wave drag component is dominant.

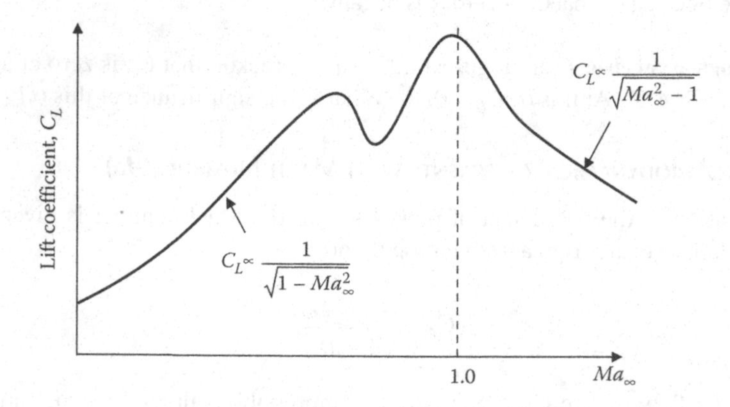

FIGURE 1.23 Typical variation of airplane C_L with Mach number.

Note: Sometimes you can see the expressions for lift, drag, or pitching moment written in the following manner:

$$L = \frac{1}{2}\gamma p(Ma)^2 SC_L \tag{1.17}$$

This is obtained by using the relation:

$$a^2 = \frac{\gamma p}{\rho} \tag{1.18}$$

for the speed of sound, and the fact that $Ma = (V/a)$.

Homework Exercise: Determine the typical variation of the forces—lift L, drag D and pitching moment M—not the coefficients, with the Mach number for subsonic and supersonic flow.

1.11 WING–BODY TRIM

As seen in Equation 1.12, to get an airplane into a state of equilibrium or trim, it is not enough to merely balance the forces, you also need to balance all the contributions to the pitching moment about the CG. Let us see how this works out for an airplane with only a wing and a body (fuselage), and no horizontal tail.

Figure 1.24 shows the various forces and moments acting on a wing–body combination in level flight ($\gamma = 0$) at an angle of attack α. The wing–body lift L^{wb} and drag D^{wb} are placed at the wing–body aerodynamic centre with the pitching moment M_{AC}^{wb}. As seen earlier, the aerodynamic centre is the point at which the wing–body pitching moment M_{AC}^{wb} is not a function of angle of attack, so it is a convenient point for the analysis. Since M_{AC}^{wb} is not a function of angle of attack, it has the same value

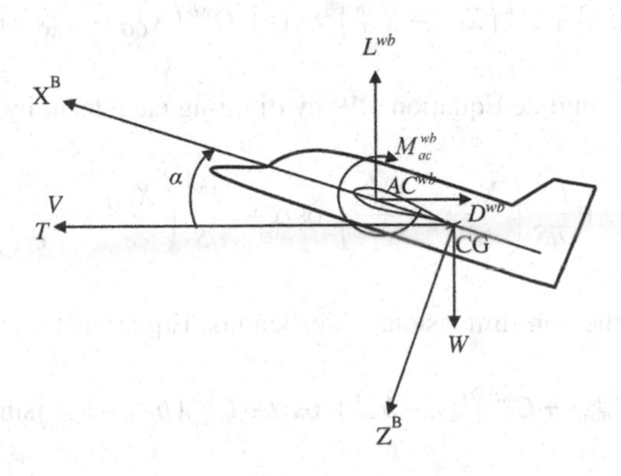

FIGURE 1.24 Various forces and moments acting on a wing–body combination in level flight.

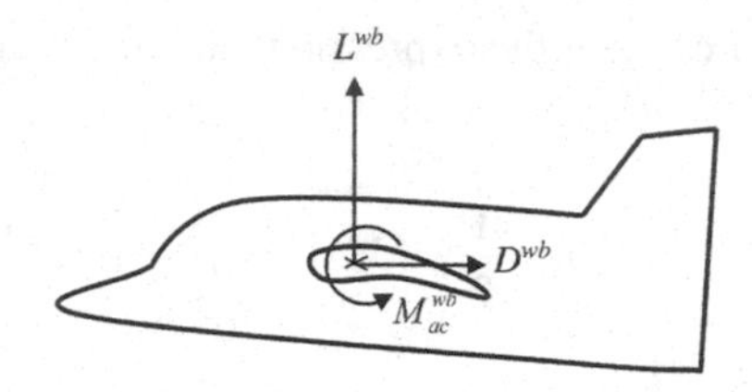

FIGURE 1.25 An airplane with positively cambered wing at zero lift showing the sense of M^{wb}_{AC}.

at zero α as at all other α, and this includes the value of α for which the lift L^{wb} is zero. For wings whose cross section is made up of airfoils with positive camber, the moment about the aerodynamic centre M^{wb}_{AC} is generally negative, and acts in the sense pictured in Figure 1.25. The body (fuselage) does not provide any significant lift, but it does contribute to drag and a pitching moment, which is also usually in the negative sense. Rather than placing these at a different point on the wing–body, it is easier to locate the resultant wing–body quantities (L^{wb}, D^{wb} and M^{wb}_{AC}) at the wing–body aerodynamic centre AC^{wb}. This is how it is pictured in Figure 1.25. In fact, the wing–body aerodynamic centre is usually not too different from the wing-alone aerodynamic centre.

Coming back to Figure 1.24, the weight acts at the *CG* along the Z^E axis. For simplicity, let us assume the thrust to also act at the *CG* and along the velocity vector. If the thrust line is at a vertical offset distance from the *CG*, it will also cause a moment. Also, for simplicity, one can assume that the *CG* and AC^{wb} both lie along the X^B axis. That is, there is no separation between them along Z^B. Distance X along the length of the aircraft can be measured from the aircraft nose tip or from wing leading edge or from any other point and that is really not an important issue; it is the distance between a point and the *CG* of the aircraft that is important.

Summing up the moments about the *CG* (refer to Figure 1.24), we get:

$$M_{CG} = M^{wb}_{AC} + L^{wb}\left(X_{CG} - X^{wb}_{AC}\right)\cos\alpha + D^{wb}\left(X_{CG} - X^{wb}_{AC}\right)\sin\alpha \tag{1.19}$$

Let us non-dimensionalize Equation 1.19 by dividing each term by $\overline{q}Sc$:

$$\frac{M_{CG}}{\overline{q}Sc} = \frac{M^{wb}_{AC}}{\overline{q}Sc} + \frac{L^{wb}}{\overline{q}S}\left(\frac{X_{CG}}{c} - \frac{X^{wb}_{AC}}{c}\right)\cos\alpha + \frac{D^{wb}}{\overline{q}S}\left(\frac{X_{CG}}{c} - \frac{X^{wb}_{AC}}{c}\right)\sin\alpha \tag{1.20}$$

Then, in terms of the non-dimensional coefficients, Equation 1.20 reads as:

$$C_{m,CG} = C^{wb}_{m,AC} + C^{wb}_{L}\left(h_{CG} - h^{wb}_{AC}\right)\cos\alpha + C^{wb}_{D}\left(h_{CG} - h^{wb}_{AC}\right)\sin\alpha \tag{1.21}$$

where *h* stands for distance *X* non-dimensionalized with the mean aerodynamic chord, *c*.

To trim the airplane (in this case, the wing–body) in pitch, one must adjust the terms in Equation 1.21 such that the sum is zero, that is:

$$C_{m,CG} = 0$$

for trim. For a small α, the term $C_D^{wb}\left(h_{CG} - h_{AC}^{wb}\right)sin\alpha$ is usually much smaller than the previous term, so it can be ignored, and $cos\alpha \approx 1$. Then, we need to examine:

$$\underbrace{C_{m,AC}^{wb}}_{(-)} + \underbrace{C_L^{wb}}_{(+)} \underbrace{\left(h_{CG} - h_{AC}^{wb}\right)}_{(?)} = 0 \tag{1.22}$$

to determine how a wing–body can be put into trim. The usual signs of the various terms in Equation 1.22 for conventional airplanes are indicated below each of them. Clearly, for trim to be possible:

$$\left(h_{CG} - h_{AC}^{wb}\right) > 0 \tag{1.23}$$

In other words, the CG must be aft of the wing–body AC. Figure 1.26 shows the same conclusion from the physical viewpoint. M_{AC}^{wb} is shown at the AC^{wb} in the usual negative sense. Positive L^{wb} acts at the AC^{wb}. As long as AC^{wb} is ahead of the CG, the moment due to L^{wb} about the CG can be positive and can possibly cancel out the negative M_{AC}^{wb} moment. Do convince yourself that if AC^{wb} were aft of the CG, trim would not be possible.

In fact, for trim to happen in this case, AC^{wb} should be exactly a distance:

$$\left(h_{CG} - h_{AC}^{wb}\right) = -\frac{C_{m,AC}^{wb}}{C_L^{wb}} \tag{1.24}$$

ahead of the CG, in which case the airplane (wing–body) would trim at an angle of attack and velocity corresponding to the C_L^{wb}.

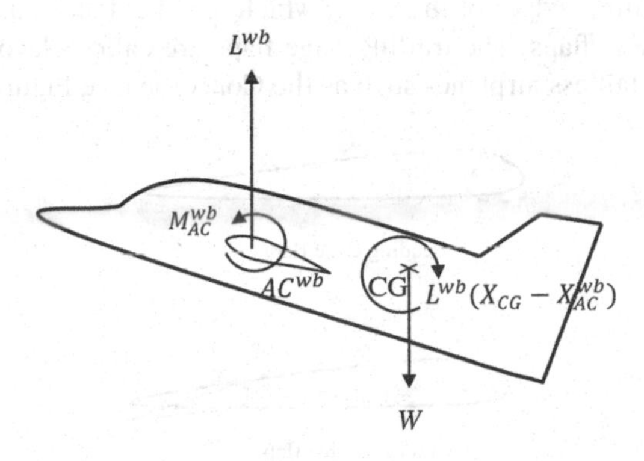

FIGURE 1.26 Trim condition with moment due to wing lift $L^{wb}\left(X_{CG} - X_{AC}^{wb}\right)$ exactly balanced by M_{AC}^{wb} at the CG.

Example 1.15

An airplane (wing–body) is flying level in trim at an altitude where the atmospheric density, $\rho = 1.225$ kg/m^3. Other airplane data are as follows:

Weight, $W = 2.27 \times 10^4$ N
Reference (wing planform) area, $S = 19$ m^2
$C^{wb}_{m,AC} = -0.016$; $h_{CG} - h^{wb}_{AC} = 0.11$

Can you find the velocity at which it is flying?

Find trim C^{wb}_L from:

$$C^{wb}_L = -\frac{C^{wb}_{m,AC}}{\left(h_{CG} - h^{wb}_{AC}\right)} = \frac{0.016}{0.11} = 0.145$$

Then, find trim velocity V^* from:

$$V^* = \sqrt{\frac{2}{\rho}\frac{W}{S}\frac{1}{C^{wb}_L}} = \sqrt{\frac{2 \times 2.27 \times 10^4}{1.225 \times 19 \times 0.145}} = 115.98 \text{ m/s}$$

(*Note*: These numerical problems are not meant to test your math skills, but to give you a feel for the kind of numbers you would encounter for each of these quantities. As an engineer, one should have a feel for the kind of values each quantity would normally take to make a reasonable first-cut judgment in various situations.)

What if we wanted to fly the airplane (wing–body) faster or slower, that is, trim it at a different velocity (and angle of attack)? That calls for a different trim C^{wb}_L, which is possible only if we could change either C^{wb}_{mAC} or $\left(h_{CG} - h^{wb}_{AC}\right)$. One could change C^{wb}_{mAC} by changing the effective camber of the wing, or $\left(h_{CG} - h^{wb}_{AC}\right)$ by moving the CG with respect to the fixed AC^{wb}. The second option is favoured on hang gliders.

Homework Exercise: Read up on how hang gliders are trimmed by shifting the flyer's weight, thus moving h_{CG}.

One way of realizing the first option of changing C^{wb}_{mAC} is by attaching flaps to the leading and/or trailing edges of the wing, which may be deflected. See Figure 1.27 for a sketch of these flaps. The trailing edge flaps are called elevons and are quite popular on many tailless airplanes such as the Concorde (see Figure 1.28). One way

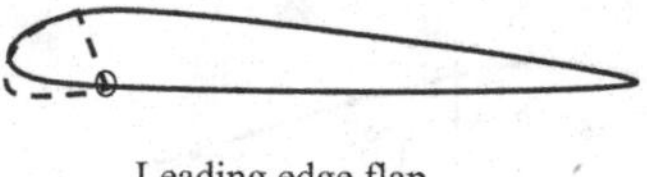

Leading edge flap

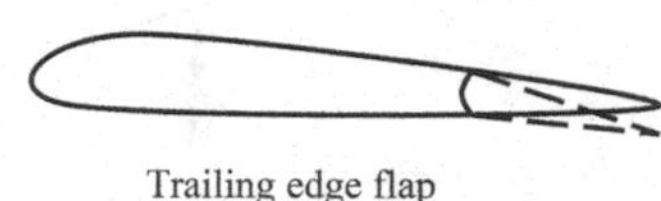

Trailing edge flap

FIGURE 1.27 Leading edge flaps and trailing edge flaps (also called elevons) on a wing.

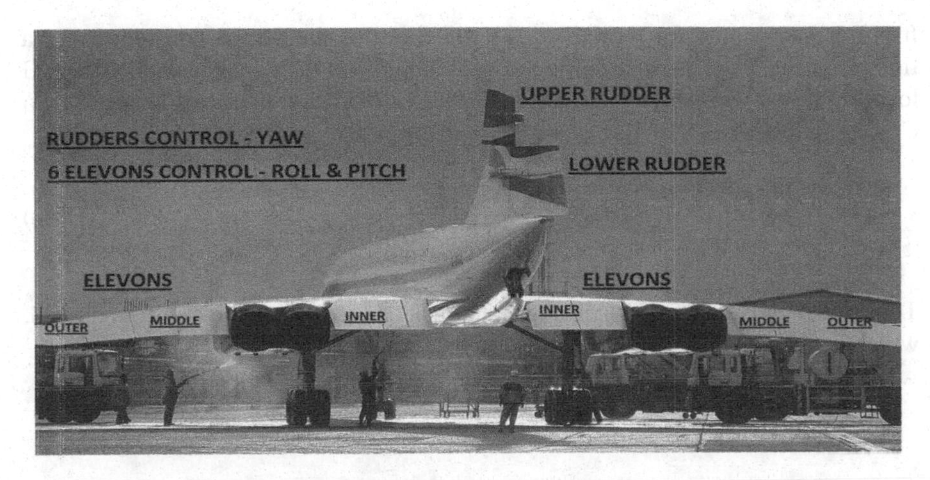

FIGURE 1.28 Elevons on the Concorde—picture shows inner, middle and outer set of elevons labelled (heritageconcorde.com; http://heritageconcorde.com/wp-content/uploads/2012/02/elevon-and-rudders-USE-FOR-WEBSITE1.jpg).

to understand their action is to imagine the elevons to be an integral part of the wing and that their deflection changes the effective camber of the wing. A sketch of how this can be visualized is shown in Figure 1.29. The downward flap deflection increases the effective camber, so this makes C_{mAC}^{wb} more negative and increases C_L^{wb} as well. For an unchanged CG position, this means trim at a higher value of C_L^{wb}, which implies a higher angle of attack and a lower velocity.

Example 1.16

For the airplane in Example 1.15, let us assume that a downward elevon deflection changes C_{mAC}^{wb} by −0.001. The new C_{mAC}^{wb} is thus $C_{mAC}^{wb} = -0.016 + (-0.001) = -0.017$.

New $C_L^{wb} = -\left(\frac{-0.017}{0.11}\right) = 0.154$ and the new trim velocity,

$$V^* = \sqrt{2(W/S)\left(1/\rho C_L^{wb}\right)} = \sqrt{2\left(\frac{2.27 \times 10^4}{19 \times 1.225 \times 0.154}\right)} = 112.5 \text{ m/s}$$

down from the previous value of 115.98 m/s.

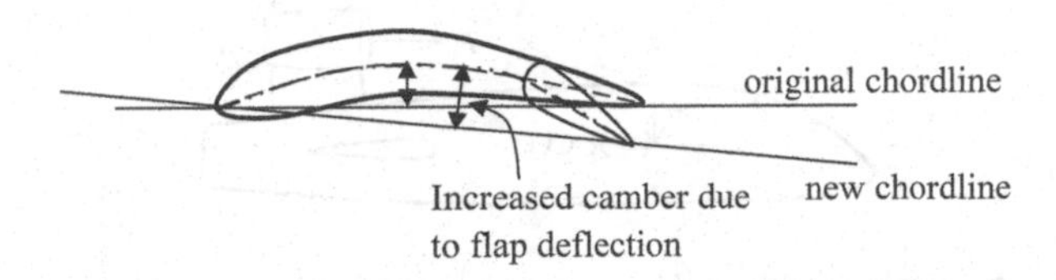

FIGURE 1.29 Schematic representation of an airfoil with increased camber due to trailing edge flap deflection.

It is not uncommon for the CG to shift in flight as the fuel is consumed. On a light airplane, even a redistribution of passengers and their seats can alter the CG location. For a wing–body system, a changed CG location implies a changed trim C_L^{wb}, as follows:

$$C_L^{wb} = -\frac{C_{m,AC}^{wb}}{\left(h_{CG} - h_{AC}^{wb}\right)} \tag{1.24a}$$

Thus, the flight velocity and trim angle of attack can change due to a shift in the CG, which is quite inconvenient, and may even be dangerous.

Example 1.17

Assume the lower limit on the velocity to be $1.2V_{stall}$, where V_{stall} is the level flight velocity corresponding to the stall C_L, C_{Lmax}. We can now determine the corresponding limit on the CG position for a wing–body to be in trim as follows:

Following the relation for trim velocity, $V^* = \sqrt{2(W/S)\left(1/\rho C_L^{wb}\right)}$, the expression for trim velocity at stall is given by $V_{stall}^* = \sqrt{2(W/S)\left(1/\rho C_{Lmax}^{wb}\right)}$.

One can determine the trim C_L^{wb} corresponding to $1.2V_{stall}$ using the above expressions as $C_L^{wb} = \left(C_{Lmax}^{wb}/1.44\right)$. Corresponding to this C_L^{wb} a limit on the CG position can be found using $C_L^{wb} = -\left(C_{m,AC}^{wb}/\left(h_{CG} - h_{AC}^{wb}\right)\right)$ as $h_{CGmax} = h_{AC}^{wb} - \left(1.44C_{mAC}^{wb}/C_{Lmax}^{wb}\right)$. This defines a point which is the forward-most limit of the CG position and requires that the CG remain aft of this point at all times. This requirement is aided, for instance, by placing the engines that are bulky at the rear of the airplane.

Example 1.18

In Figure 1.30, an airplane with a rear-mounted engine on the vertical tail is shown. The thrust line is at a height h above the centre of gravity, which causes a pitch down moment about the CG, given by the expression $M_{CG} = -T \times h$. The negative sign is following the sign convention that nose-down moment is negative. Seaplanes (see Figure 1.31) often have engines mounted high above the CG line to keep them out of contact with the water. In these cases, the contribution to pitching moment due to the thrust could be quite significant.

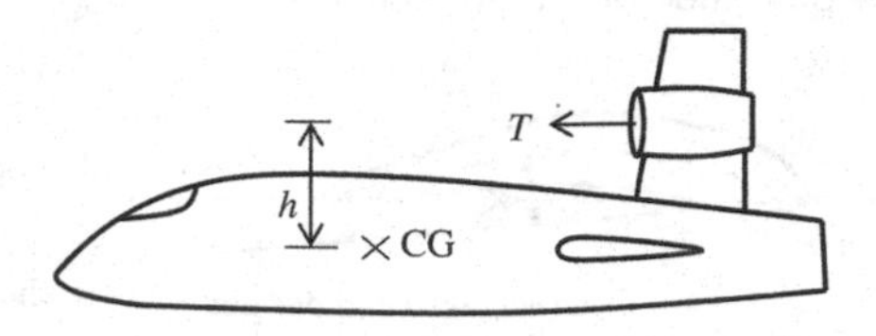

FIGURE 1.30 An airplane with engine mounted on the vertical tail showing moment due to thrust vertically displaced above the CG line.

FIGURE 1.31 Seaplane with engine mounted high above the CG line (https://en.wikipedia.org/wiki/Amphibious_aircraft#/media/File:CL-215T_43-21_(29733827710).jpg).

EXERCISE PROBLEMS

1.1 A rigid airship is an oblong ellipsoid of revolution with volume $V = 1.84 \times 10^5$ m^3. The airship is completely filled with helium. Determine the force of buoyancy on the airship at an altitude of 1.5 km above the ground. Given data: density of air (at 1.5 km altitude) ρ_{air} = 1.0581 kg/m^3 and density of helium ρ_{helium} = 0.18 kg/m^3.

1.2 A supersonic transport airplane weighing 8.89×10^5 N is flying at freestream Mach number, $Ma_\infty = 2.0$ at 11 km altitude above sea level. Calculate the lift coefficient C_L required to maintain cruise at this condition. Given data: atmospheric pressure (at 11 km altitude), $p = 2.27 \times 10^4$ N/m^2, wing planform area of the airplane, $S = 363$ m^2.

1.3 Derive the relations giving the variation of density and pressure with altitude in Box 1.4. Determine the pressure, temperature and density at an altitude of 12 km from the surface of Earth (geometric altitude), given $g_0 = 9.8$ m/s^2.

1.4 An airplane is flying at an altitude of 11 km. At this altitude, the Pitot tube fitted on the airplane measures an impact pressure of 7296 N/m^2. Determine V_{CAS}, V_{EAS} and V_{TAS} (see Box 1.7). Static pressure and temperature at this altitude are given as $p = 2.25 \times 10^4$ N/m^2 and $T = 216.78$ K.

1.5 Given $C_L = -0.027 + 0.0865\alpha$, $C_m = 0.06 - 0.0133\alpha$ for an airplane, where α is in degree, determine C_L trim.

BOX 1.7 VELOCITY MEASUREMENT IN FLIGHT

A Pitot static tube is normally used on airplanes for velocity measurement in flight. The instrument mounted on airplanes reads what is known as 'instrument' or 'indicated' airspeed (IAS or V_{IAS}). This speed is not a true measure of airplane speed. Airspeed indicators are calibrated under International Standard Atmospheric (ISA) conditions. An atmospheric parameter correction is therefore required to convert the indicated airspeed to true airspeed (TAS or V_{TAS}). Definition of other airspeeds and important relations to arrive at the true airspeed are as follows:

$$\text{TAS}: V_{TAS} = \sqrt{\frac{2\gamma RT}{(\gamma-1)}\left[\left(\frac{P_0 - p}{p}+1\right)^{\gamma-1/\gamma} - 1\right]}$$

$(P_0 - p)$ is the difference in stagnation and static pressures, also known as impact pressure, measured by the Pitot tube. γ is the ratio of specific heats; for air, $\gamma = 1.4$. R is the gas constant and T is the static temperature. p and T are the true static pressure and temperature, respectively, in the above relation.

Equivalent airspeed (EAS):

$$V_{EAS} = \sqrt{\frac{2\gamma RT_{SL}}{(\gamma-1)}\left[\left(\frac{P_0 - p}{p}+1\right)^{\gamma-1/\gamma} - 1\right]}$$

Calibrated airspeed (CAS):

$$V_{CAS} = \sqrt{\frac{2\gamma RT_{SL}}{(\gamma-1)}\left[\left(\frac{P_0 - p}{p_{SL}}+1\right)^{\gamma-1/\gamma} - 1\right]}$$

Subscript '*SL*' is used for sea-level conditions.

V_{TAS} and V_{EAS} are therefore related via the relation $V_{EAS} = V_{TAS}\sqrt{T_{SL}/T}$, which in terms of density ratio at different altitudes is $V_{EAS} = V_{TAS}\sqrt{\rho/\rho_{SL}}$.

V_{IAS} is V_{CAS} modified for instrument position and installation errors, which is airplane specific and a chart is usually available relating them. Airplane 'ground speed' is obtained finally by suitably adjusting true airspeed V_{TAS} to account for wind speed.

1.6 For the airplane in Example 1.15, determine the change in trim C_L required in the presence of a head wind of 20 m/s if it were to fly at the same velocity. How should the CG be shifted to make this trim at the new value of C_L possible?

1.7 In Figure 1.4, the locations of CG and AC for the Concorde are plotted as a function of the freestream Mach number. Study the plots carefully. One could possibly extend the curves below Mach 0.5. Can you reason (based on the information from these plots) why, during landing, the angle of attack of the airplane is large. Is this the reason why the Concorde nose needs to be swivelled down during landing?

1.8 What is the physical significance of the following timescales in aircraft flight dynamics: (i) V/g and (ii) $c/2V$?

1.9 Use the level flight trim conditions and the two timescales in the previous question to derive a new non-dimensional parameter, $\mu = 4m/(\rho Sc)$. How will you physically interpret this parameter?

REFERENCES

1. Anderson, John D., Jr., *Fundamentals of aerodynamics*, Fourth Edition, McGraw Hill Publication, 2007, p. 77.
2. Rech, Jean and Leyman, Clive S., *A case study by Aerospatiale and British Aerospace on the Concorde*, AIAA Professional Study Series.

2 Stability Concept

Associated with an equilibrium (or trim) state is the notion of its stability. At the equilibrium state, of course, all the forces and moments balance each other. So, an object in equilibrium can, in principle, remain in a state of equilibrium forever. However, in practice—and this is true for all natural systems—there are disturbances. For an airplane in flight, this may, for example, be due to wind or due to pilot input. Disturbances displace a system from its equilibrium state. The question then arises, what happens to the system after it is disturbed? Does it regain its equilibrium state or not? This leads us to the notion of stability.

Rather than giving an abstract mathematical definition of stability for an equilibrium state and then trying to reduce it to a useful form that can be applied to real-life systems, such as an airplane in flight, let us instead start with simple systems and evolve a useful definition of stability. Incidentally, the definition of stability that we will arrive at is identical to the one that is widely used in engineering analysis.

2.1 LINEAR FIRST-ORDER SYSTEM

A linear first-order dynamics can be represented in the form:

$$\dot{x} + ax = 0, \ (a \neq 0) \tag{2.1}$$

where x is the variable of interest and a is a system parameter. The equilibrium state of the system in Equation 2.1 is $x = 0$. Remember, the equilibrium state is where the variable (in this case, x) remains unchanged forever. That is, $\dot{x} = 0$ at the equilibrium state. Then, as long as $a \neq 0$, solving $ax = 0$ gives the equilibrium state to be $x = 0$.

Let there be a disturbance at an initial time, $t = 0$, which takes the variable x away from equilibrium (i.e., to a non-zero value). Call this initially disturbed value of x as $x(t = 0)$, or $x(0)$ for short. Now, this $x(0)$ is not an equilibrium state, so the system will not remain there. Instead, it should start changing (evolving) with time. In case of this simple system, it is easy to see that the time evolution of x from its disturbed state is given by:

$$x(t) = x(0)e^{-at} \tag{2.2}$$

So, if $a > 0$, then the evolution $x(t)$ 'eventually' brings the system back to its equilibrium state, $x = 0$, no matter how small or big the initial disturbance, $x(0)$, is. Likewise, if $a < 0$, then the evolution $x(t)$ blows up 'to infinity'. Of course, real-life physical systems hardly 'go away to infinity'. But, for our purposes, what matters is that for $a < 0$, the system does not regain its equilibrium state. A sketch of the time evolution for the two cases, $a > 0$ and $a < 0$, for the same initial disturbance $x(0)$ is shown in Figure 2.1.

DOI: 10.1201/9781003096801-2

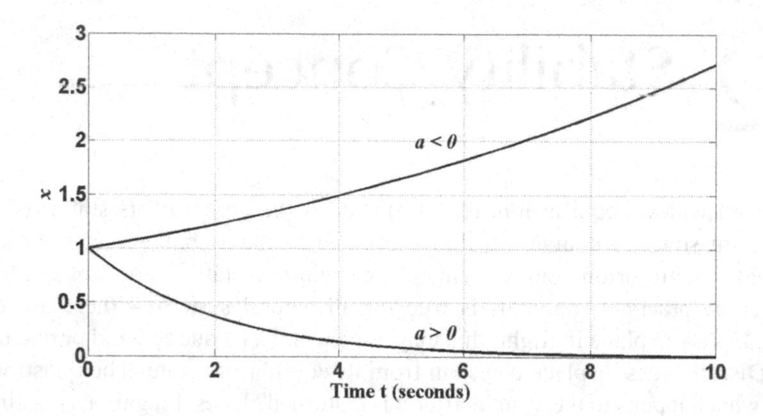

FIGURE 2.1 First-order system response for $a > 0$ and $a < 0$ from the same initial condition.

A word about the usage 'eventually' above: one may wonder how long does it take for a disturbed system to regain equilibrium when $a > 0$. Theoretically, it takes *infinite* time, no matter how small the disturbance is and how close the disturbed state $x(0)$ is to the equilibrium state, $x = 0$. But, in practice, as long as it approaches 'close enough' to the equilibrium state, it is good enough. How 'close' is 'close enough'? There is no firm answer to this question that works generally for all systems. And, it does not always matter because the next disturbance usually comes up and displaces the system before the effect of the previous disturbance has fully died down. In any case, in engineering systems, both the dynamics and the measurement (which tells you whether x has regained its equilibrium value or not) are *noisy*, so it is hard to be precise about the state of the system and its equilibrium.

So, as long as $a > 0$, we know that the system is heading in the right direction, that is, back towards the equilibrium state, and is eventually going to get there. That is usually good enough. Contrariwise, when $a < 0$, we can be sure it is headed the wrong way. In summary, we can conclude that:

For the dynamical system defined by the mathematical model $\dot{x} + ax = 0$, the equilibrium state given by $x = 0$ is

- Stable, provided $a > 0$
- Unstable, when $a < 0$

One may study Figure 2.1 and figure out that when the initial tendency of the evolution is to move in the direction of the equilibrium state, then it actually does eventually regain the equilibrium state. And when the initial tendency is in the direction away from the equilibrium state, then the evolution does diverge from the equilibrium state. Tempting as this correlation may seem and while it is valid for the first-order system here, we shall soon see that a similar deduction does not hold true for second-order (and higher-order) systems. Unfortunately, a notion of stability based on the 'initial tendency of evolution of the disturbed system' has been in vogue in flight dynamics for years and in our opinion has been the cause of some confusion. This has been called as 'static stability' to distinguish it from the more common notion of stability, which is then labelled 'dynamic stability'. In our opinion, the notion of 'static stability' is not useful enough to deserve the attention it usually receives and we shall conscientiously avoid using it.

Often, a first-order time response is characterized by time-to-halve the amplitude $t_{1/2}$ (in case of a stable response with $a > 0$) or time-to-double the amplitude t_2 (in case of an unstable response with $a < 0$). Another parameter of interest is the time constant τ. The expression for time to half (or double) the amplitude is given by:

$$t_{1/2} \text{ or } t_2 = \frac{0.693}{|a|}$$

and the time constant is defined as:

$$\tau = \frac{1}{a}$$

$t_{1/2}$ and t_2 may be used as a measure of stability and instability, respectively, for a first-order system.

Example 2.1

In Figure 2.2, three time–response curves starting from the same initial condition ($x(0) = 1$) for the first-order system Equation 2.1 are plotted together. The three curves corresponding to $a = 0.05$, 0.1 and 0.5 show stable behaviour (response converging to the equilibrium state, $x = 0$), but the rates at which the transients decay are different. Time constant τ decides the initial speed of the response; the smaller the time constant, the faster the system response. Table 2.1 summarizes the results for the first-order time response in Figure 2.2.

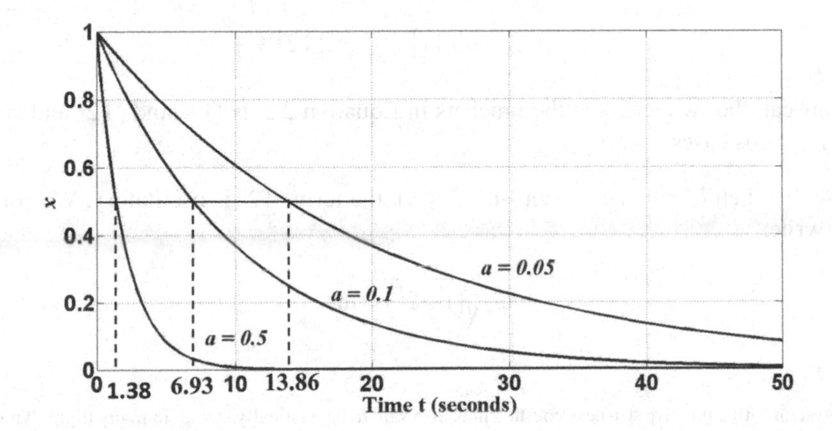

FIGURE 2.2 First-order system response for different positive values of 'a' starting from same initial condition.

TABLE 2.1
First-Order System Response

a	τ (s)	$t_{1/2}$ (s)
0.05	20	13.86
0.1	10	6.93
0.5	2	1.38

2.2 LINEAR SECOND-ORDER SYSTEM

A standard representation of a linear second-order system is:

$$\ddot{x} + d\dot{x} + kx = 0, \quad k \neq 0 \tag{2.3}$$

where d is called the damping coefficient and k is the stiffness coefficient.

Often it is more useful to write the second-order system in the following equivalent representation, which implicitly assumes that the stiffness coefficient k is positive*:

$$\ddot{x} + 2\zeta\omega_n\dot{x} + \omega_n^2 x = 0, \quad \omega_n \neq 0 \tag{2.4}$$

In this format, ζ is called the damping ratio and ω_n is the natural frequency.

The equilibrium (or trim) state of this system is again given by $x = 0$ (since at equilibrium $\dot{x}$ and $\ddot{x}$ are required to be zero). The issue of stability of this equilibrium state is as before related to the evolution of the system from a disturbed state, $x(t = 0)$ or $x(0)$ for short, with velocity $\dot{x}(0)$. Luckily, in this case too, one can explicitly solve for the time evolution as follows:

$$x(t) = x(0)\underbrace{\left[e^{-\zeta\omega_n t}\right]}_{[\mathrm{T1}]}\underbrace{\left[e^{\pm i\omega_n\sqrt{(1-\zeta^2)}\,t}\right]}_{[\mathrm{T2}]} \tag{2.5}$$

Let us call the two terms in the brackets in Equation 2.5 as 'T1' and 'T2' and consider various cases.

Case 1: When $\zeta^2 < 1$ (i.e., when $-1 < \zeta < 1$), the term 'T2' is oscillatory. We could just write:

$$\omega_n\sqrt{(1-\zeta^2)} = \omega$$

* A system with a positive stiffness coefficient is also said to be 'statically stable' in many flight dynamics references. As said earlier, we prefer not to use the concept of 'static stability'. Instead, where necessary, it is more appropriate and less confusing to just say that the system has a positive stiffness coefficient.

So

$$x(t) = \left[x(0)e^{-\zeta\omega_n t}\right]\left[e^{(\pm i\omega t)}\right] \tag{2.6}$$

and ω is called the 'damped natural frequency'. So, to judge the stability of the equilibrium state, $x = 0$, we need to investigate the term 'T1'. But, this looks exactly the same as the right-hand side of Equation 2.2 for the first-order system with $\zeta\omega_n$ in place of 'a' leading to the same conclusion:

- Stable, when $\zeta\omega_n > 0$
- Unstable, when $\zeta\omega_n < 0$

Of course, since ω_n is always positive, we get the following condition on ζ for stability:

For the dynamical system defined by the mathematical model, $\ddot{x} + 2\zeta\omega_n\dot{x} + \omega_n^2 x = 0, \omega_n \neq 0$, when $\zeta^2 < 1$ (i.e., when $-1 < \zeta < 1$), *the equilibrium state at* $x = 0$ *is*

- Stable, provided $\zeta > 0$
- Unstable, when $\zeta < 0$

We consider the case of $\zeta = 0$ separately.

Example 2.2

In Figure 2.3, time–response curves for a second-order system of the form in Equation 2.4 are shown. Positive damping ratio ($\zeta > 0$) results in a stable response as shown in Figure 2.3a, that is, response converges to the equilibrium state $x = 0$ starting from an initial condition, $\left[x(0), \dot{x}(0)\right] = [1, 0.1]$ which is a little way off from the equilibrium state. Negative damping ratio, $\zeta < 0$, results in unstable (divergent) response starting from the same initial condition, $\left[x(0), \dot{x}(0)\right] = [1, 0.1]$, as shown in Figure 2.3b, that is, response diverges from the equilibrium state $x = 0$.

A careful look at the response curves suggests that the trace of peaks or troughs in Figure 2.3 follows a first-order system response, as concluded from the analysis above. The difference between a first-order response (shown in Figure 2.1) and a second-order response (shown in Figure 2.3) is because of the appearance of the oscillatory term 'T2' in the solution of the second-order system, Equation 2.5, and this is also observed in the time response.

This is the case when the system is 'underdamped', that is, $-1 < \zeta < 1$.

Notice that in each case of Figure 2.3, the initial tendency of the system response after the disturbance is to head back towards the equilibrium point. However, only for the case of $\zeta > 0$ do the oscillations eventually die out and the system returns to the equilibrium point. In the case of $\zeta < 0$, the oscillations grow about the equilibrium point and their amplitude diverges with time. So, the concept of 'initial tendency of the system response' (aka static stability) is not at all useful here.

Let us next look at what happens when $\zeta^2 > 1$.

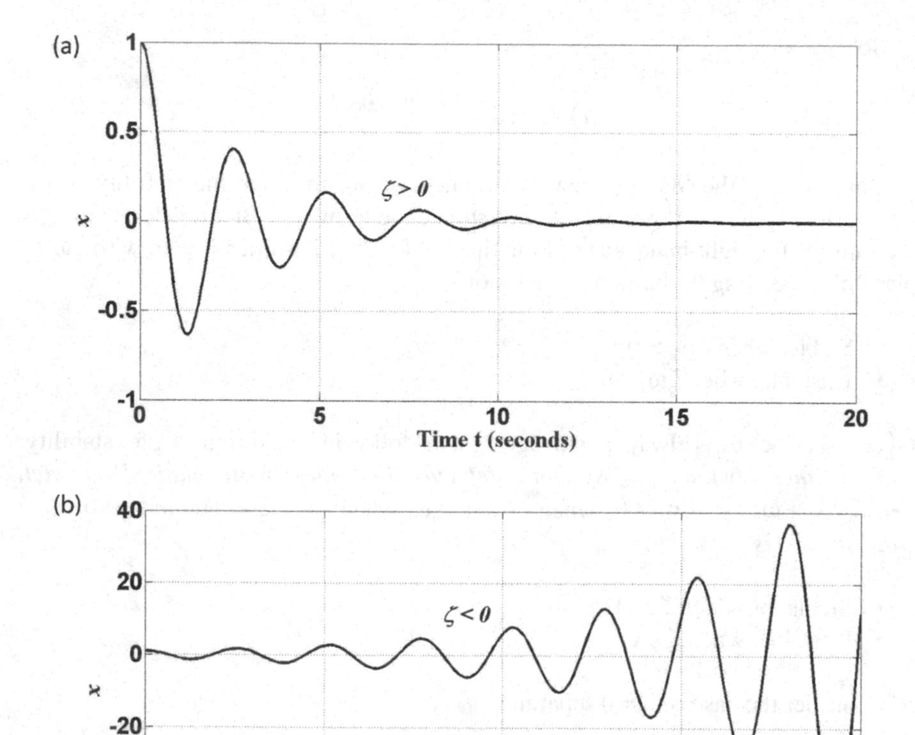

FIGURE 2.3 Second-order system response: (a) stable and (b) unstable.

Case 2: When $\zeta^2 > 1$, we can rewrite the evolution as:

$$x(t) = \left[x(0)e^{-\zeta\omega_n t}\right]\left[e^{\pm\omega_n\sqrt{(\zeta^2-1)}t}\right] \tag{2.7}$$

where the argument of both the exponential terms are real, not imaginary. So, combining the two exponential terms together, we get:

$$x(t) = x(0)e^{\left[\left(-\zeta\pm\sqrt{(\zeta^2-1)}\right)\omega_n\right]t} \tag{2.8}$$

Or to put it in the same form as Equation 2.2

$$x(t) = x(0)e^{\left[-\left(\zeta\mp\sqrt{(\zeta^2-1)}\right)\omega_n\right]t} \tag{2.9}$$

So, as for the first-order system in Equation 2.1 or the second-order system in Equation 2.4, stability depends on the coefficients $\left\{\zeta \mp \sqrt{(\zeta^2 - 1)}\right\}\omega_n$. And, as ω_n is positive, what matters is the sign of the coefficients $\left\{\zeta \mp \sqrt{(\zeta^2 - 1)}\right\}$.

- For $\zeta > 1$, both the coefficients turn out to be positive, hence the equilibrium point is stable.
- For $\zeta < -1$, both the coefficients turn out to be negative, hence the equilibrium point is unstable.

Example 2.3

Figure 2.4 shows the typical time response of the second-order dynamics after a disturbance from the equilibrium point $x = 0$ for the two cases: $\zeta > 1$ and $\zeta < -1$. Since both the exponents in Equation 2.8 (or Equation 2.9) are real, we observe that the oscillatory component of the response has disappeared. The response now decays or grows exponentially (similar to a first-order system response) depending on the sign of the damping ratio, ζ. Effectively, the system is like two first-order systems together.

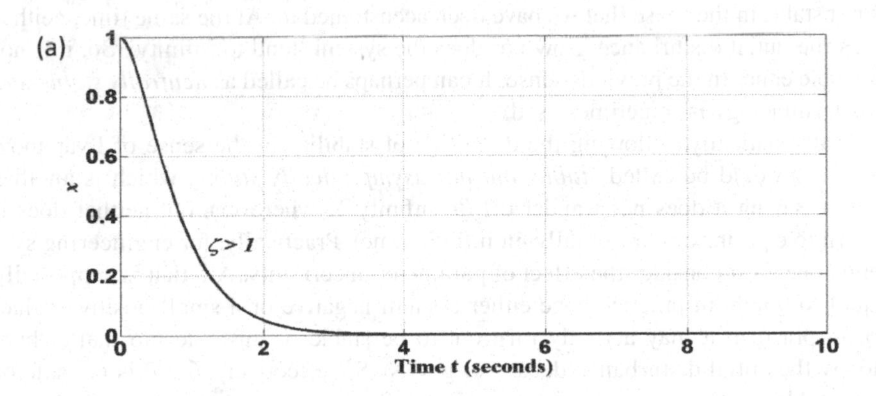

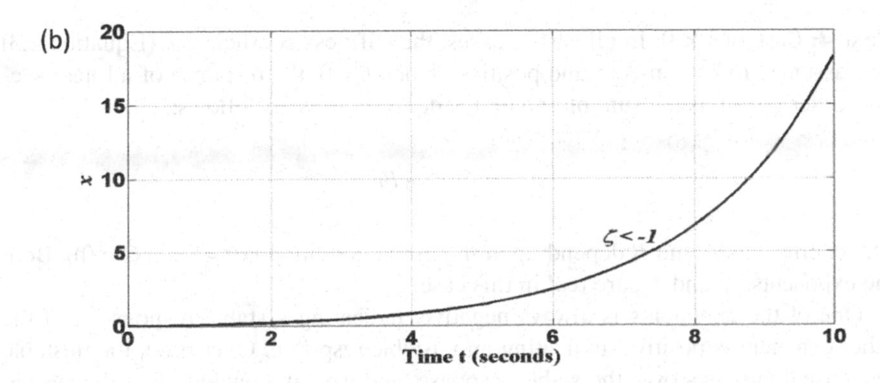

FIGURE 2.4 Second-order response for: (a) $\zeta > 1$ and (b) $\zeta < -1$.

Homework Exercise: Can you work out what happens when $|\zeta| = 1$?

Putting Cases 1 and 2 (and the Homework Exercise) together, one can state jointly that

Result 2.1: *For the dynamical system defined by the mathematical model,* $\ddot{x} + 2\zeta\omega_n\dot{x} + \omega_n^2 x = 0, \omega_n \neq 0$, *the equilibrium state given by* $x = 0$ *is*

- Stable, provided $\zeta > 0$
- Unstable, when $\zeta < 0$

Case 3: Case of $\zeta = 0$. In this case, the solution of the second-order dynamics given by Equation 2.4 is simply:

$$x(t) = x(0)e^{\pm i\omega_n t} \tag{2.10}$$

That is, a purely oscillatory response with the oscillation amplitude fixed at $x(0)$. Is this response to be called *stable* or *unstable*? Since the initial disturbance $x(0)$ does not die down and the system does not return to the equilibrium state, $x = 0$, it is not stable in the sense that we have been accustomed to. At the same time, neither does the initial disturbance grow nor does the system 'tend to infinity'. So, it is not unstable either in the previous sense. It can perhaps be called as *neutrally stable* and this terminology is sometimes used.

Mathematically, following the definition of stability in the sense of Lyapunov, this case would be called *stable, but not asymptotically stable*, which is another way of saying it does not wander off (to infinity or wherever), but neither does it return to equilibrium (eventually, in infinite time). Practically, for engineering systems, one must consider the effect of parameter uncertainty. A ζ that is supposedly equal to 0 may in practice have either a small negative or a small positive value, so, in practice, it may actually turn out to be stable or unstable (no matter how slowly the initial disturbance decays or grows). So, effectively, $\zeta = 0$ is not safe or acceptable.

Case 4: Case of $k < 0$. In all earlier cases, the stiffness coefficient k (Equation 2.3) was assumed to be non-zero and positive. When $k < 0$, the response of a linear second-order system is the sum of two first-order responses, as follows:

$$x(t) = Ae^{\lambda_1 t} + Be^{\lambda_2 t} \tag{2.11}$$

The coefficients A and B depend upon the initial condition ($x(t = 0)$, $\dot{x}(t = 0)$). Both the exponents, λ_1 and λ_2, are real in this case.

One of the exponents is always negative, indicating a stable response, and the other exponent is positive, indicating an unstable response. Over time, the unstable response dominates over the stable response, and we can conclude that the system is unstable.

Example 2.4

Consider a second-order system modelled by the equation: $\ddot{x} + 5\dot{x} - 6x = 0$. Comparing with Equation 2.3, $k = -6(<0)$ and $d = 5$. Substituting the general form of the solution, $x(t) = Ce^{\lambda t}$ for linear systems in the equation and solving for λ results in:

$$Ce^{\lambda t}\left(\lambda^2 + 5\lambda - 6\right) = 0$$

or

$$x(t)\left(\lambda^2 + 5\lambda - 6\right) = 0 \tag{2.12}$$

Since $x(t) = 0$ (the equilibrium state) is the trivial solution of the above differential equation, for any non-trivial solution $x(t) \neq 0$, we need to solve:

$$\lambda^2 + 5\lambda - 6 = 0 \tag{2.13}$$

The solution of Equation 2.13 can be determined to be $\lambda_1 = -6$ and $\lambda_2 = 1$. Thus, one arrives at the general form of solution of a second-order system or a second-order response, as given by Equation 2.11.

Finally, in terms of the initial conditions, the exact solution for our system can be written as:

$$x(t) = \left(\frac{x(0) - \dot{x}(0)}{7}\right)e^{-6t} + \left(\frac{6x(0) - \dot{x}(0)}{7}\right)e^{t} \tag{2.14}$$

One can notice from this solution that when the initial condition is the equilibrium state $\left(x(0) = 0,\ \dot{x}(0) = 0\right)$ itself, then the response of course shows $x(t) = 0$ for all time $t > 0$. For other initial conditions, the response looks as shown in Figure 2.5. Figure 2.5a is a plot showing the evolution of $x(t)$ and $\dot{x}(t)$ (on the y axis in the figure) with time, also known as *phase portrait,* where the arrows indicate increasing time and Figure 2.5b shows the response $x(t)$ as a function of time for one particular choice of $x(t)$.

We do not worry presently about special cases such as $\lambda_1 = \lambda_2$.

Homework Exercise: Do you see any connection between the sign of the stiffness coefficient, k, and the stability of the system?

The discussion above can be stated in a slightly more mathematical manner as summarized in Box 2.1.

What about third- and higher-order systems? It turns out that they can all be, in general, reduced to combinations of first- and second-order systems. So, there is no need for any special analysis of third- and higher-order systems.

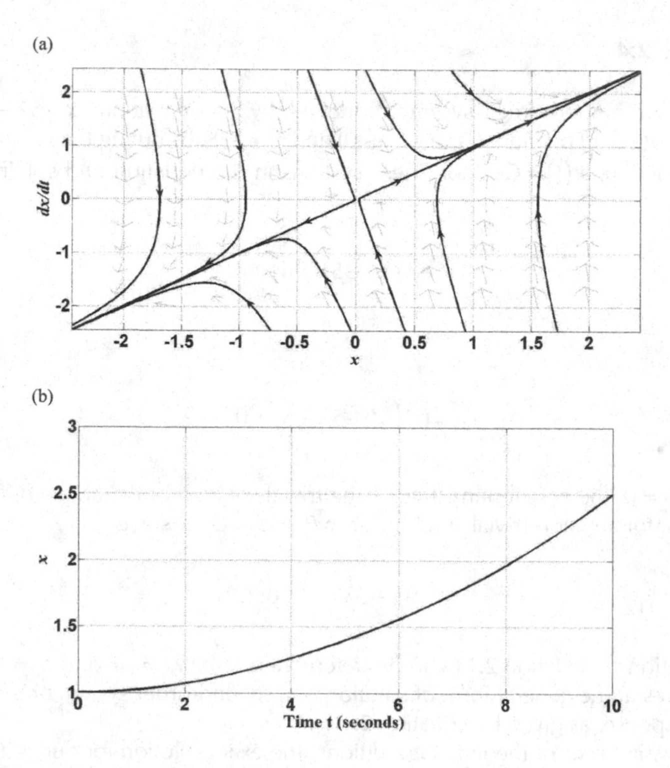

FIGURE 2.5 (a) Plot showing the evolution of $x(t)$ and $\dot{x}(t)$ (on the y axis in the figure) with time and (b) time response for $k < 0$.

BOX 2.1 A SUMMARY OF FIRST- AND SECOND-ORDER SYSTEM DYNAMICS AND STABILITY

Once we accept that all linear systems always respond in the following manner:

$$x(t) = x(0)e^{\lambda t}$$

the rest is fairly straightforward. The first derivative (velocity) is then:

$$\dot{x}(t) = x(0)\lambda e^{\lambda t} = \lambda x(t)$$

and the second derivative (acceleration) is given by:

$$\ddot{x}(t) = x(0)\lambda^2 e^{\lambda t} = \lambda^2 x(t)$$

First-order system: The solution then of the first-order system

$$\dot{x} + ax = 0$$

must satisfy

$$(\lambda + a)x(t) = 0$$

and unless it is the equilibrium solution, $x(t) = 0$, any other solution must satisfy

$$(\lambda + a) = 0 \text{ or } \lambda = -a$$

λ, which in this case is a real number, is called the *eigenvalue* of the system and stability requires λ to be negative—then a non-zero initial state, $x(0)$, dies down with time following the exponential law, $x(0)e^{\lambda t}$.

Second-order system: The solution of the second-order system

$$\ddot{x} + d\dot{x} + kx = 0$$

must satisfy

$$(\lambda^2 + d\lambda + k)x(t) = 0$$

and, as before, ignoring the equilibrium solution, $x(t) = 0$, leads us to

$$(\lambda^2 + d\lambda + k) = 0$$

whose solutions are:

$$\lambda = \frac{-d \pm \sqrt{d^2 - 4k}}{2}$$

As before, λ is called the *eigenvalue* of the system except that now λ can be a complex number. Stability requires that the real part of λ be negative. That is:

$$\text{Re}\{\lambda\} < 0$$

A concise way of depicting the various stability cases for first- and second-order systems is by plotting the possible eigenvalue combinations on a graph of $\text{Re}\{\lambda\}$ versus $\text{Im}\{\lambda\}$, as in Figure 2.6. With reference to Figure 2.6:

i. In case of a first-order system, λ must be real, so the eigenvalue must lie on the $\text{Re}\{\lambda\}$ axis. So there are only three distinct choices (these are marked with black circles):
   a. Either λ is on the negative real axis—system is stable.
   b. λ is on the positive real axis—system is unstable.
   c. λ is at the origin—we called this case *neutrally stable.*

ii. In case of a second-order system, we show six distinct cases (marked with star) in Figure 2.6:
   a. A pair of complex conjugate eigenvalues in the left-half complex plane—stable.
   b. A pair of complex conjugate eigenvalues in the right-half complex plane—unstable.
   c. A pair of complex conjugate eigenvalues on the imaginary axis—*neutrally* stable.
   d. Two real eigenvalues on the negative real axis—stable.
   e. Two real eigenvalues on the positive real axis—unstable.
   f. Two real eigenvalues, one each on the positive and negative real axis—unstable.

There could be other cases with at least one eigenvalue at the origin—these are set aside as Homework Exercise.

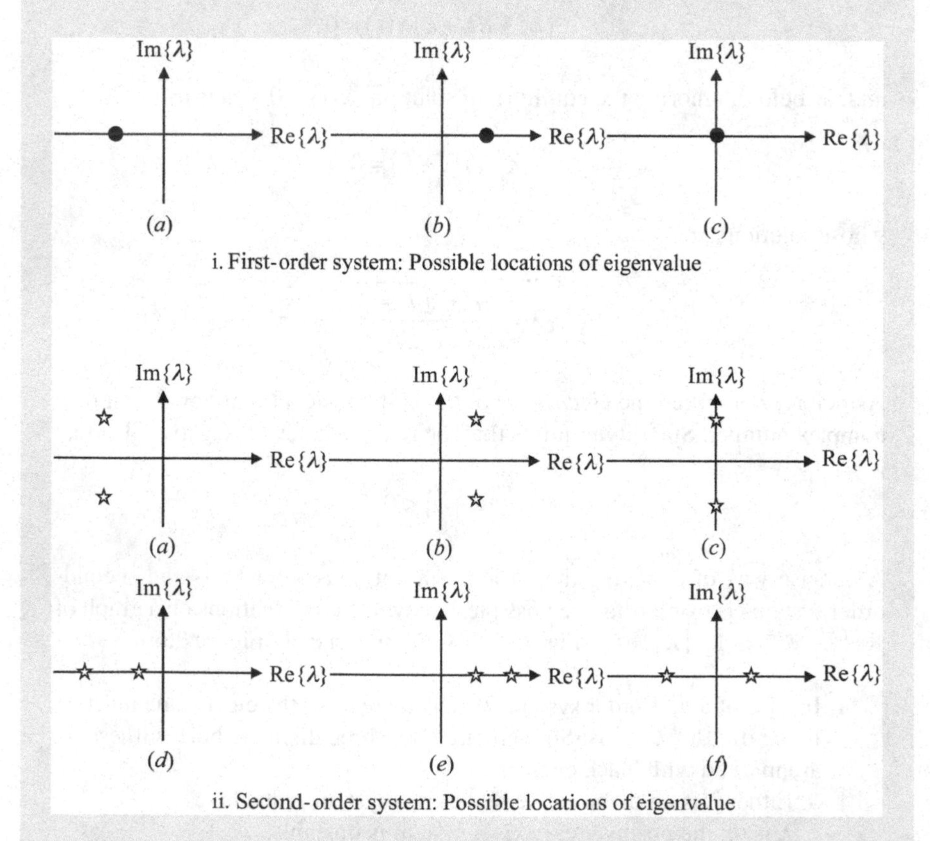

FIGURE 2.6 Locations of eigenvalues for first- and second-order systems.

2.3 NON-LINEAR SECOND-ORDER SYSTEM

Let us consider a non-linear second-order system now. One difference from a linear system is that a non-linear system may have more than one equilibrium point and the stability of each equilibrium point has to be determined separately.

As an example, consider the non-linear second-order system:

$$\ddot{x}+2\dot{x}-x+x^3=0 \tag{2.15}$$

Equilibrium points for this system are obtained by setting $\ddot{x}=\dot{x}=0$, which amounts to solving the algebraic equation $-x+x^3=0$ for its roots. When solved, the equilibrium points for the system turn out to be $x^*=0, -1, 1$. Now to judge the stability of each equilibrium point, we need to study the behaviour of the system when disturbed from that equilibrium point. To do this, we can redefine the variable x as:

$$x = x^* + y$$

about each equilibrium point x^*, where y is the perturbation variable about that equilibrium point. For instance, when this is done for the system in Equation 2.15 for any of its equilibrium points, it can be rewritten as:

$$\left(\ddot{x}^*+\ddot{y}\right)+2\left(\dot{x}^*+\dot{y}\right)-\left(x^*+y\right)+\left(x^*+y\right)^3=0 \tag{2.16}$$

Expanding each term in Equation 2.16 and re-arranging them results in:

$$\underbrace{\left(\ddot{x}^*+2\dot{x}^*-x^*+x^{*3}\right)}_{\text{'I'}}+\underbrace{\left(\ddot{y}+2\dot{y}-y+3x^{*2}y\right)}_{\text{'L'}}+\underbrace{\left(3x^*y^2+y^3\right)}_{\text{'NL'}}=0 \tag{2.17}$$

Of the three terms in Equation 2.17, the first labelled 'I' consists only of the equilibrium value of the variable x^* and can be seen to identically satisfy Equation 2.15. Therefore, that term is always equal to zero. The second term 'L' is linear in 'y' and its derivatives, while the third term 'NL' is non-linear as it contains higher powers of 'y'. Also, one would notice that both 'L' and 'NL' depend on x^* and therefore they are evaluated differently for each equilibrium point.

To proceed further, we need to assume that the disturbance y is 'small' in the sense that higher powers of y and its derivatives are negligible as compared to terms linear in y and its derivatives. This is formally called the small-perturbation approach in stability theory. In essence, to establish the stability of an equilibrium point, it is adequate to check for the system response to small perturbations (disturbances) from that equilibrium point.

In that case, one can ignore the term 'NL' in Equation 2.17, and effectively write it as:

$$\ddot{y}+2\dot{y}-y+3x^{*2}y=0 \tag{2.18}$$

where x^* is a fixed value, corresponding to one of the equilibrium points. Since Equation 2.18 appears to be a linear system (of second-order), one may use our previous experience in Sections 2.1 and 2.2 to determine the stability of the equilibrium point.

Corresponding to three equilibrium points, $x^* = 0, -1, 1$, the linear systems of interest are:

$$\ddot{y}+2\dot{y}-y=0 \quad \text{at } x^*=0$$

$$\ddot{y}+2\dot{y}+2y=0 \quad \text{at } x^*=-1, 1$$

For the equilibrium points at $x^* = -1$ and $x^* = 1$, the linear systems are identical and have $\zeta\left(=\left(1/\sqrt{2}\right)\right)>0$ and $\omega_n(=\sqrt{2}>0)$; hence, by our conclusions in Case 1 of Section 2.2, these equilibrium points are stable. For $x^* = 0$, the coefficient $k = -1 < 0$, hence following Case 4 of Section 2.2, we find $\lambda_1 = -2.414 < 0$ and $\lambda_2 = 0.414 > 0$, and we can conclude that this equilibrium point is unstable.

In this manner, one can analyse the equilibrium (or trim) state of any system and determine its stability. One must remember that stability is a property of the equilibrium state and not of the dynamical system per se. As seen in the example above, one dynamical system may have multiple equilibrium states, some stable and others unstable.

Next, we shall see how to apply the concept of stability to the equilibrium (trim) state of an airplane. The flight dynamics equations for an airplane generally constitute a non-linear system and hence must be analysed as described in this section.

2.4 PITCH DYNAMICS ABOUT LEVEL FLIGHT TRIM

Consider an airplane in straight and level flight ($\gamma^*=0$) and at a constant speed (V^* = constant). As seen earlier in Chapter 1 (Section 1.9), this is a trim state with thrust T^* and lift coefficient C_L^* given by Equation 1.13. The corresponding angle of attack is α^*. Knowing α^* and γ^*, one can deduce θ in steady, level flight as $\theta^*=\alpha^*+\gamma^*=\alpha^*+0=\alpha^*$.

We could disturb the airplane from its equilibrium (trim) state by a slight perturbation in velocity, that is, by either slowing it down or speeding it up, just a little bit. We could also disturb the airplane from its equilibrium state by a slight perturbation in its flight path, that is, by changing γ a little so as to set it into a very shallow climbing flight or descending flight. In either case, we have seen that the airplane responds with the characteristic timescale T_2, which is of the order of 10 seconds. In other words, it takes a perturbation in V or γ a time period of tens of seconds to either complete one oscillation cycle or double/halve the initial amplitude of perturbation.

In contrast, a small perturbation in the nose attitude θ elicits a response at the timescale T_1, which is of the order of 1 second. That is, the dynamics in θ happens so much quicker than that in V and γ that we could effectively assume V and γ to be constant at their trim values (V^*, γ^*) and examine the change in the small perturbation θ independently.

Since V and γ are now held constant at their trim values (V^*, γ^*), we only need to consider the pitch dynamics (Equation 1.11c) reproduced below:

$$\ddot{\theta} = \left(\frac{\bar{q}^* Sc}{I_{yy}}\right) C_m \tag{2.19}$$

Of course, at equilibrium (trim), θ^* is constant, and so $C_m^* = 0$ (as in Equation 1.12c). Let $\Delta\theta$ be a small perturbation in θ and the corresponding change in C_m would then be written as ΔC_m. So, in terms of these perturbation variables:

$$\ddot{\theta}^* + \Delta\ddot{\theta} = \left(\frac{\bar{q}^* Sc}{I_{yy}}\right)\left(C_m^* + \Delta C_m\right) \tag{2.20}$$

Since $\ddot{\theta}^* = 0$ and $C_m^* = 0$ at the trim state, we are left with:

$$\Delta\ddot{\theta} = \left(\frac{\bar{q}^* Sc}{I_{yy}}\right)\Delta C_m \tag{2.21}$$

And because $\theta = \gamma + \alpha$ in longitudinal flight (see Figure 1.15), we can write:

$$\theta^* + \Delta\theta = \alpha^* + \Delta\alpha + \gamma^* + \Delta\gamma$$

And since $\gamma^* = 0$ for the present case, $\theta^* = \alpha^*$ and $\Delta\gamma = 0$ as well, we get $\Delta\theta = \Delta\alpha$. So, we can write Equation 2.21 as:

$$\Delta\ddot{\alpha} = \left(\frac{\bar{q}^* Sc}{I_{yy}}\right)\Delta C_m \tag{2.22}$$

where $\bar{q}^* = \frac{1}{2}\rho V^{*2}$.

This is almost in the form of a second-order system in the variable $\Delta\alpha$; but, before we proceed further, we need to express ΔC_m in terms of aerodynamic variables. In fact, this is such an important aspect of flight dynamics that we need to take a break to discuss this, and then return to the analysis of pitch dynamics in level flight trim.

2.5 MODELLING SMALL-PERTURBATION AERODYNAMICS

In this section, we will limit ourselves to modelling the incremental pitching moment coefficient ΔC_m in case of an airplane perturbed from a straight and level flight trim, but which maintains V and γ at the trim value (V^*, γ^*).

The aerodynamic forces (and moments) acting on an airplane arise due to its interaction with the relative wind. Generally, two kinds of effects are modelled: (i) *static*—due to the orientation of the airplane with respect to the wind and (ii) *dynamic*—due to the relative angular motion between the airplane and the wind. It is important to note that the aerodynamic forces do not depend on the orientation of the airplane with respect to Earth (inertial axis) as given by the angles such as θ. Equally, they do not depend on the components of the airplane angular rates with respect to Earth, given by p, q, r. For a historical note on this, please see Box 2.2.

In the present instance, since the velocity V and flight path angle γ are fixed, there are two other major influences on the pitching moment—a static effect due to the angle of attack α and a dynamic effect due to the airplane angular (pitch) rate relative to the wind, $q_b - q_w$, where the subscript '*b*' refers to the body and '*w*' refers

George Hartley Bryan (1864–1928). (Source: http://en.wikipedia.org/wiki/George_H._Bryan)

BOX 2.2 A HISTORY OF THE DYNAMIC DERIVATIVES

The development of mathematical equations for studying flight vehicle motion was first put forth by G.H. Bryan (work published in 1911). In his work, he proposed the treatment of flight vehicles as a rigid body with six degrees of freedom in motion. He further proposed a mathematical technique for dynamic stability analysis of vehicle motion. His mathematical model introduced the concept of aerodynamic stability derivatives.

Surprisingly, the analysis of airplane dynamics and stability as proposed by Bryan has remained essentially unchanged all these years (but for a change in notation).

A point to note is that Bryan defined the aerodynamic derivatives in his work purely from mathematical considerations—not from an understanding of the physics or aerodynamics. Since the equations of motion were written in terms of three translational velocity components (u, v, w) and three rotational velocity components (p, q, r), Bryan simply wrote out each aerodynamic force or moment as a function of these six variables, in the form of a Taylor series. Thus arose derivatives such as M_q(later written as the non-dimensional C_{mq}), which have been unquestioningly accepted and carried over until recently.

to the wind. Just as the airplane's angular motion with respect to Earth is reckoned by judging the motion of the body-fixed axis with respect to the inertial axis, one can define a set of 'wind axes' whose motion with respect to the inertial axes lets us define terms such as q_w. The airplane's angular motion with respect to the relative wind can then be defined in terms of the motion of the body axis with respect to Earth and that of the wind axis with respect to Earth, such as $q_b - q_w$, for instance. (Refer Box 2.3 for more on 'wind axes'.)

Therefore, the perturbed pitching moment coefficient ΔC_m may be modelled as a function of:

- The perturbation in angle of attack, $\Delta\alpha$—*static effect.*
- The difference between the perturbations in body-axis and wind-axis pitch rates $(\Delta q_b - \Delta q_w)$—*dynamic effect.*

In mathematical form, the expression for the perturbed pitching moment ΔC_m appears as below:

$$\Delta C_m = C_{m\alpha}\Delta\alpha + C_{mq1}\left(\Delta q_b - \Delta q_w\right)\left(c/2V^*\right) \tag{2.23}$$

BOX 2.3 DEFINITION OF WIND AXES

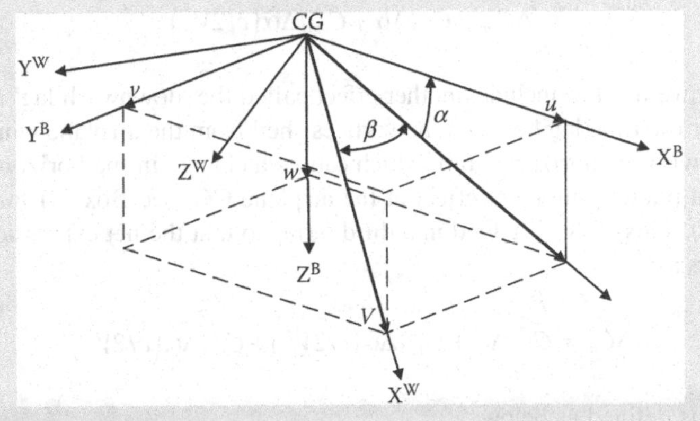

The figure shows the body axis (superscript 'B') and the wind axis (superscript 'W') placed at the airplane centre of gravity (CG). The x axis of the wind axes system is aligned with the velocity vector of the aircraft. The angle of attack (α) and the sideslip angle (β) define the orientation of the velocity vector with respect to the body-fixed axes of the aircraft. Thus, the velocity components along body-fixed axes are $u = V\cos\beta\cos\alpha$; $v = V\sin\beta$; $w = V\cos\beta\sin\alpha$.

where $C_{m\alpha}$ is defined as the partial derivative of C_m with α (all other variables being held fixed) at the equilibrium point (denoted by '*').

$$C_{m\alpha} = \left.\frac{\partial C_m}{\partial \alpha}\right|_* \tag{2.24}$$

Likewise, C_{mq1} is defined as:

$$C_{mq1} = \left.\frac{\partial C_m}{\partial\left[(q_b - q_w)(c/2V)\right]}\right|_* \tag{2.25}$$

where the factor $(c/2V^*)$ is introduced to make the term $(q_b - q_w)(c/2V^*)$ into a non-dimensional quantity.

Terms such as $C_{m\alpha}$ and C_{mq1} are usually called 'aerodynamic derivatives' or 'stability derivatives'.

Note: The derivative in Equation 2.25 is labelled 'C_{mq1}' to distinguish it from the traditional pitch damping derivative 'C_{mq}'. For more about how the traditional aerodynamic derivatives came to be defined, please see Box 2.2.

It will be shown later in Chapter 8 that:

$$\Delta q_b - \Delta q_w = \Delta\dot{\alpha} \tag{2.26}$$

For the moment, let us just accept that this is so and proceed further. Then, the expression for the incremental pitching moment coefficient works out to be:

$$\Delta C_m = C_{m\alpha}\Delta\alpha + C_{mq1}\Delta\dot{\alpha}\left(c/2V^*\right) \tag{2.27}$$

In addition, we need to include another effect called the 'downwash lag' effect. This accounts for the time lag between the vortices shed from the airplane wing and their interaction with the horizontal tail, which causes a change in the horizontal tail lift, and hence a pitching moment effect at the airplane CG (see Box 2.4 for a detailed explanation). This effect is added in a third term, so that the net expression for ΔC_m now appears as:

$$\Delta C_m = C_{m\alpha}\Delta\alpha + C_{mq1}\Delta\dot{\alpha}\left(c/2V^*\right) + C_{m\dot{\alpha}}\Delta\dot{\alpha}\left(c/2V^*\right) \tag{2.28}$$

where $C_{m\dot{\alpha}}$ is defined as below:

$$C_{m\dot{\alpha}} = \left.\frac{\partial C_m}{\partial\left[\dot{\alpha}(c/2V)\right]}\right|_* \tag{2.29}$$

Rearranging terms, Equation 2.28 can be written as:

$$\Delta C_m = C_{m\alpha}\Delta\alpha + \left(C_{mq1} + C_{m\dot{\alpha}}\right)\Delta\dot{\alpha}\left(c/2V^*\right) \tag{2.30}$$

BOX 2.4 DOWNWASH LAG EFFECT

Owing to trailing edge vortex shedding from the wings, the angle of attack seen by the horizontal tail surface behind the wing is different from the wing angle of attack by an angle called the downwash angle. Hence, the tail angle of attack is given by $\alpha_t = \alpha_w - \varepsilon(t)$, where subscripts '*t*' and '*w*' refer to the tail and the wing, respectively. An expression for the instantaneous downwash angle at the tail is obtained by considering the delay due to the travel time of vortices from wing to tail, estimated as $\Delta t = l_t / V^*$ (l_t is the distance between the wing and the tail and V^* is the aircraft speed), as $\varepsilon(t) = (d\varepsilon/d\alpha)\alpha(t - \Delta t)$. The downwash lag effect modelled by this time delay thus affects the instantaneous tail angle of attack and consequently the tail lift. A change in the angle of attack at the tail due to the downwash lag effect can be approximated to be $\Delta\alpha_t = (d\varepsilon/d\alpha) \times \dot{\alpha} \times (l_t / V^*)$. An increment in lift coefficient at the tail due to this effect is given by $\Delta C_{Lt} = C_{L\alpha_t} \Delta\alpha_t = C_{L\alpha_t}(d\varepsilon/d\alpha)\,\dot{\alpha}(l_t/V^*)$ and a change in the pitching moment coefficient due to this effect can be determined to be:

$$\Delta C_{m,cg} = -(S_t l_t / Sc) C_{L\alpha_t} (d\varepsilon/d\alpha)\,\dot{\alpha}(l_t/V^*) = -V_H C_{L\alpha_t} (d\varepsilon/d\alpha)\,\dot{\alpha}(l_t/V^*),$$

which is proportional to $\dot{\alpha}$, hence it acts as a damping term. This is how the downwash lag effect introduces a damping effect in the pitching motion defined by a non-dimensional derivative, $C_{m\dot{\alpha}} = \left(dC_m / d(\dot{\alpha}c/2V^*)\right) = -(2l_t/c) V_H C_{L\alpha_t} (d\varepsilon/d\alpha)$. The damping effect in the direction of normal acceleration is defined by the derivative, $C_{L\dot{\alpha}} = \left(dC_L / d(\dot{\alpha}c/2V^*)\right) = 2V_H C_{L\alpha_t} (d\varepsilon/d\alpha)$. In the above expressions, $V_H = (S_t l_t / Sc)$ is known as the tail volume ratio and $C_{L\alpha_t}$ is the lift curve slope of the tail. Details given in Chapter 3.

The derivative $C_{m\alpha}$ can easily be obtained at any trim angle of attack α^* by computing the local slope on a plot of C_m versus α.

Example 2.5

Estimating $C_{m\alpha}$ from $C_{m\alpha}$ versus α curve for the F-18/high-angle-of-attack research vehicle (HARV).

From Figure 1.21 in Chapter 1, we can find that:

$$C_m(@\alpha = 0°) = 0.0152 \quad \text{and} \quad C_m(@\alpha = 5°) = -0.0038.$$

Thus, the slope:

$$C_{m\alpha}(@\alpha = 0°) = \frac{0.0152 - (-0.0038)}{0 - 5} /° = -0.0038/°$$

$$= -0.0038 \times (180/3.14)/\text{rad} = -0.218/\text{rad}$$

The term $(C_{mq1} + C_{m\dot{\alpha}})$ is usually called the pitch damping derivative and can be estimated for an airplane configuration from a dynamic wind tunnel test. We shall study the pitching moment derivatives, $C_{m\alpha}$, C_{mq1} and $C_{m\dot{\alpha}}$ in greater detail in Chapter 3.

2.6 PITCH DYNAMICS ABOUT LEVEL FLIGHT TRIM (CONTD.)

Now, putting together the pitch dynamics in Equation 2.22 and the model for the perturbed pitching moment in Equation 2.30 yields:

$$\Delta\ddot{\alpha} = \left(\frac{\bar{q}^* Sc}{I_{yy}}\right)\left\{C_{m\alpha}\Delta\alpha + \left(C_{mq1} + C_{m\dot{\alpha}}\right)\Delta\dot{\alpha}\left(c/2V^*\right)\right\} \tag{2.31}$$

which, on rearranging the terms, appears as:

$$\Delta\ddot{\alpha} - \left\{\left(\frac{\bar{q}^* Sc}{I_{yy}}\right)\left(c/2V^*\right)\left(C_{mq1} + C_{m\dot{\alpha}}\right)\right\}\Delta\dot{\alpha} - \left\{\left(\frac{\bar{q}^* Sc}{I_{yy}}\right)C_{m\alpha}\right\}\Delta\alpha = 0 \tag{2.32}$$

This is a linear second-order system in the perturbed angle of attack variable, $\Delta\alpha$, and is exactly of the form in Equation 2.4, provided $C_{m\alpha} < 0$, which we shall assume for the moment. This dynamics is usually called the *short-period mode.*

Comparing Equation 2.32 with the form in Equations 2.3 and 2.4, we can write the stiffness and damping coefficients as below:

$$k = \left(\omega_n^2\right)_{SP} = -\left(\frac{\bar{q}^* Sc}{I_{yy}}\right)C_{m\alpha} \tag{2.33}$$

$$d = \left(2\zeta\omega_n\right)_{SP} = -\left(\frac{\bar{q}^* Sc}{I_{yy}}\right)\left(c/2V^*\right)\left(C_{mq1} + C_{m\dot{\alpha}}\right) \tag{2.34}$$

where V^* is the steady, level flight velocity at the trim (equilibrium) state.

From Result 2.1, we can immediately conclude that for the straight and level flight trim state (V^* = constant, $\gamma^* = 0$) to be stable to perturbations in α, the damping coefficient $(d = 2\zeta\omega_n)$ must be positive, which implies that:

$$\left(C_{mq1} + C_{m\dot{\alpha}}\right) < 0, \text{ provided already } C_{m\alpha} < 0$$

Example 2.6

The trim $(C_m = 0)$ angle of attack corresponding to zero elevator setting for F-18/HARV (Figure 1.21) can be found to be $\alpha^* = 4°$.

$$@\alpha = 4°: C_{m\alpha} = -0.0026/°; \quad C_{mq1} = -0.0834/°; \; C_{m\dot{\alpha}} = 0; \quad C_{Ltrim} = 0.3567$$

Reading the signs of the derivatives: $C_{m\alpha} < 0, \left(C_{mq1} + C_{m\dot{\alpha}}\right) < 0$, it is immediately concluded that aircraft is stable in pitch (short-period mode) at the trim angle of attack, $\alpha = 4°$.

2.6.1 Numerical Example

Let us construct Equation 2.32 for the F-18/HARV model and examine the stability result through numerical simulation. To do this, we need to choose a flight condition and then determine the aircraft trim speed at that condition. Let us choose the trim condition as a straight and level flight ($V^* =$ constant, $\gamma^* = 0$). Using the relation:

$$L = W = \frac{1}{2}\rho V^{*2} S C_L^*$$

in straight and level flight, one can determine the trim velocity as:

$$V^* = \sqrt{\frac{2}{\rho_{air}} \frac{W}{S} \frac{1}{C_L^*}}$$

Other useful data for F-18/HARV are:

$$m = 15{,}118.35 \text{ kg}, \quad S = 37.16 \text{ m}^2, \quad I_{yy} = 205113.07 \text{ kg} - \text{m}^2, \quad c = 3.511 \text{ m}$$

Further, using $g = 9.81\ m/s^2$ and $\rho_{air} = 1.225\ kg/m^3$ at sea level, we can determine the speed as:

$$V^* = \sqrt{\frac{2}{\rho_{air}} \frac{W}{S} \frac{1}{C_L^*}} = \sqrt{\frac{2}{1.225} \times \frac{15118.35 \times 9.81}{37.16} \times \frac{1}{0.3567}}$$
$$= 135.16 \text{ m/s}$$

$$\left(\frac{\bar{q}^* Sc}{I_{yy}}\right) = \frac{1}{2}\rho V^{*2} \frac{Sc}{I_{yy}} = 0.5 \times 1.225 \times 135.16^2 \times \frac{37.16 \times 3.511}{205113.07} = 7.117/\text{s}^2$$

$$c/2V^* = \frac{3.511}{2 \times 135.16} = 0.013 \text{ s}$$

$$C_{m\alpha} = -0.0026/° = -0.0026 \times \frac{180}{\pi} /\text{rad} = -0.149/\text{rad}$$

$$C_{mq1} = -0.0834/° = -0.0834 \times \frac{180}{\pi} /\text{rad} = -4.78/\text{rad}$$

Finally, one arrives at the pitch dynamics equation using:

$$\Delta\ddot{\alpha} - \frac{\bar{q}^* Sc}{I_{yy}} \left\{ \left(C_{mq1} + \underbrace{C_{m\dot{\alpha}}}_{0} \right) \left(\frac{c}{2V^*} \right) \Delta\dot{\alpha} + C_{m\alpha}\Delta\alpha \right\} = 0 \tag{2.35}$$
$$\Rightarrow \Delta\ddot{\alpha} + 0.442\Delta\dot{\alpha} + 1.046\Delta\alpha = 0$$

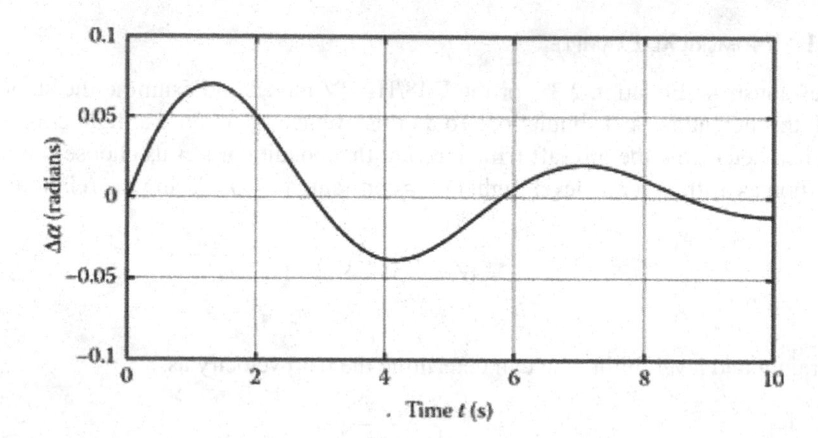

FIGURE 2.7 Short-period time response for Example 2.6.

From Equation 2.35:

$$2\zeta\omega_n = 0.442 \; rad/s$$
$$\omega_n^2 = 1.06 \; rad^2/s^2$$

which gives $\omega_n = 1.03 \; rad/s$, $\zeta = 0.214$.

The damped time period or the actual response time period T_d can be determined as:

$$T_d = \frac{2\pi}{\omega_d} = \frac{2\pi}{\omega_n\sqrt{1-\zeta^2}} = \frac{2\pi}{1.03 \times \sqrt{1-0.214^2}} = 6.247 \; s$$

The simulation result for Equation 2.35 is plotted in Figure 2.7.

From the simulation plot, note that the response is oscillatory and lightly damped. The time period of the response T_d is roughly 6.2 *seconds* (reading the time difference between two successive peaks).

Note: What if $C_{m\alpha}$ is positive? Then, comparing with the form of the second-order dynamics in Equation 2.3, we have $k < 0$, and so by Case 4 of Section 2.2, one may conclude that the equilibrium point in this case is unstable, irrespective of the value or sign of the term $(C_{mq1} + C_{m\dot{\alpha}})$. Actually, in this case, as seen in Case 4, Section 2.2, the dynamic response is no longer oscillatory and one needs to be careful in using the concept of a difference in timescales between the pitch dynamics and the dynamics in V, γ.

Example 2.7

In a similar manner as in Example 2.6, one can arrive at the following values of the response parameters for AFTI/F-16 on landing approach at $V^* = 139$ *knots*

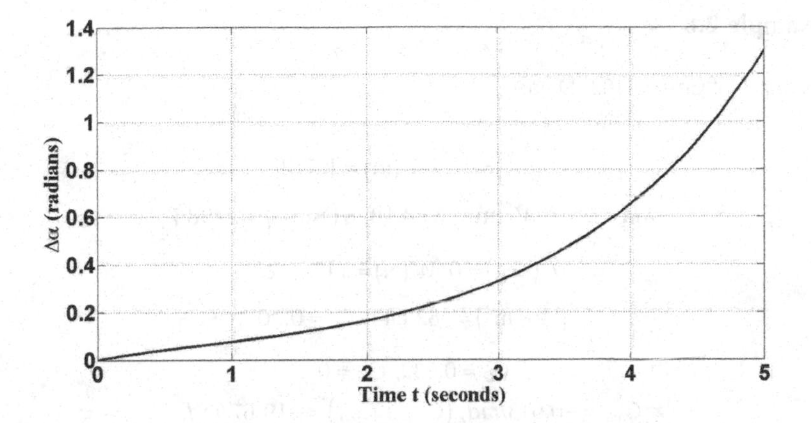

FIGURE 2.8 Exponential short-period time response for Example 2.7.

(1 *knot* = 0.515 *m/s*) (from *Control System Design* by Bernard Friedland, McGraw-Hill Publication, 1986, p. 128):

$$k = -\left(\frac{\bar{q}Sc}{I_{yy}}\right)C_{m\alpha} = -1.1621$$

$$d = -\left(\frac{\bar{q}Sc}{I_{yy}}\right)(c/2V^*)\left(C_{mq1} + C_{m\dot{\alpha}}\right) = 1.01$$

Note $k < 0$.

The exponents λ_1 and λ_2 in Equation 2.11 corresponding to the above parameter values are:

$$\lambda_{1,2} = \frac{-d \pm \sqrt{d^2 - 4k}}{2} = \frac{-1.01 \pm \sqrt{1.01^2 - 4\times(-1.1621)}}{2} = 0.685,\ -1.695$$

The positive exponent λ_1 implies instability. The response is indeed unstable and diverging exponentially as seen from the plot in Figure 2.8.

In summary, one can state that for an airplane to be stable in pitch, or for the short-period dynamics to be stable, one should ensure that:

$$C_{m\alpha} < 0 \quad \text{and} \quad \left(C_{mq1} + C_{m\dot{\alpha}}\right) < 0$$

Both the conditions must be satisfied simultaneously.

Homework Exercise: How stable is stable enough, and is there any such thing as too much stability?

Example 2.8

Example of Cessna 182. Data:

$$S\left(\text{m}^2\right)=16.7,\ c(\text{m})=1.518,$$
$$Ma=0.201,\ V^*(\text{m/s})=68.18,\ \bar{q}\left(\text{N/m}^2\right)=2298.7$$
$$\alpha^*(\text{deg})=0,\ W(\text{N})=11787.2$$
$$I_{yy}\left(\text{kg}-\text{m}^2\right)=1824.44,\ C_L^*=0.307,$$
$$C_D^*=0.032,\ C_m^*=0$$
$$C_{m\alpha}=-0.613/rad,\ \left(C_{mq1}+C_{m\dot{\alpha}}\right)=-19.67/rad.$$

The airplane is stable in pitch because the conditions $C_{m\alpha}<0, \left(C_{mq1}+C_{m\dot{\alpha}}\right)<0$ are satisfied. Let us calculate the important response parameters for this case:

$$\omega_n^2=-\left(\frac{\bar{q}Sc}{I_{yy}}\right)C_{m\alpha}$$
$$=-\frac{2298.7\times16.7\times1.518}{1824.44}\times-0.613$$
$$=19.579/s^2$$
$$\omega_n=4.425\ \text{rad/s}$$

$$2\zeta\omega_n=-\left(\frac{\bar{q}Sc}{I_{yy}}\right)(c/2V)\left(C_{mq1}+C_{m\dot{\alpha}}\right)$$
$$=-\frac{2298.7\times16.7\times1.518}{1824.44}\times\frac{1.518}{2\times68.18}\times-19.67$$
$$=6.994/s^2$$
$$\zeta=\frac{6.994}{2\omega_n}=\frac{6.994}{2\times4.425}=0.791$$

Compared to the previous case of F-18/HARV pitch dynamics, the dynamics of Cessna in pitch is highly damped with shorter damped time period:

$$T_d=\frac{2\pi}{\omega_d}=\frac{2\pi}{\omega_n\sqrt{1-\zeta^2}}$$
$$=\frac{2\times3.14}{4.425\times\sqrt{1-0.791^2}}=2.32\ \text{s}$$

A typical time history showing the variation in perturbed angle of attack is shown in Figure 2.9.

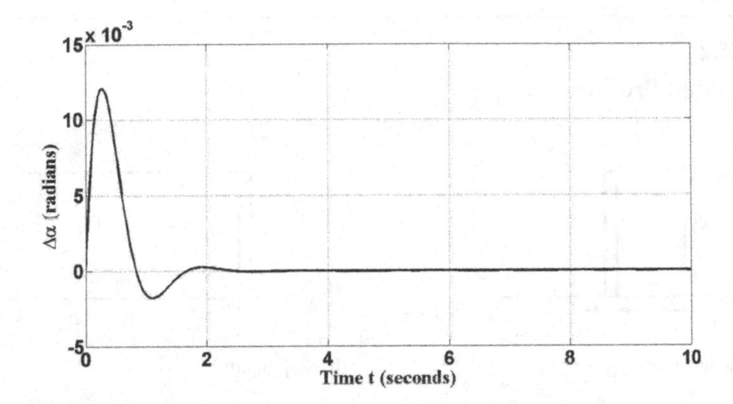

FIGURE 2.9 Short-period response of Cessna 182 (Example 2.8).

2.7 SHORT-PERIOD FREQUENCY AND DAMPING

From Equation 2.32, one can write the following relations for the short-period frequency and damping parameters:

$$\left(\omega_n^2\right)_{SP} = -\left(\frac{\bar{q}^* Sc}{I_{yy}}\right) C_{m\alpha} \tag{2.36}$$

$$\zeta_{SP} = -\left(\frac{\bar{q}^* Sc}{I_{yy}}\right) \frac{\left(c/2V^*\right)\left(C_{mq1} + C_{m\dot{\alpha}}\right)}{2\omega_n} \tag{2.37}$$

where, '*' is specifically written to indicate that those values are evaluated at the trim state. For instance, one can see from Equation 2.36 that the short-period frequency varies as the square root of the dynamic pressure $\bar{q}^*$. Thus, at higher altitudes where the air density is lesser, the short-period frequency may be expected to be lower. The same holds for the damping parameter ζ in Equation 2.37.

It is more interesting to judge the effect of 'size' of the airplane on the short-period frequency and damping, assuming $\bar{q}^*$ to be fixed. Note that the 'Sc' term in the numerator scales as L^3, and if the airplane mass is taken to scale with its volume, then the 'I_{yy}' term in the denominator scales as L^5. So, the short-period frequency ω_n scales as $1/L$, which implies that smaller airplanes tend to have higher natural response frequencies in pitch, and hence smaller timescales of response. That is, they respond much quicker, which may make it harder to control them when flown manually. This fact is of much importance for mini- and micro-aerial vehicles.

2.8 FORCED RESPONSE

A physical system is subjected to inputs in many ways. These inputs can be internal or external. For example, a pilot effecting a control deflection may be called an internal input, while wind can be referred to as an external input. System response to these inputs is known as forced response. Various kinds of inputs are shown in Table 2.2.

TABLE 2.2
Typical Input Profiles

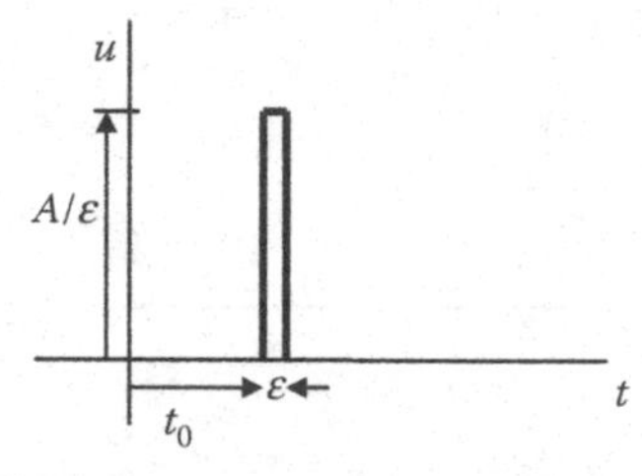

(a) Impulse input:

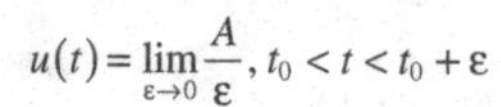

$$u(t)=\lim_{\varepsilon\to 0}\frac{A}{\varepsilon},\ t_0<t<t_0+\varepsilon$$

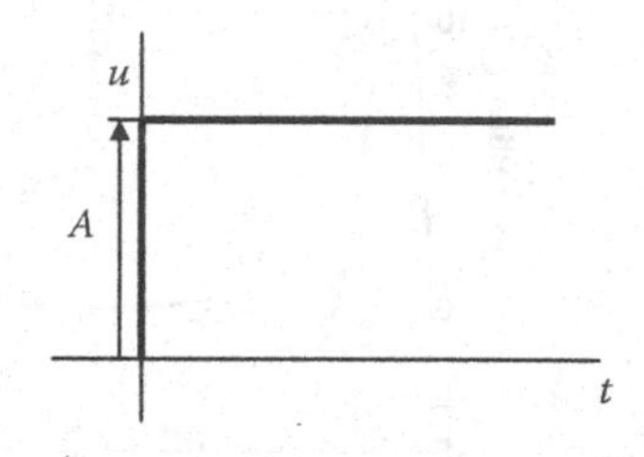

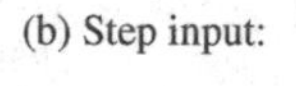

(b) Step input:

$$u(t)=A \quad t\geq 0$$

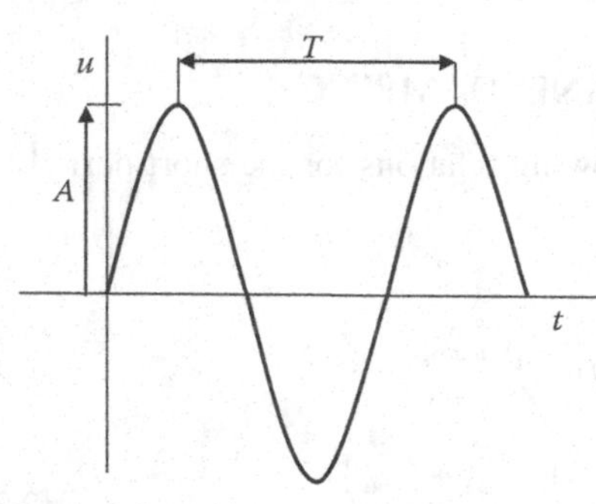

(c) Harmonic input:

$$u(t)=A\sin\omega t \quad t\geq 0;\quad \omega=\frac{2\pi}{T}$$

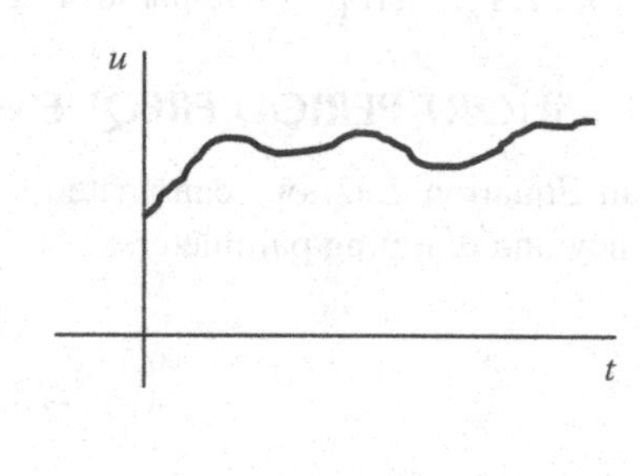

(d) Arbitrary input:

$$u(t)=f(t) \quad t\geq 0$$

In the following, we look at the forced response of first- and second-order linear systems to impulse, step and harmonic inputs; the response of a higher-order linear system to inputs will generally be a combination of first- and second-order responses.

2.8.1 First-Order System

A first-order linear system forced by an input $u(t)$ can be represented by the following equation:

$$\dot{x}+ax=u(t) \tag{2.38}$$

The forced response of a linear system consists of a homogeneous part of the solution, $x_H(t)$ (corresponding to $u(t)=0$), and a solution to the particular input called particular solution, $x_P(t)$ (corresponding to $u(t)\neq 0$). Thus,

$$x(t)=x_H(t)+x_P(t) \tag{2.39}$$

where $x_H(t) = x(0)e^{-at}$, as previously arrived at. This response is governed by the parameter a. When the system is stable, that is, $a > 0$, the homogeneous part of the solution decays and what remains is only the particular part of the solution eventually ($t \rightarrow \infty$; steady state). When the system is unstable, that is, $a < 0$, the homogeneous part of the solution $x_H(t)$ persists and the transient response itself diverges. If the initial condition $x(0) = 0$, then only the particular solution $x_P(t)$ remains. In the following, we assume $a > 0$. [**Homework Exercise**: Verify the above statements for the case when $a < 0$.]

a. **Impulse response**: A particular response of a first-order system to impulse input, $x_P(t)$, at time $t = 0$is given by $x(t) = e^{-(t/\tau)}$ for $t > 0$. Note that the particular solution does not depend upon the initial condition. The impulse response for time constant $\tau = (1/a) = \dfrac{1}{0.5} = 2\ s$ is shown in Figure 2.10. After the initial impulse, the system response dies down as per $x_H(t) = x(0)e^{-at}$.
b. **Unit step response**: In this case, $u(t) = 1$ for $t > 0$. The response of a first-order system to unit step response input, $x_P(t)$, is given by $x(t) = 1 - e^{-(t/\tau)}$ for $t > 0$. The unit step response for time constant $\tau = (1/a) = \dfrac{1}{0.5} = 2\ s$ in Figure 2.11 shows an exponential build-up to the steady-state value of $x = 1$.
c. **Harmonic response**: The particular response of a first-order system to sinusoidal input, $x_P(t)$, can be determined to be $x(t) = Csin(\omega t - \phi)$, where:

$$C = \left(A/\sqrt{a^2 + \omega^2}\right),\ \phi = \tan^{-1}(\omega/a)$$

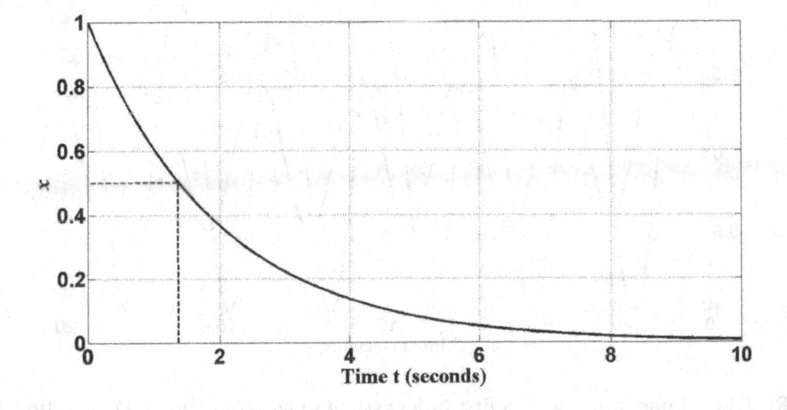

FIGURE 2.10 Impulse response of a first-order system.

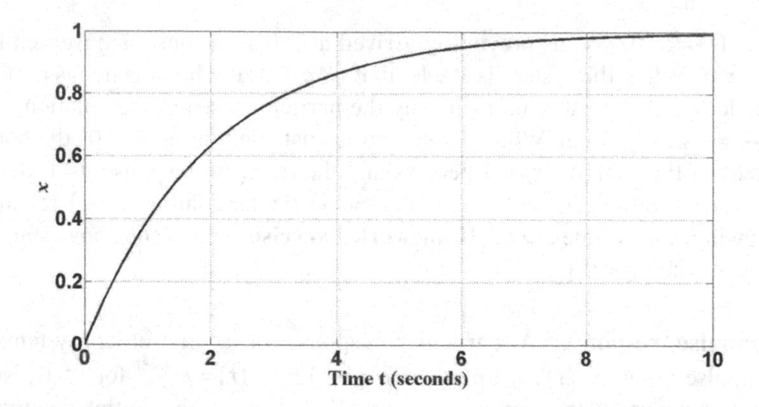

FIGURE 2.11 Step response of a first-order system.

The following conclusions can be drawn from the response of a first-order system to sinusoidal input:

1. The response has the same frequency as the input.
2. The gain in amplitude of the response, $C/A = \left(1/\sqrt{a^2+\omega^2}\right)$, is a function of the time constant ($\tau = 1/a$) of a first-order system and the frequency (ω) of the forcing function.
3. The response lags behind the input by the time difference $t = (\phi/\omega) = (1/\omega)tan^{-1}(\omega/a)$.

The harmonic response of a first-order system is shown in Figure 2.12. The harmonic input in the plot is $u(t) = 1\ sin2t$ for $t > 0$. One can notice from the figure that the response, after some initial transients due to $x_H(t)$ that takes it to a higher amplitude, settles down to an amplitude $\approx$ 0.485 and the response lags behind the input

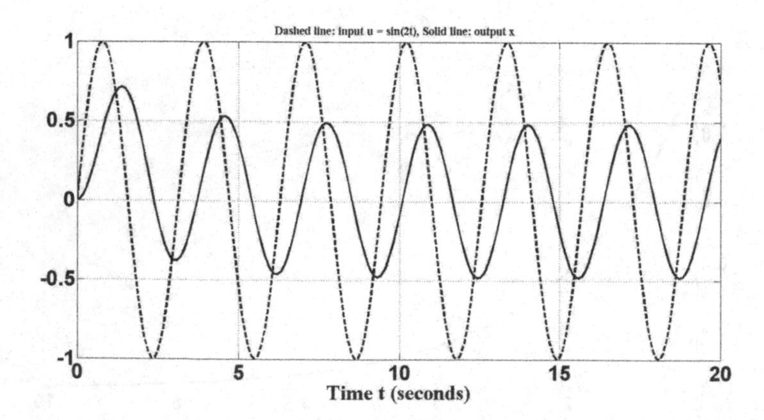

FIGURE 2.12 Time response of a first-order system to harmonic input. Dashed line: Input $u(t) = sin2t$, solid line: Output x.

by the time difference ≈ 0.66 seconds. Using the formulas above, the amplitude and time lag may be estimated as:

$$\frac{C}{A} = \frac{1}{\sqrt{0.5^2 + 2^2}} = 0.485 \quad \text{and} \quad t = \frac{1}{\omega}\tan^{-1}\left(\frac{\omega}{a}\right) = \frac{1}{2}\tan^{-1}\left(\frac{2}{0.5}\right) = 0.66 \text{ s}$$

2.8.2 Second-Order System

A forced second-order linear system can be represented by the following equation:

$$\ddot{x} + 2\zeta\omega_n\dot{x} + \omega_n^2 x = u(t) \tag{2.40}$$

where $u(t)$ is the forcing function. Similar to the case of first-order systems, forced response of a second-order system consists of a homogeneous part, $x_H(t)$ (solution of Equation 2.40 when $u(t) = 0$), and a particular solution, $x_P(t)$ (solution of Equation 2.40 due to a particular $u(t) \neq 0$). Therefore, $x(t) = x_H(t) + x_P(t)$, where $x_H(t) = Ae^{\lambda_1 t} + Be^{\lambda_2 t}$, A and B being constants which depend upon the initial conditions, as previously shown. Forced response of a second-order system thus depends upon whether the system is stable or not, very much like first-order systems. If the system is stable, the forced response is dominated by the forcing input and the particular solution $x_P(t)$; whereas, if the system is unstable, the homogeneous solution, $x_H(t)$ persists and grows without bound. In the following, we assume that the second-order system is stable and show the response $x_P(t)$ to various inputs.

For the following numerical simulation results, let us take the example of the second-order system; $\ddot{x} + 1.2\dot{x} + 9x = u(t)$. The natural frequency for this system is $\omega_n = \sqrt{9} = 3$ rad/s, and the damping ratio $\zeta = \frac{1.2}{2\omega_n} = \frac{1.2}{6} = 0.2$.

a. **Unit impulse response:** The response $x_P(t)$ of a second-order system to a unit impulse at $t = 0$ can be determined analytically to be $x(t) = \left(e^{-\zeta\omega_n t}/\omega_d\right)\sin(\omega_d t)$; $\omega_d = \omega_n\sqrt{1-\zeta^2}$. This response is shown in Figure 2.13. After the initial impulse, $x(t)$ dies down as per $x_H(t)$.

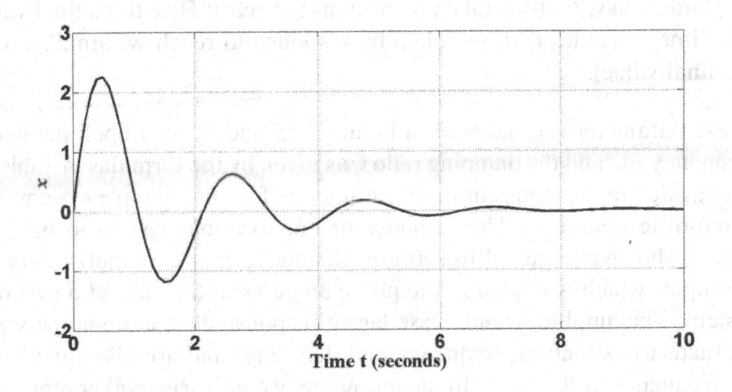

FIGURE 2.13 Impulse response of a second-order system.

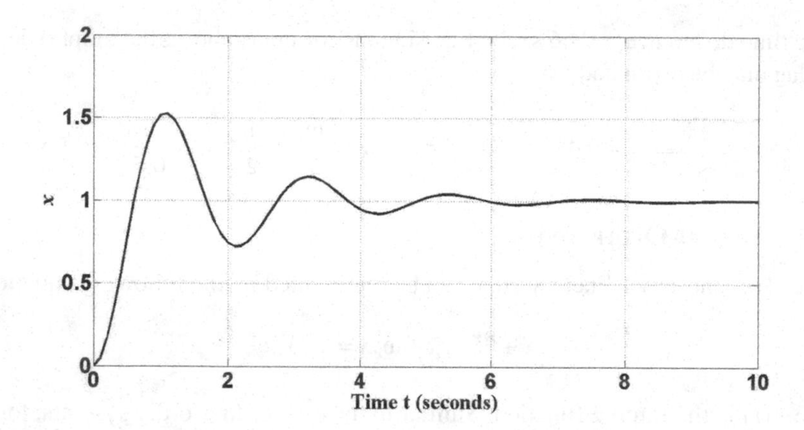

FIGURE 2.14 Response of a second-order system to unit step input.

b. **Unit step response:** The response of a second-order system to unit step input is:

$$x(t) = 1 - \frac{e^{-\zeta\omega_n t}}{\sqrt{1-\zeta^2}} \sin\left(\omega_d t + \tan^{-1}\frac{\sqrt{1-\zeta^2}}{\zeta}\right) \quad \text{for } t \geq 0$$

For our example system, the simulation response is shown in Figure 2.14. The response settles down at the steady-state value of $x(t) = 1$.

Characteristic parameters associated with unit step response of a second-order system are:

- Maximum overshoot, M_p (maximum peak value measured from unity, in %).
- Time to peak, t_p (time taken by response to reach the first peak of overshoot).
- Rise time, t_r (time taken by response to reach from 0% to 100% (for under-damped systems) and from 10% to 90% (for over-damped systems) of the final value).
- Time delay, t_d (time taken by response to reach 50% of its final value).
- Time to settle, t_s (time taken by response to reach within ±2% of its final value).

These parameters are marked in Figure 2.15 and depend upon the natural frequency ω_n and the damping ratio ζ as given by the formulas in Table 2.3. Alongside are the values of these parameters for our example system.

c. **Harmonic response:** The response of our example system to harmonic input is shown in Figure 2.16. Observe that the response (output) lags behind the input, which is related to the phase angle as in the case of a first-order system. The amplitude and phase lag of response depend upon the system characteristics (natural frequency and damping) and are also functions of the frequency of the input. In the following, we will see another representation of harmonic response in a more compact manner.

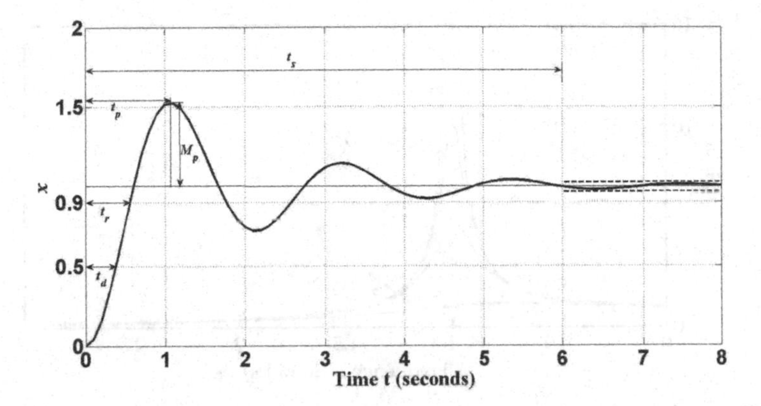

FIGURE 2.15 A typical second-order system response to unit step input with characteristic parameters.

TABLE 2.3
Second-Order Response Characteristic Parameters

M_p (%)	t_p (s)	t_r (s)	t_s (s)
$e^{-\left(\frac{\zeta}{\sqrt{1-\zeta^2}}\pi\right)} \times 100$	$\frac{\pi}{\omega_n\sqrt{1-\zeta^2}}$	$\frac{\pi - \tan^{-1}\frac{\sqrt{1-\zeta^2}}{\zeta}}{\omega_n\sqrt{1-\zeta^2}}$	$\frac{4}{\zeta\omega_n}$ (±2% criterion)
52.66	1.068	0.602	6.667

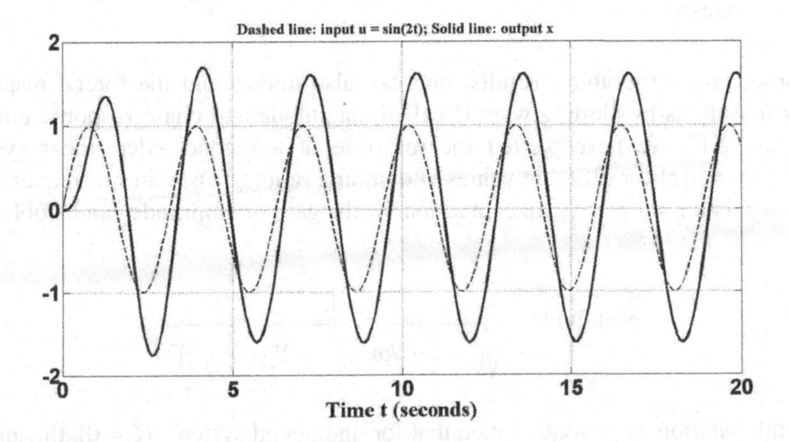

FIGURE 2.16 Time response of a second-order system to harmonic input. Dashed line: Input $u = \sin(2t)$, solid line: Output x.

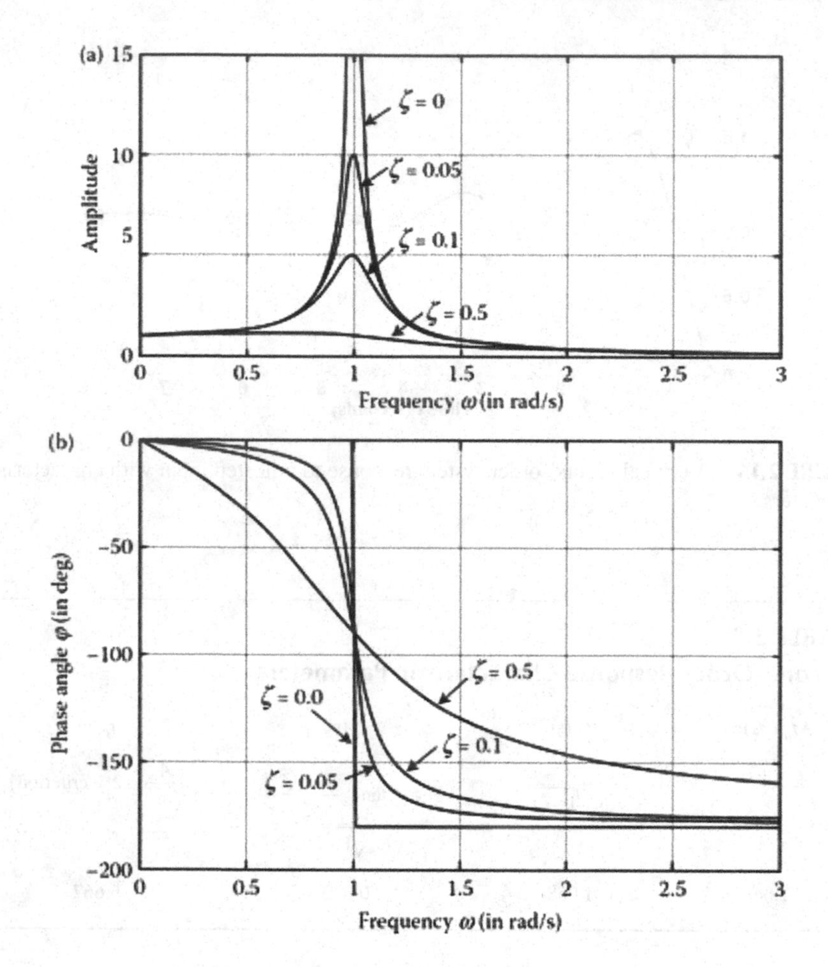

FIGURE 2.17 (a) Amplitude and (b) phase angle versus frequency plots showing a second-order system response.

Consolidating the above results, one can also understand the forced response of linear systems by plotting what is called magnitude and phase response curves. In Figure 2.17, we have plotted the response of a second-order linear system $\ddot{x} + 2\zeta\dot{x} + x = u(t)$ for different values of damping ratio, ζ. The natural frequency of the system is $\omega_n = 1\ rad/s$. An expression for the gain or amplitude can be obtained to be:

$$\text{Amplitude} = \frac{1}{\sqrt{\left(1-\left(\omega^2/\omega_n^2\right)\right)^2 + \left(2\zeta\left(\omega/\omega_n\right)\right)^2}}$$

From this relation, one would notice that for undamped systems ($\zeta = 0$), the amplitude is infinity at $\omega = \omega_n$, what is known as frequency of resonance or resonant frequency. For damped second-order system, the frequency at which the amplitude peaks, which may be called the 'shifted resonant frequency', can be obtained as $\omega_r = \omega_n\sqrt{1-2\zeta^2}$, which tells us that neither a 'shifted resonant frequency' nor a peak in response amplitude exists for $\zeta \geq \left(1/\sqrt{2}\right)$. Also, the 'shifted resonant frequency' and hence the peak in response amplitude shifts to the left in Figure 2.17a with increasing damping ratio in the range $0 < \zeta < \left(1/\sqrt{2}\right)$.

The expression for phase angle is given by:

$$\varphi = -\tan^{-1}\left(\frac{2\zeta\frac{\omega}{\omega_n}}{1-\left(\frac{\omega}{\omega_n}\right)^2}\right)$$

Therefore, at $\omega = \omega_n$, phase $\varphi = -90°$, for all $\zeta \neq 0$.

The following observations can be made from Figures 2.17a and 2.17b:

- For $\zeta = 0$, the amplitude of the response increases with increasing frequency of the harmonic input until the input frequency matches the natural frequency of the system, $\omega = \omega_n$ or $\frac{\omega}{\omega_n} = 1$. At this frequency, 'resonance' occurs, which is marked by the response reaching its peak value. For a second-order system with no damping ($\zeta = 0$), the peak value of the response is infinite, which is only a mathematical result. In reality, a system may be expected to have some amount of damping and the peak response at the resonance frequency will be finite as shown in Figure 2.17a for non-zero values of the damping ratio. As the damping ratio increases, the amplitude of the response peak decreases and shifts to the left of the resonant frequency. For significantly large values of the damping ratio, the amplification (or the gain) of the response is negligible. In fact, when $\zeta = 0.707$, the gain (ratio of the response amplitude to the input amplitude) is negligible and for a damping ratio higher than this value, there is no peak observed in the response. When the frequency of the forcing input function is increased further, the response gain actually decreases.
- The phase angle versus frequency plot in Figure 2.17b indicates the time lag between the input and the response. The time lag increases with increasing frequency of the forcing input. At frequencies below resonant frequency $\omega = \omega_n$, increasing the damping ratio has increased time delay effect in the response, which tells us that in a more viscous medium, the response of the system is going to be sluggish. [**Homework Exercise**: Think of a spring-mass-damper system under harmonic excitation and try to justify what is concluded above.]

2.9 RESPONSE TO PITCH CONTROL

The most common way for the pilot of an airplane to induce a pitching motion is to deflect the elevators. The elevators are flaps placed at the trailing edge of the horizontal tail, as pictured in Figure 2.18. The action of the elevators in producing a

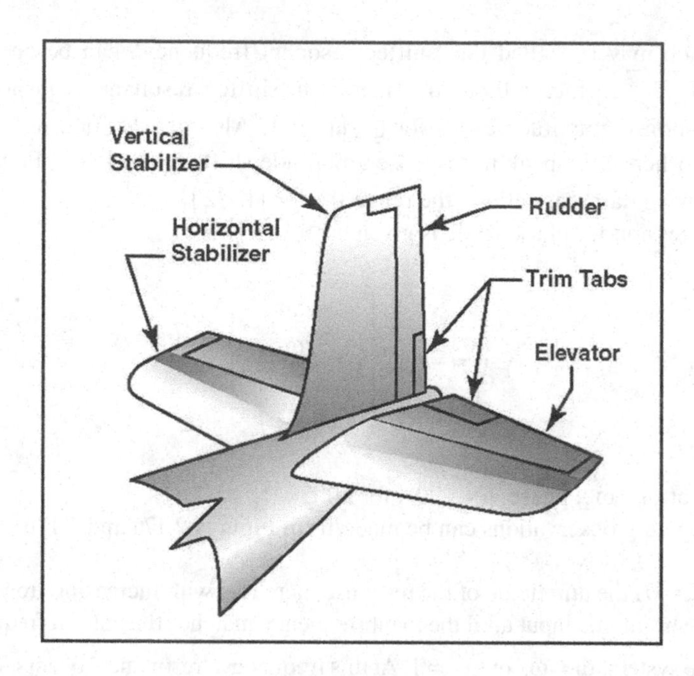

FIGURE 2.18 Elevator at the trailing edge of the horizontal tail. (www.americanflyers.net; http://www.americanflyers.net/aviationlibrary/pilots_handbook/images/chapter_1_img_32.jpg).

pitching moment is sketched in Figure 2.19, where the positive elevator is marked by a downward deflection, indicated by the symbol, δe.

- For reasonably small downward deflections (positive δe), the elevators induce a net additional positive lift (directed upwards) on the horizontal tail. This additional lift usually acts way behind the airplane CG; hence, it creates a negative (nose-down) pitching moment about the CG. Thus, a downward elevator deflection tends to decrease the angle of attack of the airplane.
- Likewise, a reasonably small upward deflection of the elevators (negative δe) creates an additional down-lift on the horizontal tail, which produces a nose-up (positive) pitching moment at the airplane CG. In this manner, an upward elevator deflection tends to increase the angle of attack of the airplane.
- Either case, upward or downward elevator deflection, usually contributes to an additional source of drag on the airplane. Fortunately, as long as the

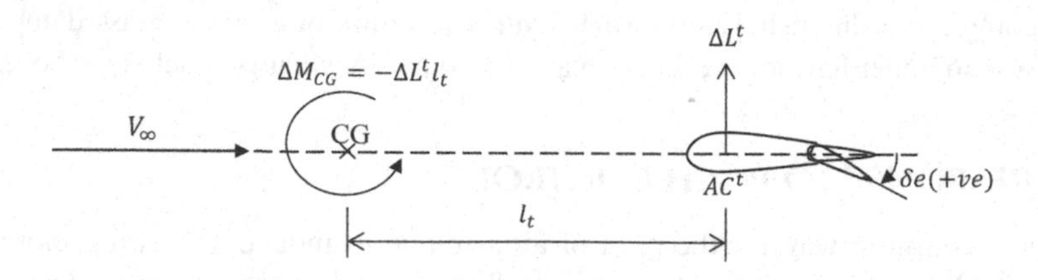

FIGURE 2.19 Action of elevator producing pitching moment.

moment arm between the horizontal tail aerodynamic centre (AC^t) and the airplane CG is large enough, a small change in horizontal tail lift (and hence a small elevator deflection) is adequate to produce a reasonably large change in the pitching moment. Thus, the additional drag due to elevator deflection will also be usually small, but note that this may not hold true in supersonic flight for some airplane configurations.

2.9.1 Pitch Dynamics about Level Flight Trim with Elevator Control

Consider a level flight trim as before with V^* and $\gamma^* = 0$, both held constant. The trim angle of attack is α^* and the elevator is set at a deflection δe^* at trim.

We can write the incremental pitching moment ΔC_m due to a small change in elevator deflection from trim, denoted by $\Delta\delta e$, as:

$$\Delta C_m = C_{m\delta e}\Delta\delta e$$

where,

$$C_{m\delta e} = \left.\frac{\partial C_m}{\partial \delta e}\right|_* \tag{2.41}$$

This is called the *pitch control derivative* or the *elevator effectiveness*.

With the elevator deflection as an additional parameter, the incremental pitching moment model in Equation 2.30 is modified and appears as below:

$$\Delta C_m = C_{m\alpha}\Delta\alpha + \left(C_{mq1} + C_{m\dot{\alpha}}\right)\Delta\dot{\alpha}\left(c/2V^*\right) + C_{m\delta e}\Delta\delta e \tag{2.42}$$

The pitch dynamics Equation 2.32 for the short-period dynamics mode now has an extra term due to the elevator deflection, and it looks like this:

$$\Delta\ddot{\alpha} - \left(\frac{\overline{q}^* Sc}{I_{yy}}\right)\left(c/2V^*\right)\left(C_{mq1} + C_{m\dot{\alpha}}\right)\Delta\dot{\alpha} - \left(\frac{\overline{q}^* Sc}{I_{yy}}\right)C_{m\alpha}\Delta\alpha = \left(\frac{\overline{q}^* Sc}{I_{yy}}\right)C_{m\delta e}\Delta\delta e \tag{2.43}$$

Note that the previous expressions for stiffness and damping and the condition for stability—called the *natural* response of the system—remain unchanged, even with the additional term on the right-hand side. However, when the pilot applies an elevator input (over and above the trim value δe^*), the airplane responds in pitch to the 'forcing' as per Equation 2.43.

Example 2.9

Calculations and numerical simulation showing short-period response to harmonic forcing.

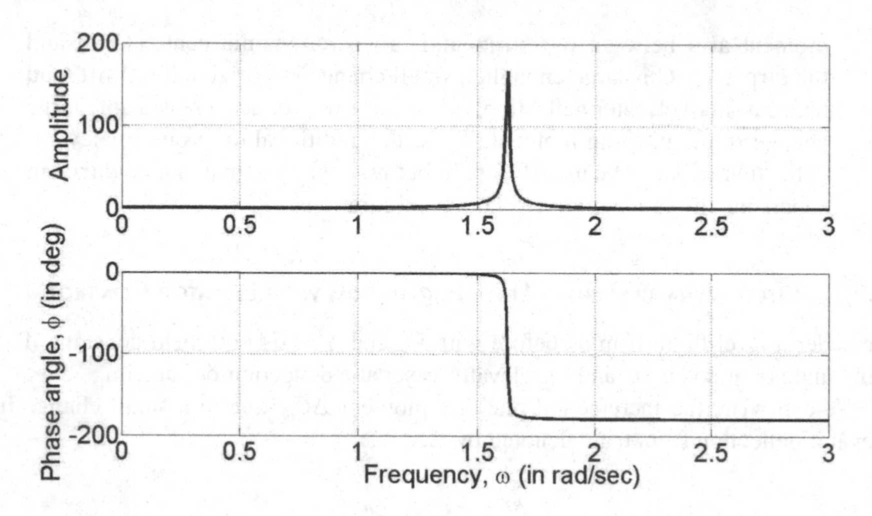

FIGURE 2.20 Amplitude and phase plots versus frequency showing pitch response to elevator deflection.

In this example, the equation of motion, similar to Equation 2.35 at a trim state of an aircraft is now augmented by the elevator forcing term on the right-hand side and appears as:

$$\Delta\ddot{\alpha} + 0.0077\Delta\dot{\alpha} + 2.6473\Delta\alpha = -9.3489\Delta\delta e \tag{2.44}$$

Consider a harmonic elevator input, $\Delta\delta e = A\sin\omega t$, where $A = -2.6473/9.3489$. Thus, Equation 2.44 can be rewritten as:

$$\Delta\ddot{\alpha} + 0.0077\Delta\dot{\alpha} + 2.6473\Delta\alpha = 2.6473\ \sin\omega t \tag{2.45}$$

Conventionally, a positive (down) elevator deflection results in a decrease in the angle of attack indicating a phase lag of 180°. From Equation 2.45, we can determine the natural frequency, $\omega_n = \sqrt{2.6473} = 1.6271$ rad/s, the damping ratio, ζ = 0.0024 and the shifted resonant frequency, $\omega_r = \omega_n\sqrt{1-2\zeta^2} = 1.6271\sqrt{1-2\times 0.0024^2} = 1.627$ rad/s. For Equation 2.45, the amplitude and phase plots are plotted in Figure 2.20.

Notice the magnification of the angle of attack response at the shifted resonant frequency, $\omega_r = 1.627$ rad/s and phase changing from in-phase (0°) to out-of-phase (180°). The finite value of amplitude is due to the presence of finite damping in Equation 2.45.

EXERCISE PROBLEMS

2.1 Consider a non-linear second-order system described by the equation:

$$\ddot{x} - \left(1 - x^2\right)\dot{x} + x^3 - x = 0$$

Determine fixed points (equilibrium states) of this system and comment on their stability. Also, compute the trajectories and confirm your stability results.

2.2 Consider a simple pendulum model shown in the figure below. The equations describing the undamped motion of the pendulum can be arrived at as $\ddot{\theta}+(g/l)\sin\theta=0$.

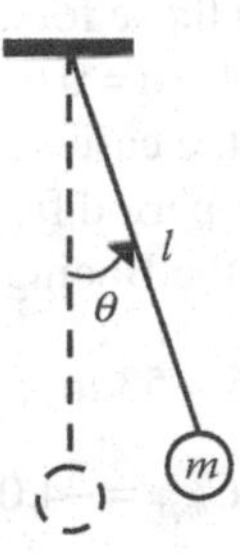

Determine the steady-state (equilibrium) positions of the pendulum. Discuss the stability of each equilibrium condition.

2.3 For a fighter aircraft, F104-A, the following data are given:

$$S=18.5\ \text{m}^2,\quad \bar{c}=2.9\ \text{m},\quad I_{yy}=79444.6\ \text{kg}-\text{m}^2,\quad C_{m0}=0.01$$
$$C_{m\delta e}=-1.46/\text{rad},\quad C_{m\alpha}=-0.64/\text{rad},\quad C_{mq1}=-5.8/\text{rad},\quad C_{m\dot{\alpha}}=0$$

Determine a trim condition (α^*) at $\delta e=-2°$ for this aircraft. The aircraft trim speed can be assumed to be fixed at $V^*=200$ m/s. Write down the equation for small perturbation pitch motion for this aircraft about the trim condition (V^*, α^*) at sea-level condition. Determine the characteristic exponents and study the stability of the trim condition. Simulate the natural/free response of the airplane in pitch motion.

2.4 Carry out forced response analysis of the dynamics of F104-A in pitch using the data given in Exercise Problem 2.3. Determine the peak overshoot, rise time, time to peak and time to settle for unit step response of the equation (developed in Problem 2.3) using the data given, analytically as in Table 2.3 and from the simulated response.

2.5 From the expressions in Equations 2.36 and 2.37 for pure pitching motion of aircraft, figure out the variation of short-period frequency and damping with Mach number.

2.6 Using the data given in Example 2.8 for Cessna 182, determine the short-period frequency and damping ratio at altitudes, $h=1500$, 3000 and 5000 m. Compare the values. Further, determine the variation of short-period frequency and damping with Mach number.

2.7 How stable is stable enough, and is there any such thing as too much stability? [**Hint**: Think in terms of the effect of stability on the controllability or the ability to manoeuvre the airplane for a highly stable airplane (Example 2.8) and a less stable airplane (Example 2.6).] Try different forms of elevator input and study the effect on the pitch dynamics of the airplanes in Example 2.6 and 2.8.

2.8 An airplane flying at a speed of 100 m/s sees the relative wind coming at angle, $\alpha = 30°$, $\beta = 10°$. Determine the components of the aircraft velocity along its body-fixed axes.

2.9 For an airplane with the following data, determine the coefficients $C_{m\dot{\alpha}}$ and $C_{L\dot{\alpha}}$. Data: c(m) = 1.6; S(m²) = 18.6; l_t(m) = 3.05; S^t(m²) = 4.65; $C_{L\alpha_t} = 2\pi$/rad. Compare these coefficients with those for a mini-aerial vehicle with the following data: c(m) = 0.25; S(m²) = 1; l_t(m) = 0.6; S_t(m²) = 0.3; $C_{L\alpha_t} = 2\pi$/rad. Assume constant $(d\varepsilon/d\alpha) = 0.15$ in both the cases.

2.10 Calculate and compare the short-period frequency and damping parameters, ω_n and ζ, for the F-4 Phantom and the Boeing 747–200, given the following data:

$$V^* = 300 \text{ m/s}, \quad S = 53 \text{ m}^2, \quad c = 5.3 \text{ m}, \quad I_{yy} = 17450 \text{ kg} - \text{m}^2$$

F4:
$$C_{m\alpha} = -0.4/\text{rad}, \quad C_{mq1} = -4.0/\text{rad}, \quad C_{m\dot{\alpha}} = 0$$

$$V^* = 275 \text{ m/s}, \quad S = 550 \text{ m}^2, \quad c = 9.1 \text{ m},$$

747 – 200:
$$I_{yy} = 4{,}950{,}000 \text{ kg} - \text{m}^2, \quad C_{m\alpha} = -1.6/\text{rad},$$
$$C_{mq1} = -34.5/\text{rad}, \quad C_{m\dot{\alpha}} = 0$$

2.11 It is said that all things being equal, the short-period damping ζ of an airplane at an altitude of 11 km (where the density of air is approximately a third of the sea-level density) will reduce to about 54% of its value at sea level. Can you verify this statement?

3 Longitudinal Trim and Stability

3.1 WING–BODY TRIM AND STABILITY

Let us recap our analysis of the wing and body alone (no horizontal tail or canard) from Chapter 1, and examine wing–body trim and stability in light of what we have learnt in Chapter 2 on stability.

Figure 3.1 gives a sketch of the wing–body with all the relevant forces and moments marked in. As before, the thrust may be assumed to act at the CG along the direction of the velocity vector and the moment due to the drag force may be neglected (see Chapter 1, Section 1.11). Then the balance of pitching moments about the CG gives:

$$M = M_{AC}^{wb} + L^{wb}\left(X_{CG} - X_{AC}^{wb}\right) \tag{3.1}$$

Distances X are measured from a reference point in the sense marked by the arrows in Figure 3.1. Dividing Equation 3.1 by $\overline{q}Sc$ gives the non-dimensional version:

$$C_m = C_{mAC}^{wb} + C_L^{wb}\left(h_{CG} - h_{AC}^{wb}\right) \tag{3.2}$$

where h is the distance X non-dimensionalized by the wing mean aerodynamic chord c. Trim requires the net pitching moment about the CG to be zero, so that gives:

$$\underbrace{C_{mAC}^{wb}}_{(-)} + \underbrace{C_L^{wb}}_{(+)}\underbrace{\left(h_{CG} - h_{AC}^{wb}\right)}_{(+)} = 0 \tag{3.3}$$

where the usual signs of the various terms are marked below each of them. $\left(h_{CG} - h_{AC}^{wb}\right) > 0$ implies that the CG must lie aft of (behind) the wing–body aerodynamic centre AC^{wb} for trim to be possible. This is a recap of Section 1.11.

Next, let us examine the question of stability. To find the derivative $C_{m\alpha}$, which was defined in Chapter 2, Section 2.6, as $\partial C_m / \partial \alpha|_*$, where * indicates that the derivative is to be evaluated at the trim point, we differentiate each term of Equation 3.2 with respect to α:

$$C_{m\alpha} = \underbrace{C_{L\alpha}^{wb}}_{(+)}\underbrace{\left(h_{CG} - h_{AC}^{wb}\right)}_{(+)} \tag{3.4}$$

Remember that C_{mAC}^{wb} is not a function of α. The signs below the terms on the right-hand side of Equation 3.4 come from Equation 3.3. Together, they imply that $C_{m\alpha} > 0$ for a wing–body in trim. As we saw in Chapter 2, Section 2.6, when $C_{m\alpha} > 0$, the pitch dynamics (short-period mode) is unstable, irrespective of the sign of the damping

DOI: 10.1201/9781003096801-3

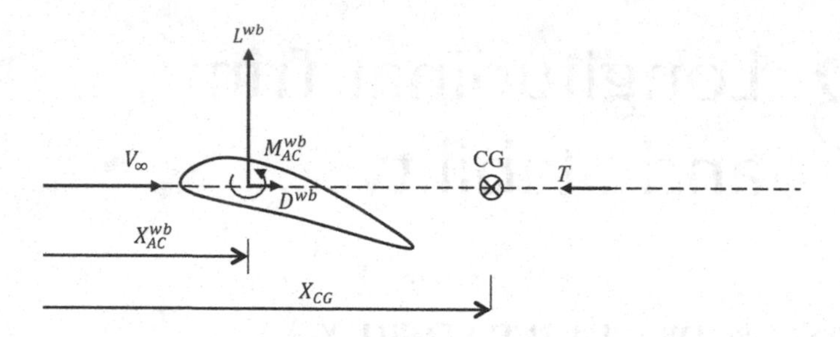

FIGURE 3.1 Sketch of wing–body showing forces, moments and distances.

term $(C_{mq1} + C_{m\dot{\alpha}})$. Thus, a wing–body in trim cannot be stable in pitch—a small perturbation from trim will build up and take the wing–body away from its trim state. Or, in other words, its short-period dynamics is divergent, unstable.

Example 3.1

Let us consider the wing and body alone of a general aviation airplane. The wing–body of the airplane has the following properties:

$$C_{mAC}^{wb} = -0.04, \quad h_{AC}^{wb} = 0.25, \quad C_{L\alpha}^{wb} = 4.5/\text{rad}$$

The CG of the airplane is located at a distance $h_{CG} = 0.4$ (all distances are from the leading edge of the wing).

C_{Ltrim} for the airplane can be found using Equation 3.3, which gives:

$$C_{Ltrim} = \frac{-C_{mAC}^{wb}}{h_{CG} - h_{AC}^{wb}} = -\frac{-0.04}{0.4 - 0.25} = 0.267$$

The pitch stability derivative $C_{m\alpha}$ can be calculated using Equation 3.4, which gives:

$$C_{m\alpha} = C_{L\alpha}^{wb}\left(h_{CG} - h_{AC}^{wb}\right) = 4.5 \times (0.4 - 0.25) = 0.675/\text{rad} > 0$$

Thus, it is possible to trim this airplane at a positive angle of attack corresponding to C_{Ltrim}, but this trim is unstable in pitch.

Example 3.2 Tailless Airplanes

There are several examples of tailless airplanes, such as the one shown in Figure 3.2. How are they flown?

Well, one way would be to trim the airplane by locating the CG behind the aerodynamic centre (AC^{wb}), as we just discussed. This would of course make the airplane unstable in pitch. Then, an automatic control system can be used, which could stabilize the airplane at every instant. Alternatively, if C_{mAC}^{wb} could be made positive, then the airplane would trim with the CG ahead of the AC^{wb}. In that case, you can verify (*Homework Exercise!*) that the airplane would indeed be stable in pitch, provided $\left(C_{mq1} + C_{m\dot{\alpha}}\right) < 0$.

FIGURE 3.2 The Indian Light Combat Aircraft (LCA)—a tailless wing–body configuration (http://ajaishukla.blogspot.com/2016/01/tejas-fighter-to-make-international.html). (Photo courtesy: Ajai Shukla.)

But, how could you arrange for C_{mAC}^{wb} to be positive? One way would be to use what is called an airfoil with *reflex camber* at the trailing edge, as shown in Figure 3.3a. The trailing edge section contributes sufficient positive C_{mAC}^{wb} to drive the sum to a positive quantity. A better way, perhaps, would be to use wing sweepback combined with twist variation along the span such that the outer wing segments contribute a positive C_{mAC}^{wb}. The latter arrangement is shown in Figure 3.3b. The inward positively cambered wing segment results in positive lift and negative C_{mAC}^{wb}, while the outward wing segment results in down-lift and positive C_{mAC}^{wb}, which can together be made to result in a positive C_{mAC}^{wb}. The lift from the outer segment would subtract from the total wing lift, but if you remember the typical spanwise lift distribution over a wing from a course on *aerodynamics*, then you will know that the lift distribution is dominant over the inboard wing segment and falls off over the outer wing segment, so it is not so bad. There is a lift loss, but it is not terrible.

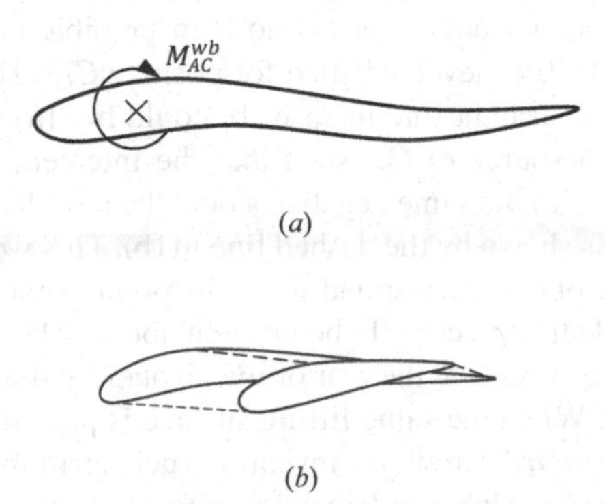

FIGURE 3.3 (a) An airfoil with reflex camber at the trailing edge and (b) a wing sweep and twist arrangement.

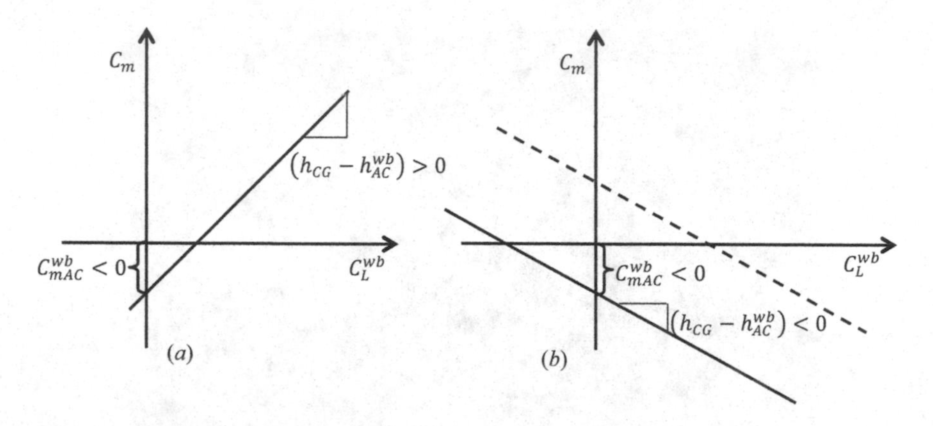

FIGURE 3.4 Graphs of C_m plotted against C_L^{wb} for two cases: (a) CG behind AC^{wb} and (b) CG ahead of AC^{wb}.

What would happen if the CG of a tailless airplane with $C_{mAC}^{wb} < 0$ were placed ahead of its AC^{wb}? Could one say that the airplane then has stability, but has no trim? Strictly speaking, as we saw in Chapter 2, the concept of stability has meaning only when it is associated with a trim (or equilibrium) state. Stability is a property of the trim state, not of the airplane. However, for the sake of making a point, let us go ahead with the discussion below.

Figure 3.4 shows two graphs—in each case, C_m is plotted against C_L^{wb}. The one in (a) is for the case of *CG* behind the AC^{wb}, and the one in (b) for the case of CG ahead of AC^{wb}.

As can be seen in (a), the intercept on the C_m axis, which is C_{mAC}^{wb}, happens to be negative and since $C_{L\alpha}^{wb}\left(h_{CG} - h_{AC}^{wb}\right)$ is positive in this case (see Equation 3.4), the straight line slopes upward with slope $\left(h_{CG} - h_{AC}^{wb}\right) > 0$. The graph intersects the C_L^{wb} axis where the condition $C_{mAC}^{wb} + C_L^{wb}\left(h_{CG} - h_{AC}^{wb}\right) = 0$ is satisfied. The intercept is at positive C_L^{wb}, so trim is possible; but, due to $(h_{CG} - h_{AC}^{wb})$ being positive, the trim is not stable.

For case (b), the intercept on the C_m axis is still the same because C_{mAC}^{wb} continues to be negative, but with the CG ahead of the AC^{wb} now, the slope $\left(h_{CG} - h_{AC}^{wb}\right) < 0$. Hence, the straight line in (b) slopes downward and never intersects the C_L^{wb} axis (for positive C_L^{wb}). In other words, there is no trim possible because the condition $C_{mAC}^{wb} + C_L^{wb}\left(h_{CG} - h_{AC}^{wb}\right) = 0$ is never satisfied for positive C_L^{wb}. However, if there was some way by which the straight line in case (b) could be 'lifted up', as it were, by providing an additional source of C_m, such that the intercept on the C_m axis now became positive, then, with the same negative slope, the modified straight line would intersect the C_L^{wb} axis as shown by the dashed line in (b). This would indicate a stable trim at a positive value of C_L^{wb} corresponding to the point of intercept.

How could the stable trim in case (b) be brought about? The answer is: by providing an additional lifting surface at the rear of the airplane, called a *horizontal tail* or a *horizontal stabilizer.* When the same lifting surface is placed near the nose of the airplane, it is called a *canard*. Credit for inventing such a trimming device is usually given to the French aviator Alphonse Pénaud (see Box 3.1).

Note: The usage *horizontal stabilizer* may be considered unfortunate because as we observe from Figure 3.4b, a horizontal tail (or canard) is not really required for

BOX 3.1 ALPHONSE PÉNAUD (1850–1880)

A nineteenth-century French aviation pioneer, also known as the father of rubber-powered airplane models, first discovered the use of tail on an airplane. One of his famous model airplanes *Planophore* is shown below.

(Source: http://alain.vassel.pages-perso-orange.fr/siecle19.html)

(Source: http://commons.wikimedia.org/File:AlphonsePenaudPlenophore.jpg)

stability, but for trim. If a trim is possible, stability can be arranged by placing the CG ahead of a particular point (such as an aerodynamic centre). Perhaps, it would have been more appropriate to call it the *horizontal trimmer.*

3.2 WING–BODY PLUS TAIL: PHYSICAL ARGUMENTS

We figure out how the trim and stability riddle can be solved by introducing a horizontal tail—first, without math, just by physical arguments.

Consider the schematic diagram in Figure 3.5, which shows a wing–body plus a horizontal tail. The wing–body lift acts at the aerodynamic centre (AC^{wb}) as before, and there is a negative (nose-down) pitching moment also at AC^{wb}, as expected for a typical wing–body combination. The CG is located at a distance ahead of the AC^{wb}. Clearly, the net moment at the CG due to the wing–body alone is nose-down (negative) and no trim is possible.

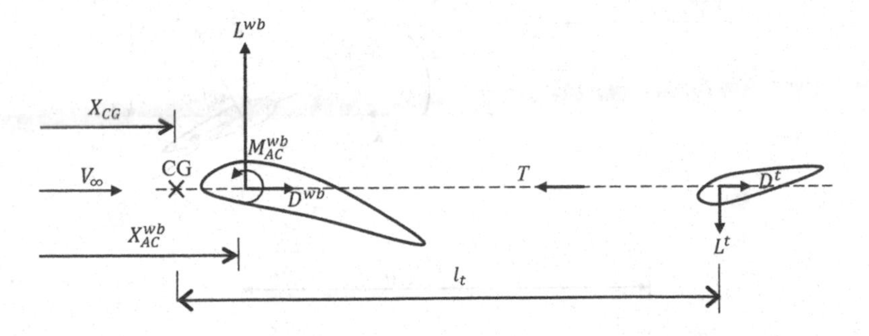

FIGURE 3.5 Schematic representation of the wing–body plus tail trim and stability.

To counter this nose-down moment, an equal and opposite nose-up moment must be generated at the CG by the horizontal tail. This is possible if the tail creates a down-lift L^t at its aerodynamic centre such that the tail lift times the moment arm between the tail aerodynamic centre AC^t and the overall CG is exactly equal to the desired nose-up moment at the *CG*.

In the following section, we establish the same concept by mathematically modelling the wing–body plus tail system.

Note that the horizontal tail is normally required to provide a down-lift to make the wing–body plus tail system trim, so this subtracts from the wing lift, reducing the overall airplane lift. In practice, as long as the tail arm from the CG is large enough, the tail down-lift required to trim is quite small and the loss in airplane lift due to tail down-lift is not considerable. We shall illustrate this with a numerical example shortly.

3.3 WING–BODY PLUS TAIL: MATH MODEL

Let us now formally consider a conventional airplane configuration consisting of a wing–body and a horizontal tail as sketched in Figure 3.6.

The wing–body aerodynamic forces and the pitching moment are marked with the correct sign convention at the wing–body aerodynamic centre. Wing–body angles are reckoned with respect to the wing–body zero-lift line (ZLL^{wb})—that is, if the relative velocity acts along this line, the wing–body produces zero lift.

The weight acts at the CG and the thrust is also assumed to act through the CG for convenience. The body-fixed X^B axis is marked in the figure shifted from the CG

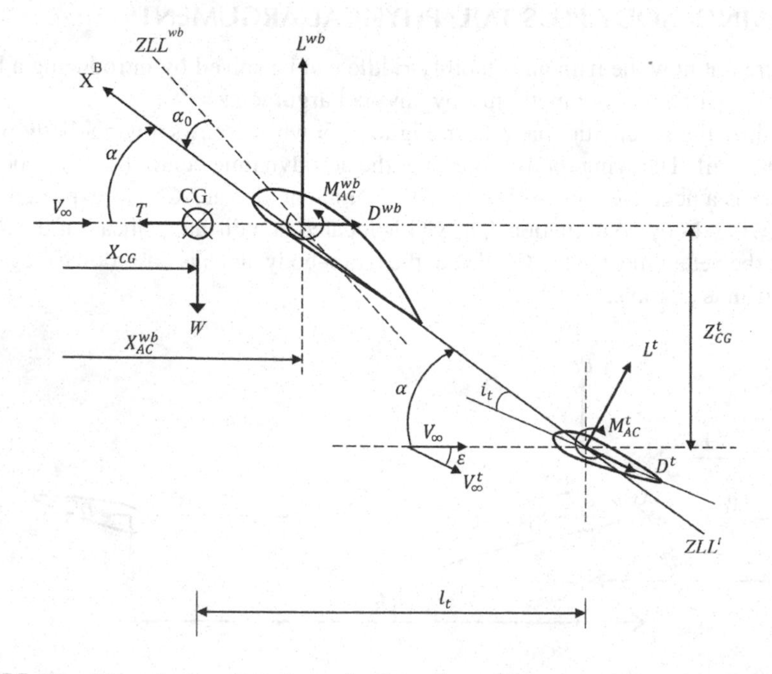

FIGURE 3.6 Wing–body and horizontal tail configuration.

only for ease of representation. The angle between the velocity V_∞ and the X^B axis is, as always, the angle of attack, α. The wing–body ZLL^{wb} is at an angle α_0 from the X^B axis. Note that the angles in Figure 3.6 are exaggerated for clarity.

Additionally, a horizontal tail is located aft of the wing–body, the tail aerodynamic centre being located at a distance l_t behind the CG. The tail airfoil section is usually a symmetric one, so its chord line is also its zero-lift line (ZLL^t). The tail is set at an angle i_t such that the ZLL^t is inclined nose-down with respect to the X^B axis by this angle i_t, called the 'tail setting angle'. This is done to facilitate a down-lift on the tail so that it creates a nose-up pitching moment about the airplane CG, as required. Thus, if one ignored the angle ε in Figure 3.6, the relative velocity would approach the tail at an angle $\alpha - i_t$. The tail setting angle is usually of the order of 2–4°.

The tail lift, drag and pitching moment are placed at the tail aerodynamic centre AC^t, but the tail pitching moment M_{AC}^t is usually zero if the tail airfoil section is symmetric. However, at the horizontal tail, there is an additional complication due to wing downwash, which causes a downward deflection of velocity at the tail. This effect is represented by the downwash angle ε. See Box 3.2 for a fuller explanation. The net flow incidence angle at the horizontal tail therefore is $\alpha - i_t - \varepsilon$. Also, the tail lift and drag vectors are then tilted backward by the angle ε.

BOX 3.2 MODELLING WING DOWNWASH

A complete understanding of downwash probably requires an intense course on *aerodynamics*, but for our present purposes, a simpler description is adequate.

Upwash and downwash are created due to the lifting mechanism of the wing—no lift and no up/downwash. As shown in Figure 3.7, one way of modelling the wing lift is to place a bound vortex at its quarter chord. This creates an upwash ahead of the airplane and a downwash behind the airplane, as sketched in Figure 3.7 (lower panel).

At the wing tips, the bound vortex is bent and extends downstream to an infinite distance; this is called the tip vortex. The tip vortex models the wing-tip trailing vortex observed on airplanes due to flow slippage from the high-pressure lower surface to the low-pressure upper surface of a lifting wing. The tip vortex creates an upwash outside the wing span and a downwash within the wing span, as shown in Figure 3.7 (upper panel).

Figure 3.7 shows the typical spanwise velocity distribution over a moderate-to-high aspect ratio wing. As you no doubt have learned in a course on *aerodynamics*, the spanwise lift at the wing tip is zero because it simply cannot support a pressure difference (unless there is a surface such as a winglet). Nevertheless, there is suction on the upper wing surface and a higher pressure on its lower surface, and this pressure difference sets up a flow around the wing tip, as pictured in Figure 3.8. This is usually referred to as the *wing-tip vortex* or *the wing trailing vortex*, as it trails behind the wing in flight. This is the same mechanism that is responsible for creating the induced drag.

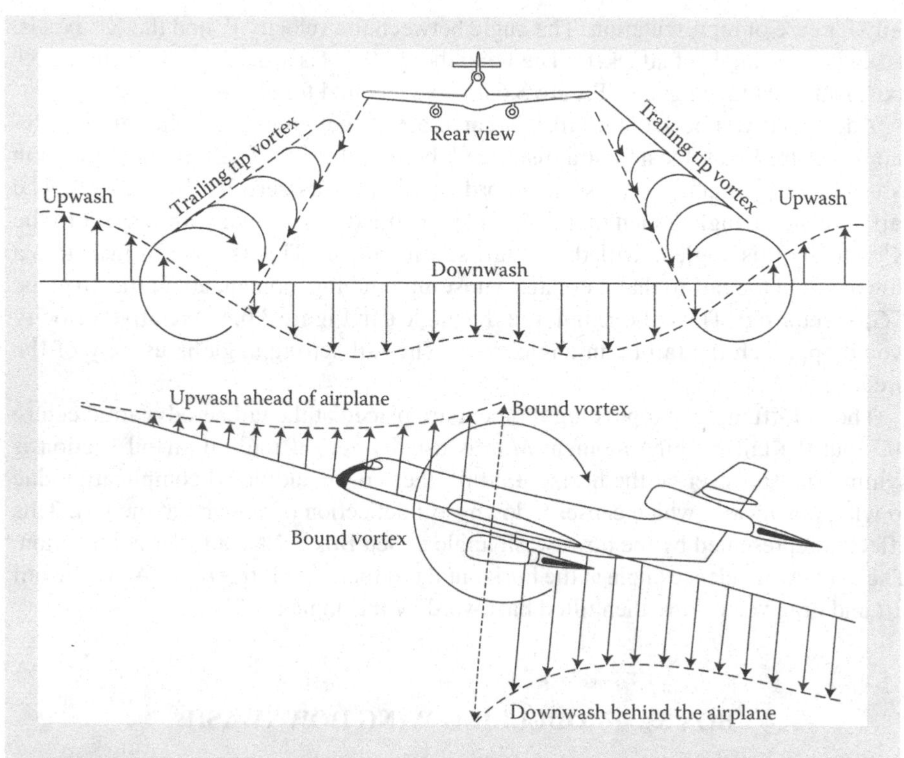

FIGURE 3.7 Upwash and downwash due to flowfield around wing. (With permission from Gustavo Corujo, http://history.nasa.gov; http://history.nasa.gov/SP-367/f54.htm.)

FIGURE 3.8 Trailing edge vortices emanating from tips of the wing of an airplane seen in the clouds (http://flyingindian.files.wordpress.com/2010/09/wingtip_vortices_lg.jpg).

When the spanwise lift distribution is an ideal elliptic one, the induced drag can be derived to be:

$$C_{Di} = \frac{C_L^2}{\pi AR_w}$$

which can be rearranged to look like:

$$C_{Di} = C_L \times \frac{C_L}{\pi AR_w}$$

AR_w is the aspect ratio of the wing. Surprisingly (or not), for an elliptical lift distribution, the downwash angle, ε, turns out to be precisely:

$$\varepsilon = \frac{C_L}{\pi AR_w}$$

So, the induced drag is:

$$C_{Di} = C_L \times \varepsilon$$

More pertinently, the downwash is a function of wing C_L, which is a function of the angle of attack, α. Hence, the downwash angle ε is a function of α.

If, hypothetically, the tail downwash angle were a function of the wing angle of attack α, as shown below:

$$\varepsilon_{tail}(t) = f(\alpha_{wing}(t))$$

then one could perhaps model ε as:

$$\varepsilon_{tail}(t) = \varepsilon_0 + \left(\frac{\partial \varepsilon}{\partial \alpha}\right)\alpha \tag{3.5}$$

where, ε_0 accounts for the downwash due to C_{L0} (any lift that may be created even when $\alpha = 0$). However, it is not so.

In fact, the downwash at the tail is a delayed function of the wing angle of attack, such as:

$$\varepsilon_{tail}(t) = f(\alpha_{wing}(t - \tau)) \tag{3.6}$$

where, τ is the time taken for the trailing vortices to travel downstream from the wing to the tail; roughly $\tau = l_t'/V_\infty$, where l_t' is the distance between wing–body AC and tail AC, and V_∞ is the freestream velocity (or, equivalently, airplane trim velocity, V^*). Note the slight difference between l_t' here and l_t in Figure 3.6, the latter being the distance from airplane CG to the tail AC.

The functional relation in Equation 3.6 can be modelled as:

$$\varepsilon_{tail}(t) = \varepsilon_0 + \left(\frac{\partial \varepsilon}{\partial \alpha}\right)\alpha + \frac{\partial \varepsilon}{\partial \dot{\alpha}\left(c/2V^*\right)}\dot{\alpha}\left(\frac{c}{2V^*}\right)$$

or, with the usual short form for the aerodynamic derivatives as:

$$\varepsilon_{tail}(t) = \varepsilon_0 + \varepsilon_\alpha \alpha + \varepsilon_{\dot{\alpha}}\dot{\alpha}\left(c/2V^*\right) \tag{3.7}$$

Comparing Equations 3.5 and 3.7, one can realize that the ε_α term is independent of the time delay factor, τ. It is the $\varepsilon_{\dot{\alpha}} = \left(\partial\varepsilon/\partial\left[\dot{\alpha}\left(c/2V^*\right)\right]\right)$ term that can be understood to capture the effect of the time lag as follows:

$$\frac{\partial \varepsilon}{\partial \dot{\alpha}} \sim \frac{\Delta \varepsilon}{\Delta \dot{\alpha}} \sim \frac{\Delta \varepsilon}{\Delta \alpha}\Delta t$$

where $\Delta t = -l_t'/V^*$.

So, $\partial\varepsilon/\partial\dot{\alpha}$ may be viewed approximately as:

$$\frac{\partial \varepsilon}{\partial \dot{\alpha}} \sim -\frac{\partial \varepsilon}{\partial \alpha}\left(l_t'/V^*\right) \tag{3.8}$$

In effect, it represents the action of a $(\partial\varepsilon/\partial\alpha)$ that existed at a time $\Delta t = l_t'/V^*$ earlier. More importantly, Equation 3.8 shows that the derivative $\left(\frac{\partial \varepsilon}{\partial \dot{\alpha}}\right)$ is negative, provided $(\partial\varepsilon/\partial\alpha)$ is positive (which it usually is).

In the formulation of the aerodynamic forces that is employed in this text, the derivative $\varepsilon_{\dot{\alpha}} = \left(\partial\varepsilon/\partial\left[\dot{\alpha}\left(c/2V^*\right)\right]\right)$ naturally merges with other like terms, and hence there is no need to further approximate it.

Let us now write out expressions for the total aircraft (wing–body plus tail) lift and pitching moment.

3.3.1 Airplane Lift

The total lift on the airplane now consists of the lift produced over the wing–body and the lift produced over the horizontal tail. The total lift can therefore be expressed as:

$$L = L^{wb} + L^t \tag{3.9}$$

where the effect of the backward tilt of the tail lift by angle ε is neglected.

Dividing each term in Equation 3.9 by $\bar{q}S$, where $\bar{q}\left(=1/2\rho V_\infty^2\right)$ is the dynamic pressure based on the freestream density and velocity, ρ and V_∞, respectively, and S is the wing planform area, gives the non-dimensional lift equation to be:

$$\frac{L}{\bar{q}S}=\frac{L^{wb}}{\bar{q}S}+\frac{S_t}{S}\frac{L^t}{\bar{q}S_t} \tag{3.10}$$

or

$$C_L=C_L^{wb}+\frac{S_t}{S}C_L^t \tag{3.11}$$

Note that we have used the freestream velocity V_∞ to non-dimensionalize both the wing–body and tail lift terms. For the wing–body, the relative velocity is very close to the freestream velocity, whereas for the tail there is a small correction due to the wing downwash. However, it is usually small enough not to make a fuss over.

Note: In the case of a canard configuration, it is the canard relative velocity that is close to the freestream velocity, whereas the relative velocity at the wing is affected by the canard downwash.

Following the usual practice, each lift coefficient (wing–body and tail) can be written as a lift-curve slope times the local incidence angle, as below:

$$C_L=C_{L\alpha}^{wb}\alpha_w+\frac{S_t}{S}C_{L\alpha}^t\alpha_t \tag{3.12}$$

where, with reference to Figure 3.6, $\alpha_w=\alpha_0+\alpha$, and $\alpha_t=\alpha-i_t-\varepsilon$, as discussed earlier. Thus:

$$C_L=C_{L\alpha}^{wb}(\alpha_0+\alpha)+\frac{S_t}{S}C_{L\alpha}^t(\alpha-i_t-\varepsilon) \tag{3.13}$$

Referring to Box 3.2, the downwash angle can be modelled as:

$$\varepsilon_{tail}(t)=\varepsilon_0+\varepsilon_\alpha\alpha+\varepsilon_{\dot{\alpha}}\dot{\alpha}\left(c/2V^*\right) \tag{3.14}$$

Inserting this expression for the downwash angle into Equation 3.13 for C_L, we can write:

$$C_L=C_{L\alpha}^{wb}(\alpha_0+\alpha)+\frac{S_t}{S}C_{L\alpha}^t\left(\alpha-i_t-\left\{\varepsilon_0+\varepsilon_\alpha\alpha+\varepsilon_{\dot{\alpha}}\dot{\alpha}\left(\frac{c}{2V^*}\right)\right\}\right) \tag{3.15}$$

Then, rearranging terms gives:

$$C_L=\left\{C_{L\alpha}^{wb}\alpha_0-\frac{S_t}{S}C_{L\alpha}^t(i_t+\varepsilon_0)\right\}+\alpha\left\{C_{L\alpha}^{wb}+\frac{S_t}{S}C_{L\alpha}^t(1-\varepsilon_\alpha)\right\} \\ -\dot{\alpha}\left(\frac{c}{2V^*}\right)\left\{\frac{S_t}{S}C_{L\alpha}^t\varepsilon_{\dot{\alpha}}\right\} \tag{3.16}$$

A generic expression for the airplane lift coefficient C_L may be written as:

$$C_L = C_{L0} + C_{L\alpha}\alpha + C_{L\dot{\alpha}}\dot{\alpha}\left(c/2V^*\right) \tag{3.17}$$

Comparing Equations 3.16 and 3.17 term by term, one can obtain the following expressions:

$$C_{L0} = C_{L\alpha}^{wb}\alpha_0 - \frac{S_t}{S}C_{L\alpha}^{t}\left(i_t + \varepsilon_0\right) \tag{3.17a}$$

$$C_{L\alpha} = C_{L\alpha}^{wb} + \frac{S_t}{S}C_{L\alpha}^{t}\left(1 - \varepsilon_\alpha\right) \tag{3.17b}$$

$$C_{L\dot{\alpha}} = -\frac{S_t}{S}C_{L\alpha}^{t}\varepsilon_{\dot{\alpha}} \tag{3.17c}$$

In Equation 3.17a, $C_{L\alpha}^{wb}\alpha_0$ may be denoted by C_{L0}^{wb}, the wing–body lift coefficient at zero angle of attack. Since both i_t and ε_0 are positive, it can be seen that the tail, at zero angle of attack, contributes a down-lift to the total airplane lift coefficient. This is to be expected as the tail was specifically set at an incidence angle i_t for this very purpose. However, the factor (S_t/S) is usually quite small, of the order of 0.1–0.2, so the tail contribution to Equation 3.17a is quite small as compared to the wing term. The term C_{L0}^{wb} is also quite small, so that often the combination of the two terms on the right-hand side of Equation 3.17a is slightly negative. This explains the negative values of C_{L0} that we encountered in the examples of Chapter 1 (*Homework Exercise* from Chapter 1).

In Equation 3.17b, the tail term (second on the right) adds a positive contribution to the lift slope, so this adds to the net lift. All together, the total tail lift at typical trim angles of attack is usually either positive or marginally negative.

The expression in Equation 3.17c is due to the tail alone and usually not of much importance as far as the airplane dynamic modes are concerned.

3.3.2 Airplane Pitching Moment

Referring to Figure 3.6, an expression for net pitching moment about the CG can be arrived at by summing up the contributions from individual components as follows:

$$\begin{aligned} M_{CG} = M_{AC}^{wb} - L^{wb}\left(X_{AC}^{wb} - X_{CG}\right) - \left(L^t \cos\varepsilon\right)l_t \\ + \left(D^t \sin\varepsilon\right)l_t - \left(D^t \cos\varepsilon\right)Z_{CG}^t - \left(L^t \sin\varepsilon\right)Z_{CG}^t \end{aligned} \tag{3.18}$$

Normally, the downwash angle is small enough so that we can assume sin ε to be small and cos ε ~ 1. Also, for many aircraft (but not all), the vertical separation of the tail from the wing, Z_{CG}^t, can be ignored. And usually, $D^t \ll L^t$. In that case, the order of the various terms in Equation 3.18 may be estimated as below:

$$M_{CG} = M_{AC}^{wb} - L^{wb}\left(X_{AC}^{wb} - X_{CG}\right) - \left(L^t \cos\varepsilon\right)l_t + \left(D^t \sin\varepsilon\right)l_t - \left(D^t \cos\varepsilon\right)Z_{CG}^t - \left(L^t \sin\varepsilon\right)Z_{CG}^t$$

$$\surd \qquad \surd \qquad \surd \qquad (s.s).\surd \qquad \left(s.\surd\right).s \qquad \left(\surd.s\right).s$$

where a '√' is a term of normal magnitude and '*s*' is a term of relatively small magnitude. Of the six terms on the right-hand side of Equation 3.18, the first three are normal and may be retained. The next three all have a product of two small quantities, and hence are of a considerably smaller magnitude; so, they may be ignored. Thus, the terms containing D^t and Z_{CG}^t drop out from Equation 3.18, leaving behind a simpler equation for the pitching moment as follows:

$$M_{CG} = M_{AC}^{wb} - L^{wb}\left(X_{AC}^{wb} - X_{CG}\right) - L^t \cdot l_t \tag{3.19}$$

Example 3.3 T-Tail Airplanes

Quite a few airplanes have their horizontal tail mounted on top of the vertical stabilizer. They are usually called 'T-tails' (Figure 3.9). This configuration is also popular for seaplanes that operate from water.

T-tails have a large vertical separation Z_{CG}^t, so they are usually spared from being immersed in the wing downwash. On the other hand, the separation Z_{CG}^t provides a large moment arm for the tail drag to contribute a nose-up (positive) pitching moment at the CG that should not be ignored.

FIGURE 3.9 Example of a T-tail configuration (https://en.wikipedia.org/wiki/Advance_Airlines_Flight_4210#/media/File:Beechcraft_b200_superkingair_zk453_arp.jpg). (By Adrian Pingstone—Own work, Public Domain, https://commons.wikimedia.org/w/index.php?curid=4427214.)

Divide each term of Equation 3.19 by $\overline{q}Sc$ to obtain a non-dimensional version, as follows:

$$C_{mCG} = C_{mAC}^{wb} - C_L^{wb}\left(h_{AC}^{wb} - h_{CG}\right) - \left(\frac{S_t}{S}\right)C_L^t\left(\frac{l_t}{c}\right) \tag{3.20}$$

where h is the distance X made non-dimensional with the mean aerodynamic chord c, as before.

The factor $(S_t/S)(l_t/c)$, or $(S_t l_t/Sc)$, appears as a ratio of two terms, each with the dimension of 'volume' and is therefore called the *horizontal tail volume ratio (HTVR)*, symbol V_H. So, Equation 3.20 can be written more cleanly as:

$$C_{mCG} = \underbrace{C_{mAC}^{wb}}\underbrace{-C_L^{wb}\left(h_{AC}^{wb} - h_{CG}\right)}\underbrace{-V_H C_L^t} \tag{3.21}$$

(−)	(−)	(+)	(if CG ahead of AC^{wb})
(−)	(+)	(+/−)	(if CG behind AC^{wb})

No doubt the terms in Equation 3.21 must add up to zero for the airplane to trim with a net zero moment about its *CG*. Let us examine the signs of the various terms marked under Equation 3.21 to see how this may happen.

- *If* CG *ahead of AC*^{wb}, then the first two terms on the right of Equation 3.21 are both negative. So, trim requires the third term to be positive, which can happen when $C_L^t < 0$, or the tail is down-lifting.
- *If* CG *behind AC*^{wb}, then the first term is negative, as always, but the second term is now positive. Depending on the relative magnitudes of these terms, the third term could be either positive or negative, that is, the tail may require to up-lift or down-lift to balance the moments at the *CG*.

Thus, the tail lift does not always have to be acting downward, though more often than not, it is so.

As before, we can write the wing–body and tail lift in terms of their respective angles of attack, thus:

$$C_{mCG} = C_{mAC}^{wb} + C_{L\alpha}^{wb}\alpha_w\left(h_{CG} - h_{AC}^{wb}\right) - V_H C_{L\alpha}^t \alpha_t \tag{3.22}$$

where $\alpha_w = \alpha_0 + \alpha$ and $\alpha_t = \alpha - i_t - \varepsilon$, as discussed earlier. The downwash angle ε is modelled as in Equation 3.14. When these are inserted in Equation 3.22 and the terms rearranged, you can derive the following expression for the airplane pitching moment coefficient (at the *CG*):

$$\begin{aligned} C_{mCG} &= \left\{C_{mAC}^{wb} + C_{L\alpha}^{wb}\alpha_0\left(h_{CG} - h_{AC}^{wb}\right) + V_H C_{L\alpha}^t\left(i_t + \varepsilon_0\right)\right\} \\ &\quad + \alpha\left\{C_{L\alpha}^{wb}\left(h_{CG} - h_{AC}^{wb}\right) - V_H C_{L\alpha}^t\left(1 - \varepsilon_\alpha\right)\right\} + \dot{\alpha}\left(c/2V^*\right)\left\{V_H C_{L\alpha}^t \varepsilon_{\dot{\alpha}}\right\} \end{aligned} \tag{3.23}$$

A generic expression for the airplane pitching moment coefficient C_m may be written as:

$$C_{mCG} = C_{m0} + C_{m\alpha}\alpha + C_{m\dot{\alpha}}\dot{\alpha}\left(c/2V^*\right) \quad (3.24)$$

Comparing the expressions in Equations 3.23 and 3.24 term by term, one can obtain the following relations:

$$C_{m0} = C_{mAC}^{wb} + C_{L0}^{wb}\left(h_{CG} - h_{AC}^{wb}\right) + V_H C_{L\alpha}^{t}\left(i_t + \varepsilon_0\right) \quad (3.24a)$$

$$C_{m\alpha} = C_{L\alpha}^{wb}\left(h_{CG} - h_{AC}^{wb}\right) - V_H C_{L\alpha}^{t}\left(1 - \varepsilon_\alpha\right) \quad (3.24b)$$

$$C_{m\dot{\alpha}} = V_H C_{L\alpha}^{t}\varepsilon_{\dot{\alpha}} \quad (3.24c)$$

where, $C_{L\alpha}^{wb}\alpha_0$ has been written as C_{L0}^{wb}, the wing–body lift at zero angle of attack.

Homework Exercise: Repeat the analysis above for the case of a T-tail.

Let us now discuss trim and stability in terms of C_{m0} and $C_{m\alpha}$. Refer back to the dashed line in Figure 3.4b, which is what we would like to have—trim at positive C_L (positive α) and, for stability, a negative slope at the point of intercept. In other words, for trim and stability, we would like:

$$C_{m0} > 0 \quad \text{and} \quad C_{m\alpha} < 0 \quad (3.25)$$

From Equation 3.24a, for C_{m0}, the first term C_{mAC}^{wb} is almost always negative, whereas the second term could be of either sign, depending on the relative locations of the CG and AC^{wb}. Usually, the sum of these two wing–body terms is negative. Thus, by manipulating the third (tail) term, particularly the tail setting angle i_t, the designer can adjust C_{m0} to almost any desired positive value.

Now, let us examine Equation 3.24b for $C_{m\alpha}$. The first (wing–body) term, as we have seen earlier, can be made negative simply by locating the CG ahead of the AC^{wb}. But, notice that the second (tail) term, $V_H C_{L\alpha}^{t}\left(1 - \varepsilon_\alpha\right)$, is always negative, no matter what. Effectively, this means that even if the first term were positive (CG behind AC^{wb}), by suitably selecting the *HTVR* parameter, V_H, one can always obtain the desired negative value of $C_{m\alpha}$. This is an added benefit of introducing the horizontal tail. Besides providing a positive (nose-up) moment in Equation 3.24a to make trim possible, the horizontal tail also naturally provides a stabilizing moment in Equation 3.24b. This allows the designer the luxury of letting the CG move behind the AC^{wb}, thus easing the constraint on CG location a little bit.

Homework Exercise: Can you judge why it may be difficult on typical commercial and military airplanes to keep the CG constrained to a location ahead of AC^{wb}? Think of the various sources of weight and also find an approximate location for the AC^{wb} assuming a trapezoidal, sweptback wing.

Equation 3.24c is discussed in the following section.

Example 3.4

Horizontal Tail Sizing: Selection of i_t and V_H

This example gives you an idea of selecting the horizontal tail design parameters i_t and V_H and a feel for the numbers.

Data:

$$C_{L0}^{wb} = 0; \quad C_{L\alpha}^{wb} = 0.08/°; \quad C_{L\alpha}^{t} = 0.1/°$$

$$\varepsilon_0 = 0; \quad \varepsilon_\alpha = 0.35$$

$$C_{mAC}^{wb} = -0.032; \quad h_{CG} - h_{AC}^{wb} = 0.11$$

This is a case where the CG is actually behind AC^{wb}.

The problem is to select i_t and V_H such that:

$$C_{m\alpha} = -0.0133/° \quad \text{and} \quad C_{m0} = 0.06.$$

SOLUTION

Just apply Equation 3.24b and then Equation 3.24a.

Plugging the numbers into Equation 3.24b yields:

$$-0.0133 = 0.08 \times 0.11 - V_H \times 0.1 \times (1 - 0.35)$$

giving $V_H = 0.34$. Note that V_H itself consists of two separate parameters (S_t/S) and (l_t/c). There is not enough information in this example to deduce these two individually.

From Equation 3.24a, then:

$$0.06 = -0.032 + 0.0 \times 0.11 + 0.34 \times 0.1 \times (i_t + 0)$$

which gives $i_t = 2.7°$.

Example 3.5

Wing–Body–Tail Trim and Stability

Carrying on from Example 3.4, assume that $(S_t/S) = 0.1$. Then, calculate the trim point and stability of that trim. Find trim C_L.

From the given requirements in Example 3.4:

$$C_{m\alpha} = -0.0133/° \quad \text{and} \quad C_{m0} = 0.06$$

one can draw a plot of C_m versus α as shown in Figure 3.10. The Y-intercept is $C_{m0} = 0.06$ and the slope is $C_{m\alpha} = -0.0133/°$. A simple calculation yields a trim angle of attack $\alpha^* = 4.5°$.

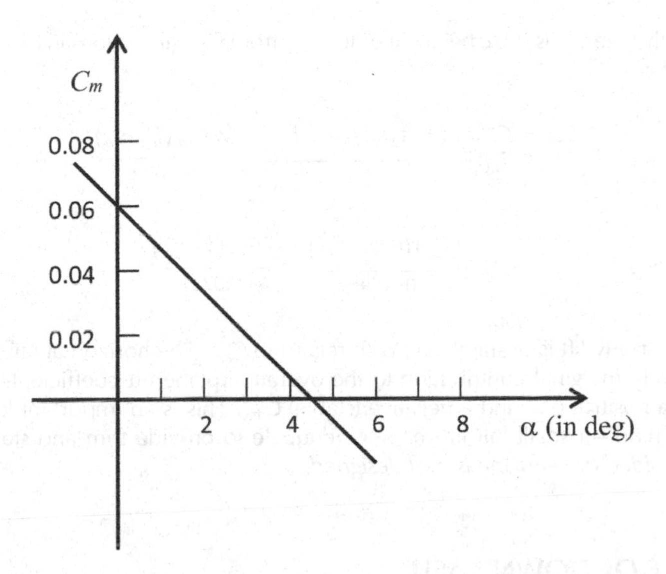

FIGURE 3.10 Plot of C_m versus α in Example 3.5.

Since $C_{m\alpha}$ is negative and provided $(C_{m\alpha} + C_{mq1})$ is also negative, we are assured that this trim is stable, that is, perturbations in the short-period dynamics about this trim state will diminish and die out.

Now, we determine the trim C_L at this point; let us call it C_L^*.

$$C_L^* = C_{L0} + C_{L\alpha}\alpha^*$$

We have expressions for C_{L0} and $C_{L\alpha}$ from Equations 3.17a and b. Plugging in the numbers into these equations, we get:

$$C_{L0} = C_{L0}^{wb} - \frac{S_t}{S} C_{L\alpha}^t (i_t + \varepsilon_0) = 0 - 0.1 \times 0.1 \times (2.7 + 0) = -0.027$$

$$C_{L\alpha} = C_{L\alpha}^{wb} + \frac{S_t}{S} C_{L\alpha}^t (1 - \varepsilon_\alpha)$$

Thus, $C_L^* = -0.027 + 4.5 \times 0.0865 = 0.362$. Notice, once again, that C_{L0} is marginally negative, and this is all due to the tail down-lift (seen more clearly below).

It is instructive to split this trim C_L^* into a wing–body contribution and a tail contribution as follows:

$$C_L^* (\text{wing–body contribution}) = 0 + 4.5 \times 0.08 = 0.36$$

$$C_L^* (\text{tail contribution}) = -0.027 + 0.0065 \times 4.5 = 0.002$$

Almost all the lift comes from the wing–body. The tail contribution is really small, but the net tail contribution to C_L^* is still positive at this trim angle of attack.

Note that the CG is 0.11c behind the AC^{wb}, so from the calculations in Example 3.4 we see that:

$$C_{m0} = \underbrace{C_{mAC}^{wb}}_{-0.032} + \underbrace{C_{L0}^{wb}\alpha_0\left(h_{CG} - h_{AC}^{wb}\right)}_{0} + \underbrace{V_H C_{L\alpha}^{t}\left(i_t + \varepsilon_0\right)}_{+0.092}$$

$$C_{m\alpha} = \underbrace{C_{L\alpha}^{wb}\left(h_{CG} - h_{AC}^{wb}\right)}_{0.0088} - \underbrace{V_H C_{L\alpha}^{t}\left(1 - \varepsilon_\alpha\right)}_{-0.0221}$$

The wing–body lift is destabilizing with regard to $C_{m\alpha}$. The horizontal tail, despite its relatively marginal contribution to the overall airplane lift coefficient, is what ensures a positive C_{m0} and a negative (stable) $C_{m\alpha}$. This is an important lesson to keep in mind—a small tail lift may be adequate to provide trim and stability in pitch *provided the airplane is well designed.*

3.4 ROLE OF DOWNWASH

As explained in Box 3.2, downwash at the horizontal tail is due to the wing trailing edge vortex adding a downward component of velocity to the freestream velocity vector. This is pictured in Figure 3.6 where the freestream velocity is marked V_∞ and the introduction of the downwash angle ε changes the freestream velocity vector to V_∞^t. While the magnitude of the velocity may not change much due to the downwash, the change in orientation by angle ε can have some interesting effects.

First, in Equation 3.24a, it contributes a term $V_H C_{L\alpha}^t$ to the pitching moment at zero angle of attack, but ε_0 is usually quite small, so it may not make much of a difference.

Second, in the term $V_H C_{L\alpha}^t\left(1 - \varepsilon_\alpha\right)$ in Equation 3.24b, it subtracts from the stabilizing pitching moment arising due to the horizontal tail. In this sense, it is detracting. Since the typical value of ε_α could be in the range 0.3–0.4, the loss in tail effectiveness due to downwash can be significant.

Finally, Equation 3.24c reveals a term $C_{m\dot{\alpha}} = V_H C_{L\alpha}^t \varepsilon_{\dot{\alpha}}$, which is negative as explained in Box 3.2. Here, the downwash term $\varepsilon_{\dot{\alpha}}$ contributes a damping effect to the pitch dynamics and in this manner $C_{m\dot{\alpha}}$ cooperates with the pitch damping derivative C_{mq1}.

Note: As formulated in this text, the terms C_{mq1} and $C_{m\dot{\alpha}}$ add up naturally to give a net pitch damping derivative that is responsible for damping the short-period dynamics in Equation 2.32. As against this, the traditional aerodynamic model employed in the past does not merge C_{mq} and $C_{m\dot{\alpha}}$ neatly, though both effects are labelled 'pitch damping'. Incidentally, typical dynamic wind tunnel tests to obtain these derivatives always seem to yield C_{mq1} and $C_{m\dot{\alpha}}$ together. With the aerodynamic model as formulated in this text, there is no need to separate these two derivatives as they always go together in forming the pitch damping term. However, in the past, much effort used to be expended to separate them and obtain independent estimates of C_{mq} and $C_{m\dot{\alpha}}$. This is now shown to be unnecessary.

3.5 NEUTRAL POINT

One big take away from the previous section is that with the introduction of the horizontal tail, the CG can stray behind (aft of) the wing–body AC and the airplane can still have stability in pitch (i.e., have a stable short-period dynamics). The obvious question is how far back can the CG go, and what limits the aft CG position?

Let us examine the expression for $C_{m\alpha}$ again.

$$C_{m\alpha} = C_{L\alpha}^{wb}\left(h_{CG} - h_{AC}^{wb}\right) - V_H C_{L\alpha}^{t}\left(1 - \varepsilon_\alpha\right)$$

Under normal circumstances, the second term (including the minus sign ahead of it) on the right (due to the tail) is negative (stabilizing). The first (wing–body) term with CG behind AC^{wb} is positive (destabilizing). As the CG moves further behind AC^{wb}, the first term becomes more and more positive. Clearly, given a certain HTVR, V_H, a point is reached where the positive first term and the negative second term cancel each other, leaving $C_{m\alpha}$ to be zero. This point can be thought of as the boundary between stability and instability. Any further rearward movement of the CG will result in a positive $C_{m\alpha}$ and an unstable short-period mode. This point is called the neutral point (NP), presumably because it offers 'neutral' stability (neither stable nor unstable) in pitch.

Note that the second (tail) term is strictly not independent of the CG position, as the parameter V_H contains a length l_t, which is equal to $(X_{AC}^{t} - X_{CG})$. It is usually a small effect, which we can ignore for now, but we will address this shortly.

So, the NP is the location of the CG, where $C_{m\alpha}$ is zero. This is easy to find, by setting $C_{m\alpha}$ in Equation 3.24b to zero—the corresponding CG position h_{CG} is denoted by h_{NP}, where '*NP*' stands for 'neutral point'.

$$0 = C_{L\alpha}^{wb}\left(h_{NP} - h_{AC}^{wb}\right) - V_H C_{L\alpha}^{t}\left(1 - \varepsilon_\alpha\right) \tag{3.26}$$

On solving for h_{NP}, we get:

$$h_{NP} = h_{AC}^{wb} + V_H \left\{\frac{C_{L\alpha}^{t}}{C_{L\alpha}^{wb}}\right\}\left(1 - \varepsilon_\alpha\right) \tag{3.27}$$

It is obvious that in the absence of the tail, the second term on the right-hand side of Equation 3.27 disappears and

$$h_{NP} = h_{AC}^{wb}.$$

In other words, for a wing–body alone, for $C_{m\alpha}$ to be negative (stable condition), the CG should remain ahead of the AC^{wb}, but then there is an issue with trimming, as we have seen previously.

The second term in Equation 3.27 is the contribution due to the horizontal tail. Being positive, it allows the CG to move behind the AC^{wb}. Since $C_{L\alpha}^{t}/C_{L\alpha}^{wb} \sim 1$, the key factor deciding the location of the NP in Equation 3.27 is the horizontal tail volume ratio, V_H. Recall that V_H is itself the product of two ratios:

$$V_H = \left(\frac{S_t}{S}\right)\left(\frac{l_t}{c}\right)$$

The first of these, (S_t/S), sets the tail size (relative to the wing planform area), and the second, (l_t/c), fixes the tail arm (relative to the wing mean aerodynamic chord). The tail size (S_t) and the tail arm from the CG (l_t) are the main parameters that the designer can select to decide on the location of the NP.

The term $(1 - \varepsilon_\alpha)$ in Equation 3.27 shows that the effect of downwash through the term ε_α is to subtract from the effectiveness of the tail. If $\varepsilon_\alpha \sim 0.3$–$0.4$, it means that the tail functions only at 60–70% efficiency.

Example 3.6

Calculation of NP

For the numerical data in Example 3.4, determine the NP.

Plugging the numbers into Equation 3.27:

$$h_{NP} = h_{AC}^{wb} + V_H\left\{\frac{C_{L\alpha}^t}{C_{L\alpha}^{wb}}\right\}(1-\varepsilon_\alpha)$$

$$= h_{AC}^{wb} + 0.34 \times \left(\frac{0.1}{0.08}\right) \times (1-0.35)$$

So, $h_{NP} = h_{AC}^{wb} + 0.276$.

Refer to the sketch in Figure 3.11, where the various positions are marked on an airplane. The mean aerodynamic chord of the wing is also sketched in the figure. Assume distances to be measured from the leading edge of the mean aerodynamic chord. We are not told the absolute location of h_{AC}^{wb} in the data of Example 3.4, but it is a safe guess to take it around quarter chord, that is, $h_{AC}^{wb} \sim 0.25$. We are given that: $h_{CG} - h_{AC}^{wb} \sim 0.11$, so $h_{CG} \sim 0.36$. And, we just calculated $h_{NP} - h_{AC}^{wb} = 0.276$, so $h_{NP} \sim 0.526$, which happens to be about half mean aerodynamic chord. If we have no information about an airplane, it is a good first guess to assume $h_{NP} \sim 0.5$ on the mean aerodynamic chord. So, now you know that the CG must lie ahead of this point!

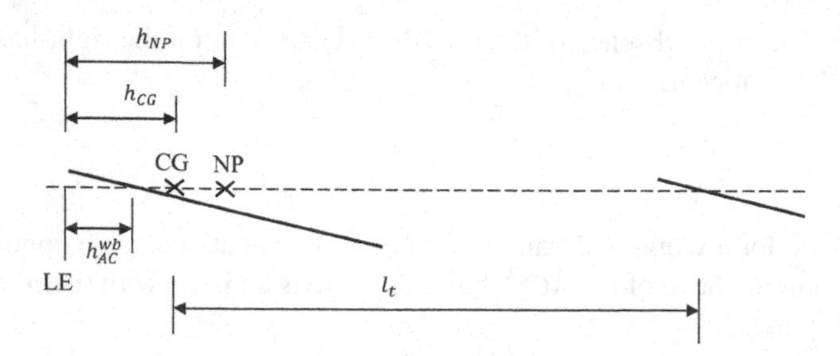

FIGURE 3.11 Sketch for Example 3.6 ($h_{CG} \sim 0.36$ and $h_{NP} \sim 0.526$).

3.5.1 Static Margin

Looking back at Equation 3.24b, we can manipulate the terms to appear as:

$$C_{m\alpha} = C_{L\alpha}^{wb}\left[h_{CG} - \underbrace{\left(h_{AC}^{wb} + V_H\left\{C_{L\alpha}^{t}/C_{L\alpha}^{wb}\right\}(1-\varepsilon_\alpha)\right)}_{h_{NP}}\right]$$

and recognize the marked term as h_{NP}. So, one can reframe Equation 3.24b as:

$$C_{m\alpha} = C_{L\alpha}^{wb}\left(h_{CG} - h_{NP}\right) \tag{3.28}$$

The non-dimensional distance $h_{NP} - h_{CG}$ is called the static margin, *SM*.

$$SM = h_{NP} - h_{CG} \tag{3.29}$$

Equation 3.28 explicitly states that as long as the CG is ahead of the NP (i.e., the *SM* is positive), $C_{m\alpha}$ will be negative and one of the conditions for the airplane to be stable in pitch (short-period mode) is met.

3.5.2 NP as Aerodynamic Centre of Entire Airplane

You will recollect from Equation 3.24 that C_{mCG} can be generally written as:

$$C_{mCG} = C_{m0} + C_{m\alpha}\alpha \tag{3.30}$$

ignoring the $\dot{\alpha}$ term that does not matter when we are talking only of trim and stiffness.

With the tail absent (wing–body alone), we can write C_{mCG} as:

$$C_{mCG} = \left\{C_{mAC}^{wb} + C_{L0}^{wb}\left(h_{CG} - h_{AC}^{wb}\right)\right\} + \left\{C_{L\alpha}^{wb}\left(h_{CG} - h_{AC}^{wb}\right)\right\}\alpha \tag{3.31}$$

Now, consider a case where the CG coincides with the AC^{wb}, that is, $h_{CG} - h_{AC}^{wb} = 0$. Then, Equation 3.31 reduces to:

$$C_{mCG} = C_{mAC}^{wb}$$

which is independent of α because the coefficient of the α term in Equation 3.30 reduces to zero. In fact, you may remember that is precisely how we first defined the aerodynamic centre—as a point about which the pitching moment (coefficient, C_m) is independent of the angle of attack.

With the tail added, and using the form of $C_{m\alpha}$ in Equation 3.28, Equation 3.30 for C_{mCG} for a wing–body–tail system appears as:

$$C_{mCG} = \left\{C_{mAC}^{wb} + C_{L0}^{wb}\left(h_{CG} - h_{AC}^{wb}\right) + V_H C_{L\alpha}^{t}\left(i_t + \varepsilon_0\right)\right\} + \left\{C_{L\alpha}^{wb}\left(h_{CG} - h_{NP}\right)\right\}\alpha \tag{3.32}$$

We get h_{AC}^{wb} from Equation 3.27 as shown below:

$$h_{AC}^{wb} = h_{NP} - V_H \left\{ \frac{C_{L\alpha}^t}{C_{L\alpha}^{wb}} \right\} (1 - \varepsilon_\alpha)$$

And using this in Equation 3.32 gives an expression for C_{mCG} in terms of h_{NP} as follows:

$$C_{mCG} = \left\{ C_{mAC}^{wb} + C_{L0}^{wb}(h_{CG} - h_{NP}) + C_{L0}^{wb} V_H \left\{ C_{L\alpha}^t / C_{L\alpha}^{wb} \right\} (1 - \varepsilon_\alpha) + V_H C_{L\alpha}^t (i_t + \varepsilon_0) \right\} + \left\{ C_{L\alpha}^{wb}(h_{CG} - h_{NP}) \right\} \alpha \qquad (3.33)$$

Now, consider the case where the CG coincides with the NP, that is, $h_{CG} - h_{NP} = 0$. Then, Equation 3.33 comes down to:

$$C_{mCG} = \left\{ C_{mAC}^{wb} + C_{L0}^{wb} V_H \left\{ \frac{C_{L\alpha}^t}{C_{L\alpha}^{wb}} \right\} (1 - \varepsilon_\alpha) + V_H C_{L\alpha}^t (i_t + \varepsilon_0) \right\} \qquad (3.34)$$

which is C_{mNP} by definition (because of our choice of CG being at NP), and once again this is independent of α. By choosing $h_{CG} = h_{NP}$, we put to zero the coefficient of the α term in Equation 3.33.

Thus, it follows that for a wing–body plus tail, the NP is the location where the pitching moment (coefficient C_m) is independent of the angle of attack. In other words, the NP is the 'aerodynamic centre' of the complete airplane (wing–body plus tail)!

Homework Exercise: The expression for C_{mCG} in Equation 3.32 is not very clean, although it serves the purpose. Better still, can you derive a cleaner expression for C_{mCG} that appears as below:

$$C_{mCG} = C_{mNP} + C_L^{wb}(h_{CG} - h_{NP}) \qquad (3.35)$$

where, C_{mNP} is the pitching moment coefficient at the NP of the airplane as in Equation 3.34 and C_L^{wb} is the lift coefficient of the wing–body alone, expressed as $C_L^{wb} = C_{L\alpha}^{wb}(\alpha + \alpha_0) = C_{L\alpha}^{wb}\alpha + C_{L0}^{wb}$.

Bigger Homework Exercise: Some of you may suspect that Equation 3.35 should logically have appeared as:

$$C_{mCG} = C_{mNP} + C_L(h_{CG} - h_{NP}) \qquad (3.36)$$

Where, C_L is the entire airplane lift coefficient and not that of the wing–body alone, and you would be right. Can you figure out the error that resulted in the slightly incorrect expression of Equation 3.35 appearing rather than the more appealing and correct version of Equation 3.36?

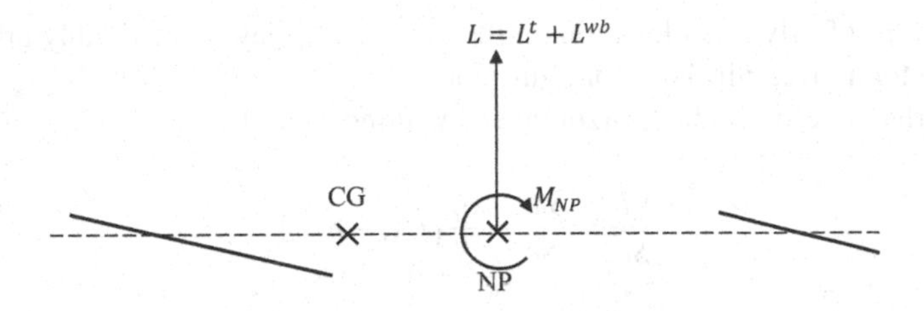

FIGURE 3.12 Sketch of forces and moment on an airplane displayed at the neutral point.

This discussion also suggests that a good way to represent the forces on the entire airplane (wing–body plus tail) is to place the pitching moment (coefficient C_{mNP}) and the lift and drag (coefficients C_L, C_D) at the NP of the airplane, as shown in Figure 3.12. Were the lift, drag and pitching moment to be placed at another point, then with a change in the angle of attack, all of them would change in such a manner that the net moment about the NP would remain unchanged. By placing them all at the NP, this condition is obtained trivially since any change in the lift and drag does not create a moment about the point at which they act, and the pitching moment there is already invariant with the angle of attack. That is so convenient!

3.6 REPLACING V_H WITH V_H'

Refer to Figure 3.13, which shows a wing–tail combination with the various locations and distances as marked. Note that l_t is the distance between the tail aerodynamic centre AC^t and the CG, whereas l_t' is the distance between AC^t and AC^{wb}. So far, we have been using l_t to define various quantities such as V_H and to derive expressions for C_L and C_m. However, the CG (as we shall soon see at greater length) is not a fixed location and can vary quite a bit in flight. So, with varying CG, l_t must also be changed, which is a bit tedious. If we ignore this change, as we have been doing so far, the error may not be severe, but it does damage the analysis a bit. The alternative is to use l_t'—the distance between the AC^t and the AC^{wb}—as the preferred variable. The two are related as:

$$l_t = l_t' - \left(X_{CG} - X_{AC}^{wb}\right) \quad \text{or} \quad \frac{l_t}{c} = \frac{l_t'}{c} - \left(h_{CG} - h_{AC}^{wb}\right) \tag{3.37}$$

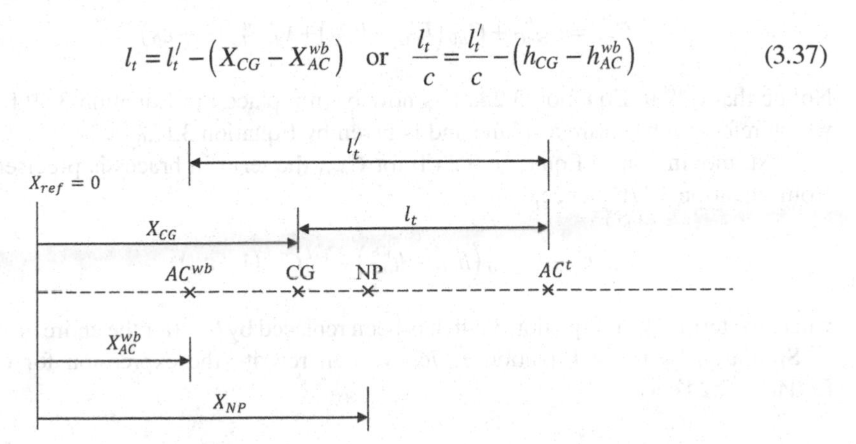

FIGURE 3.13 Sketch of various locations and distances in case of a wing–body plus tail.

Since AC^{wb} is a fairly fixed location on the aircraft, $l_t^{/}$ may be justifiably taken to be a constant for a given airplane configuration.

Similarly, in defining the horizontal tail volume ratio, V_H:

$$\frac{S_t l_t}{Sc} = \frac{S_t l_t^{/}}{Sc} - \frac{S_t}{S}\left(h_{CG} - h_{AC}^{wb}\right)$$

That is,

$$V_H = V_H^{/} - \frac{S_t}{S}\left(h_{CG} - h_{AC}^{wb}\right) \tag{3.38}$$

Now, let us go back and substitute these expressions in place of l_t and V_H in Equations 3.24 for C_m:

$$C_{m0} = C_{mAC}^{wb} + \left(h_{CG} - h_{AC}^{wb}\right)\left\{C_{L0}^{wb} - \frac{S_t}{S}C_{L\alpha}^{t}\left(i_t + \varepsilon_0\right)\right\} + V_H^{/} C_{L\alpha}^{t}\left(i_t + \varepsilon_0\right) \tag{3.24'a}$$

$$C_{m\alpha} = \left(h_{CG} - h_{AC}^{wb}\right)\left\{C_{L\alpha}^{wb} + \frac{S_t}{S}C_{L\alpha}^{t}\left(1 - \varepsilon_\alpha\right)\right\} - V_H^{/} C_{L\alpha}^{t}\left(1 - \varepsilon_\alpha\right) \tag{3.24'b}$$

$$C_{m\dot{\alpha}} = \left\{V_H^{/} - \frac{S_t}{S}\left(h_{CG} - h_{AC}^{wb}\right)\right\} C_{L\alpha}^{t}\varepsilon_{\dot{\alpha}} \tag{3.24'c}$$

Note that l_t does not figure in Equation 3.17 for C_L since the lift coefficient does not depend on the tail arm; it is the moment that is dependent on the tail arm.

The term in braces in Equation 3.24′a is precisely C_{L0} (see Equation 3.17a), so Equation 3.24′a can be written as:

$$C_{m0} = C_{mAC}^{wb} + C_{L0}\left(h_{CG} - h_{AC}^{wb}\right) + V_H^{/} C_{L\alpha}^{t}\left(i_t + \varepsilon_0\right) \tag{3.39}$$

Notice that C_{L0}^{wb} in Equation 3.24a has now been replaced in Equation 3.39 by C_{L0}, which refers to the entire airplane and is given by Equation 3.17a.

Next, moving on to Equation 3.24′b for $C_{m\alpha}$, the term in braces is precisely $C_{L\alpha}$ from Equation 3.17b; hence,

$$C_{m\alpha} = C_{L\alpha}\left(h_{CG} - h_{AC}^{wb}\right) - V_H^{/} C_{L\alpha}^{t}\left(1 - \varepsilon_\alpha\right) \tag{3.40}$$

where the term $C_{L\alpha}^{wb}$ in Equation 3.24b has been replaced by $C_{L\alpha}$ (for the entire airplane).

Similarly, by using Equation 3.17c, we can rewrite the expression for $C_{m\dot{\alpha}}$ in Equation 3.24′c as;

$$C_{m\dot{\alpha}} = \left\{V_H^{/} C_{L\alpha}^{t}\varepsilon_{\dot{\alpha}} + \left(h_{CG} - h_{AC}^{wb}\right) C_{L\dot{\alpha}}\right\}$$

3.6.1 Revised Expressions for NP

Can we obtain a formula for the NP from the form of $C_{m\alpha}$ in Equation 3.40?

As before, the NP is that location of the CG where $C_{m\alpha} = 0$. So, from Equation 3.40, we have:

$$0 = C_{L\alpha}\left(h_{NP} - h_{AC}^{wb}\right) - V_H' C_{L\alpha}^t (1 - \varepsilon_\alpha)$$

which can be solved for h_{NP} as:

$$h_{NP} = h_{AC}^{wb} + V_H'\left(C_{L\alpha}^t / C_{L\alpha}\right)(1 - \varepsilon_\alpha) \tag{3.41}$$

Since, $(1 - \varepsilon_\alpha) \sim 0.6$–$0.7$ and $\left(C_{L\alpha}^t / C_{L\alpha}\right) \sim 1$, $\left(h_{NP} - h_{AC}^{wb}\right)$ is mainly dependent on V_H'—the slightly redefined HTVR. In other words, a larger tail area placed at a larger distance aft of the wing allows the CG to stray further behind the wing–body AC while maintaining stability of the short-period mode. So, conceptually, nothing has changed. Only in place of $(V_H / C_{L\alpha}^{wb})$ in the earlier expression for h_{NP}, we have a revised factor of $(V_H' / C_{L\alpha})$ in Equation 3.41.

So, do the two expressions—one with $(V_H / C_{L\alpha}^{wb})$ and the other with $(V_H' / C_{L\alpha})$—give the same value of the NP location, or will they be different?

From the derivation below, it can be seen to be the same:

$$\begin{aligned} \frac{V_H'}{C_{L\alpha}} &= \frac{\left\{V_H + (S_t/S)\left(h_{CG} - h_{AC}^{wb}\right)\right\}}{\left\{C_{L\alpha}^{wb} + (S_t/S) C_{L\alpha}^t (1 - \varepsilon_\alpha)\right\}} \\ &= \left(\frac{V_H}{C_{L\alpha}^{wb}}\right) \frac{\left\{1 + (S_t/S)\left(h_{CG} - h_{AC}^{wb}\right)/V_H\right\}}{\left\{1 + (S_t/S)\left(C_{L\alpha}^t / C_{L\alpha}^{wb}\right)(1 - \varepsilon_\alpha)\right\}} \end{aligned} \tag{3.42}$$

From Equation 3.27:

$$\frac{C_{L\alpha}^t}{C_{L\alpha}^{wb}}(1 - \varepsilon_\alpha) = \frac{\left(h_{NP} - h_{AC}^{wb}\right)}{V_H}$$

Hence, we can write Equation 3.42 as:

$$\frac{V_H'}{C_{L\alpha}} = \left(\frac{V_H}{C_{L\alpha}^{wb}}\right) \frac{\left\{1 + (S_t/S)\left(h_{CG} - h_{AC}^{wb}\right)/V_H\right\}}{\left\{1 + (S_t/S)\left(h_{NP} - h_{AC}^{wb}\right)/V_H\right\}} \tag{3.43}$$

Note the different terms h_{CG} in the numerator and h_{NP} in the denominator. However, when defining the NP, we examine the case where the CG is at the so-called NP. Hence, $h_{CG} = h_{NP}$ in that case. Then, $\left(V_H' / C_{L\alpha}\right) = \left(V_H / C_{L\alpha}^{wb}\right)$ and both the expressions yield the same location of the NP.

3.6.2 NP as Aerodynamic Centre of the Entire Airplane

We derived the following expression in Equation 3.35:

$$C_{mCG} = C_{mNP} + C_L^{wb}\left(h_{CG} - h_{NP}\right)$$

for the pitching moment coefficient C_{mCG} in terms of the pitching moment coefficient at the NP C_{mNP}, which happens to be independent of the angle of attack. Now, we can work this a little better by using V_H^l instead of V_H.

From Equation 3.39 for C_{m0}:

$$C_{m0} = C_{mAC}^{wb} + C_{L0}\left(h_{CG} - h_{AC}^{wb}\right) + V_H^l C_{L\alpha}^t\left(i_t + \varepsilon_0\right)$$

And using Equation 3.41 to write h_{AC}^{wb} as:

$$h_{AC}^{wb} = h_{NP} - V_H^l\left(C_{L\alpha}^t / C_{L\alpha}\right)\left(1 - \varepsilon_\alpha\right)$$

we get:

$$C_{m0} = C_{mAC}^{wb} + C_{L0}\left(h_{CG} - h_{NP}\right) + C_{L0}V_H^l\left(\frac{C_{L\alpha}^t}{C_{L\alpha}}\right)\left(1 - \varepsilon_\alpha\right) + V_H^l C_{L\alpha}^t\left(i_t + \varepsilon_0\right)$$

That is,

$$C_{m0} = C_{mAC}^{wb} + C_{L0}\left(h_{CG} - h_{NP}\right) + V_H^l C_{L\alpha}^t\left\{\left(\frac{C_{L0}}{C_{L\alpha}}\right)\left(1 - \varepsilon_\alpha\right) + \left(i_t + \varepsilon_0\right)\right\} \tag{3.44a}$$

When $h_{CG} = h_{NP}$:

$$C_{m0} = C_{mNP} = C_{mAC}^{wb} + V_H^l C_{L\alpha}^t\left\{\left(\frac{C_{L0}}{C_{L\alpha}}\right)\left(1 - \varepsilon_\alpha\right) + \left(i_t + \varepsilon_0\right)\right\} \tag{3.44b}$$

So, one can as well write Equation 3.44a as below, by rearranging the terms:

$$C_{m0} = \left\{C_{mAC}^{wb} + V_H^l C_{L\alpha}^t\left(\frac{C_{L0}}{C_{L\alpha}}\right)\left(1 - \varepsilon_\alpha\right) + V_H^l C_{L\alpha}^t\left(i_t + \varepsilon_0\right)\right\} + C_{L0}\left(h_{CG} - h_{NP}\right)$$

where the term in braces can be seen to be C_{mNP} from Equation 3.44b, so:

$$C_{m0} = C_{mNP} + C_{L0}\left(h_{CG} - h_{NP}\right) \tag{3.44c}$$

Similarly, from Equation 3.40 for $C_{m\alpha}$:

$$C_{m\alpha} = C_{L\alpha}\left(h_{CG} - h_{AC}^{wb}\right) - V_H^l C_{L\alpha}^t\left(1 - \varepsilon_\alpha\right)$$

One may rearrange the terms to appear as:

$$C_{m\alpha} = C_{L\alpha}\left(h_{CG} - \left\{h_{AC}^{wb} + V_H^{l}\left(\frac{C_{L\alpha}^{t}}{C_{L\alpha}}\right) - (1-\varepsilon_{\alpha})\right\}\right) \tag{3.40a}$$

where, from Equation 3.41, it can be seen that the expression within the braces is h_{NP}. Hence,

$$C_{m\alpha} = C_{L\alpha}(h_{CG} - h_{NP}) \tag{3.45}$$

Hence, from Equations 3.44c and 3.45, we can piece together the static part (ignoring the $\dot{\alpha}$ term) of the pitching moment coefficient, C_{mCG}, as:

$$C_{mCG} = C_{m0} + C_{m\alpha}\alpha = \left[C_{mNP} + C_{L0}(h_{CG} - h_{NP})\right] + \left[C_{L\alpha}(h_{CG} - h_{NP})\right]\alpha \tag{3.46}$$

We can combine the two terms on the right-hand side of Equation 3.46 as:

$$C_{mCG} = C_{mNP} + [C_{L0} + C_{L\alpha}\alpha](h_{CG} - h_{NP})$$

But $[C_{L0} + C_{L\alpha}\alpha] = C_L$ for the entire airplane. So,

$$C_{mCG} = C_{mNP} + C_L(h_{CG} - h_{NP}) \tag{3.47}$$

What does this mean? It says that C_m at the airplane CG can be written as the sum of a pure moment C_{mNP}, not a function of the angle of attack, placed at its NP and the moment due to the lift (coefficient, C_L) located at the NP acting at the CG through a moment arm $(h_{CG} - h_{NP})$.

Just as the pure moment (C_{mAC}^{wb}, independent of the angle of attack) of a wing–body and its lift (C_L^{wb}) are placed at the wing–body aerodynamic centre, in the same manner, the pure moment and lift coefficient of the entire airplane can be placed at the NP.

In other words, the NP is the 'aerodynamic centre' for the entire airplane.

3.6.3 Trim and Stability, Again!

From Equation 3.47, we can find the trim C_L as:

$$C_L^* = -\frac{C_{mNP}}{(h_{CG} - h_{NP})} \tag{3.48}$$

Then, with $C_{mNP} > 0$, as is usually the case with a horizontal tail in place, for a positive C_L^*, we require $(h_{CG} - h_{NP}) < 0$, that is, the CG must remain ahead of the NP.

And then, from Equation 3.45, we deduce that with $(h_{CG} - h_{NP}) < 0$ and $C_{L\alpha}$ of course positive, we obtain negative $C_{m\alpha}$. Hence, this arrangement allows for stability in pitch (stable short-period dynamics).

The multiplicity of formulas in this chapter may be slightly confusing and worrying to some. So, we collect all the relevant ones in a single Table 3.1 for easy reference.

TABLE 3.1
Summary of Formulas from Chapter 3

$$C_L = C_{L0} + C_{L\alpha}\alpha + C_{L\dot{\alpha}}\dot{\alpha}(c/2V^*) \quad (3.17)$$

$$C_{L0} = C_{L0}^{wb} - \frac{S_t}{S}C_{L\alpha}^t(i_t + \varepsilon_0) \quad (3.17a)$$

$$C_{L\alpha} = C_{L\alpha}^{wb} + \frac{S_t}{S}C_{L\alpha}^t(1 - \varepsilon_\alpha) \quad (3.17b)$$

$$C_{L\dot{\alpha}} = -\frac{S_t}{S}C_{L\alpha}^t\varepsilon_{\dot{\alpha}} \quad (3.17c)$$

$$C_{mCG} = C_{m0} + C_{m\alpha}\alpha + C_{m\dot{\alpha}}\dot{\alpha}(c/2V^*) \quad (3.24)$$

$$C_{m0} = C_{mAC}^{wb} + C_{L0}^{wb}(h_{CG} - h_{AC}^{wb}) + V_H C_{L\alpha}^t(i_t + \varepsilon_0) \quad (3.24a)$$

$$C_{m0} = C_{mAC}^{wb} + C_{L0}(h_{CG} - h_{AC}^{wb}) + V_H^{\prime} C_{L\alpha}^t(i_t + \varepsilon_0) \quad (3.39)$$

$$C_{m0} = C_{mNP} + C_{L0}(h_{CG} - h_{NP}) \quad (3.44c)$$

$$C_{m\alpha} = C_{L\alpha}^{wb}(h_{CG} - h_{AC}^{wb}) - V_H C_{L\alpha}^t(1 - \varepsilon_\alpha) \quad (3.24b)$$

$$C_{m\alpha} = C_{L\alpha}(h_{CG} - h_{AC}^{wb}) - V_H^{\prime} C_{L\alpha}^t(1 - \varepsilon_\alpha) \quad (3.40)$$

$$C_{m\alpha} = C_{L\alpha}(h_{CG} - h_{NP}) \quad (3.45)^*$$

$$C_{m\dot{\alpha}} = V_H C_{L\alpha}^t\varepsilon_{\dot{\alpha}} \quad (3.24c)$$

$$C_{m\dot{\alpha}} = (h_{CG} - h_{AC}^{wb})C_{L\dot{\alpha}} + V_H^{\prime} C_{L\alpha}^t\varepsilon_{\dot{\alpha}}$$

$$C_{mCG} = C_{mNP} + C_L(h_{CG} - h_{NP}) \quad (3.47)^*$$

$$h_{NP} = h_{AC}^{wb} + V_H\left\{\frac{C_{L\alpha}^t}{C_{L\alpha}^{wb}}\right\}(1 - \varepsilon_\alpha) \quad (3.27)$$

$$h_{NP} = h_{AC}^{wb} + V_H^{\prime}(C_{L\alpha}^t / C_{L\alpha})(1 - \varepsilon_\alpha) \quad (3.41)$$

* Sometimes an approximate version of these with C_L^{wb} in place of C_L, and $C_{L\alpha}^{wb}$ in place of $C_{L\alpha}$, may be used. The $C_{m\dot{\alpha}}\dot{\alpha}(c/2V^*)$ term has not been included in Equation (3.47).

Question: Which one is more accurate—the expressions with V_H or the ones with $V_H^{/}$? Looking at Table 3.1, the formulas with V_H and the corresponding ones with $V_H^{/}$ are entirely equivalent. So, it is perfectly fine to use either of them, depending on how the data are presented.

3.7 EFFECT OF CG MOVEMENT

We have seen that the choice of the CG location for an airplane has a significant impact on its trim and stability characteristics. We have seen how to select the horizontal tail parameters, i_t and V_H, for a given CG location to obtain a desired design trim point and stability level. Having done that, however, we have to reckon with the movement of CG in flight.

The CG movement of an airplane in flight can be due to several causes; consumption of fuel and ejection of any kind of mass (stores, cargo, etc.) are the dominant reasons. In smaller airplanes, passenger weight distribution can also be a significant factor determining the CG.

To understand the effect of a shift in CG position on airplane trim and stability, refer to Figure 3.12 where the net lift (C_L) and pitching moment (C_{mNP}) of the airplane have been placed at the NP as discussed earlier. First, from physical arguments.

Assume that the CG has shifted back (aft—towards the NP in Figure 3.12) by a small distance.

- With a smaller moment arm, the nose-down moment at the CG due to C_L is decreased.
- Hence, there is an effective nose-up moment created.
- The airplane would pitch up to a higher angle of attack.
- At the higher angle of attack, generally one may expect a larger C_L.
- The increase in C_L acting through the same moment arm gives an additional nose-down moment at the CG.
- This subtracts from the nose-up moment created earlier until $C_{mNP} + C_L(h_{CG} - h_{NP}) = 0$ again, when a new trim is established. By then, the airplane has pitched up to a higher angle of attack.

Now, the same from the equations for the pitching moment in Table 3.1:

Usually, $C_{L0} \sim 0$, so Equation 3.44c with the remaining significant term looks like:

$$C_{m0} = C_{mNP} + C_{L0}(h_{CG} - h_{NP}) \sim C_{mNP}$$

and you can see that C_{m0} is pretty much unaffected by a change in h_{CG}. But, from Equation 3.45:

$$C_{m\alpha} = C_{L\alpha}(h_{CG} - h_{NP}) = -C_{L\alpha} \cdot SM$$

a rearward movement of CG, decreases the static margin, $SM = (h_{NP} - h_{CG})$, and thus $C_{m\alpha} = -C_{L\alpha} \times SM$ is also decreased in magnitude. That is, stability is reduced as the CG moves back towards the NP. We have already seen that the absolute limit of the rear CG movement where stability is lost is the location of the NP itself.

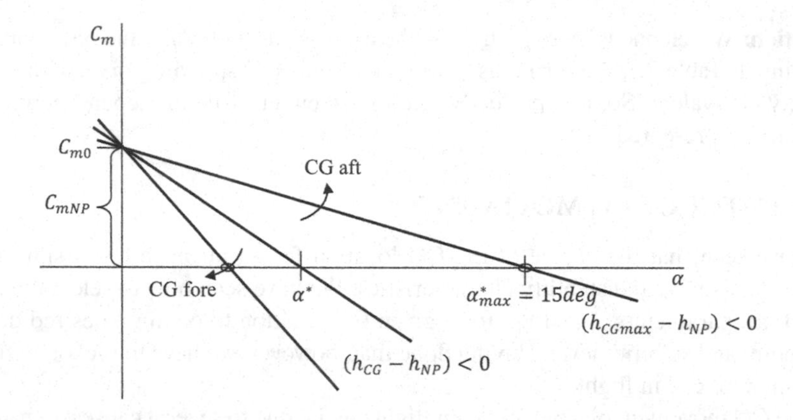

FIGURE 3.14 Graph of C_m versus α showing the effect of shift in CG position, and the practical rearmost CG position corresponding to trim at landing angle of attack, here taken to be nominally 15°.

Contrariwise, as the CG moves forward, away from the NP, the stability is increased.

Finally, let us view this graphically:

Using the data in Examples 3.4 and 3.5, let us draw the C_m versus α graph (with our present linear approximation), as shown in Figure 3.14.

To recap, $C_{m0} = 0.06$ and $C_{m\alpha} = -0.0133/°$ at the design point, which corresponds to $\alpha^* = 4.5°$ and C_L^*. This defines the full line in Figure 3.14 passing through α^*.

Remember, C_{m0} remains fairly unchanged with the CG movement.

As the CG moves aft, the slope $C_{m\alpha}$ becomes less negative; hence, the graph in Figure 3.14 swings up as shown to intersect the α axis at a larger value of the angle of attack (and hence C_L). Likewise, as the CG moves forward, the graph swings down and intersects the α axis at a smaller value as marked in Figure 3.14.

With a changed trim value of α (and C_L), comes a different trim velocity V^*, which can be quite inconvenient as we may have a desired value of trim velocity for our flight. How can we compensate for the effect of CG movement and yet retain the same trim flight condition? We shall examine this question in Chapter 4.

Example 3.7

Imagine an airplane where different trims are obtained by movement of the CG alone. We shall use the same data as in Figure 3.14. We plan to land this airplane at an angle of attack of $\alpha^* = 15°$, which is a reasonable upper limit for general aviation and commercial transport airplanes. How far back must the CG be moved for the airplane to trim at this angle of attack at landing?

To determine this limit, draw a line on Figure 3.14 from $C_{m0} = 0.06$ on the Y axis to $\alpha^* = 15°$ on the X axis. The slope of this line is $C_{m\alpha}$, and this works out to:

$$C_{m\alpha} = -\frac{0.06}{15} = -0.004/°$$

Corresponding to this value of $C_{m\alpha}$, we can find the CG location:

$$C_{m\alpha} = C_{L\alpha}^{wb}\left(h_{CG} - h_{AC}^{wb}\right) - V_H C_{L\alpha}^{t}\left(1 - \varepsilon_\alpha\right)$$

$$-0.004 = 0.08\left(h_{CG} - h_{AC}^{wb}\right) - 0.34 \times 0.1 \times (1 - 0.35)$$

$$\left(h_{CG} - h_{AC}^{wb}\right) = 0.226$$

So, at landing, the CG must be moved to a point 0.226*c* behind the wing–body aerodynamic centre.

From the data and previous examples, we know that at the design trim point $\alpha^* = 4.5°$, the CG was already 0.11*c* behind AC^{wb}. Now, to trim at $\alpha^* = 15°$, the CG must move an additional 0.116*c* rearward. Note that the NP in this problem is 0.266*c* behind the AC^{wb}.

Finally, we can determine the new C_L^* at $\alpha^* = 15°$ as:

$$C_L^* = C_{L0} + C_{L\alpha}\alpha^*$$
$$= 0.027 + 0.0865 \times 15 = 1.27$$

Observe that trimming at a new angle of attack with CG movement alone necessarily means a change in stability, which is undesirable. So, in reality, airplanes use elevator control to move from one trim state to another, with unchanged stability characteristics, as we shall see in Chapter 4.

3.8 REAR CG LIMIT DUE TO AIRPLANE LOADING AND CONFIGURATION AT TAKE-OFF

Theoretically, in flight, we have seen that the rearmost position of the CG is marked by the NP, where $C_{m\alpha}$ is zero. However, in practice, we would want a definite value of stability for the short-period mode; hence, a finite negative number is desirable for $C_{m\alpha}$. We shall see later how this number is arrived at.

For the present, let us note another possible source that may limit the rearmost position of the CG of an airplane, and this comes from its take-off ground run. During take-off, the airplane lifts up its nose wheel and swivels about the main rear wheels. We usually allow for a rotation angle of about 15°. Examine Figure 3.15, which shows an airplane at take-off with the rear wheels still on the

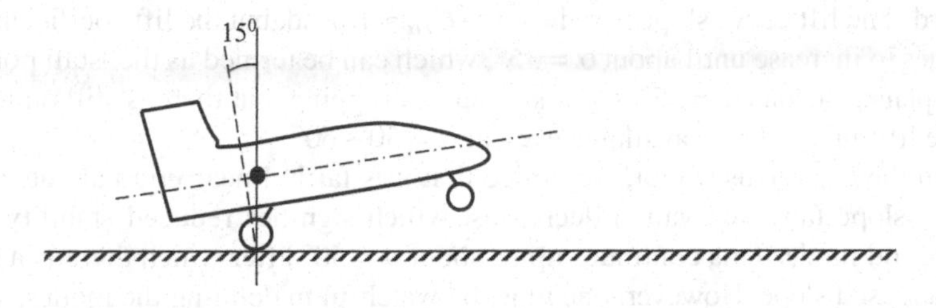

FIGURE 3.15 Sketch of airplane and rearmost allowable CG position during rotation at take-off ground run.

FIGURE 3.16 Example of a 'tail-sitting' airplane due to improper loads on the ground (http://www.aerospaceweb.org/aircraft/jetliner/b747/b747_21.jpg).

ground and the nose rotated up by an angle of 15°. With reference to the dashed line marked from the rear landing gear, the vertical with the ground makes an angle of 15°. This gives a location on the airplane reference axis, marked by a black circle in Figure 3.15. If the CG were to be behind (aft of) this location, then, in this condition, the airplane would tip over and sit on its tail on the runway. See Figure 3.16 for an example. To avoid such an embarrassment, the airplane CG should be ahead of this rearward location (black circle), such as indicated in Figure 3.15 for example.

3.9 C_m, C_L CURVES—NON-LINEARITIES

Before we close this chapter, it will do us good to take a look at what the trend of C_L and C_m for a real airplane looks like. Thus far, our analysis has been limited to the linear approximation, which is usually valid up to about 15° angle of attack (this number varies from airplane to airplane). That is also the usual limit of controlled flight for many general aviation and commercial airplanes, but many military aircraft fly beyond that limit. As an example, C_L and C_m curves for the F-18/high-angle-of-attack research vehicle (HARV) are shown in Figure 3.17.

You notice that the linear approximation is quite valid for α up to 15° or so with a lift-curve slope of $C_{L\alpha} \approx 4.77$/rad. Beyond that, there is no stall in the usual sense of the word. The lift-curve slope is reduced to $C_{L\alpha} \approx 1.7$/rad, but the lift coefficient itself continues to increase until about $\alpha = 35°$, which can be termed as the 'stall point' for this airplane. Beyond $\alpha = 35°$, C_L at last starts dropping, but there is still quite some positive lifting capability available even at $\alpha = 50 - 60°$.

From the C_m versus α plot, we notice that it is fairly linear up to about $\alpha = 10°$, then the slope (absolute value) decreases, which signifies reduced stability in the short-period mode. This continues up to about $\alpha = 50°$ after which there is a stretch with increased slope. However, one must be watchful in defining the modes, such as short period, at such high values of α; some of the assumptions we make in obtaining their equations and characteristics may not hold.

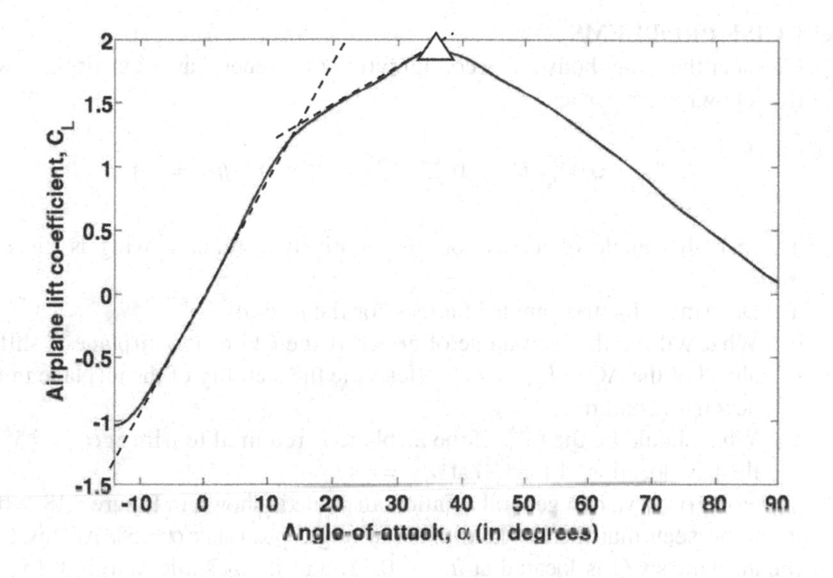

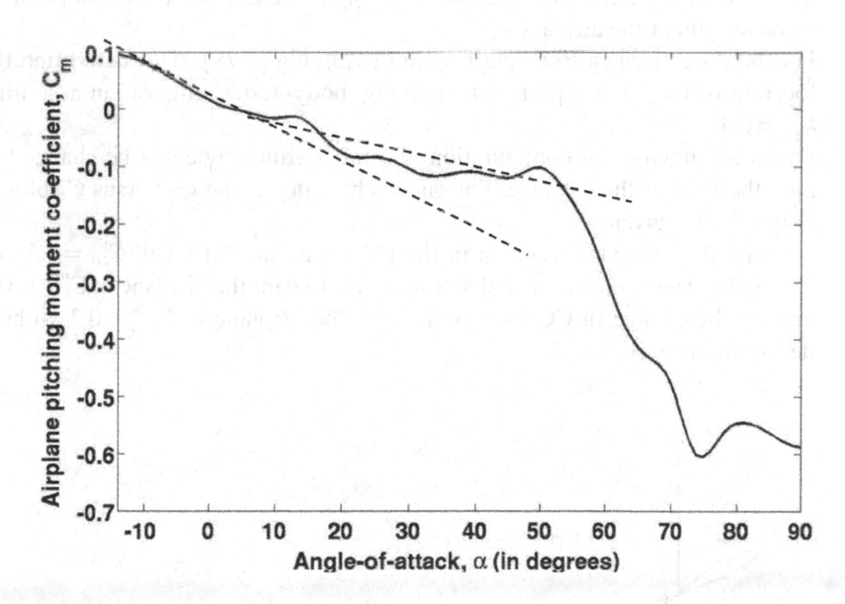

FIGURE 3.17 Variation of C_L and C_m for the F-18/HARV from −14° to +90° angle of attack (Δ-Stall).

EXERCISE PROBLEMS

3.1 Consider the wing–body alone configuration of a general aviation airplane with the following properties:

$$C_{mAc}^{wb} = -0.04, \quad h_{AC}^{wb} = 0.25, \; C_{L\alpha}^{wb} = 4.5/\text{rad}, \quad h_{CG} = 0.4$$

The zero-lift angle of attack for the positively cambered wing is given as $\alpha_0 = -2°$.

a. Determine the trim angle of attack for the aircraft.
b. What will be the trim angle of attack if the CG of the airplane is shifted ahead of the AC to $h_{CG} = 0.1$? Determine the stability of the airplane in this new trim condition.
c. What should be the C_{mAC}^{wb} if the airplane is required to trim at $\alpha_{\text{trim}} = 5°$ for the new location of the CG at $h_{CG} = 0.1$?

3.2 C_m versus α curve of a general aviation airplane is shown in Figure 3.18, where it can be seen that the airplane trims at angle of attack $\alpha = 5°$. At this trim, the airplane's CG is located at $h_{CG} = 0.25$, and it has static margin 0.15. The airplane is required to change its trim angle of attack to $\alpha = 10°$ by changing its *CG* location. Determine the new location of the CG and the corresponding static margin of the airplane.

3.3 For the data given in Example 3.4, and assuming $(S_t/S) = 0.1$, determine the location of the neutral point for the wing–body–tail configuration assuming $h_{CG} = 0.4$.

3.4 On an all-moving tail configuration, the tail setting angle can be changed to alter the trim of the airplane. For one such airplane, the C_m versus C_L plot of Figure 3.19 is given.

Using the information given in the plot in Figure 3.19, and $C_{L\alpha}^t = 4.5/\text{rad}$, $V_H = 0.25$, determine (a) the tail setting angle to trim the airplane at $C_L^{trim} = 0.3$ and (b) the change in CG location to trim the airplane at $C_L^{trim} = 0.3$ without deflecting the tail.

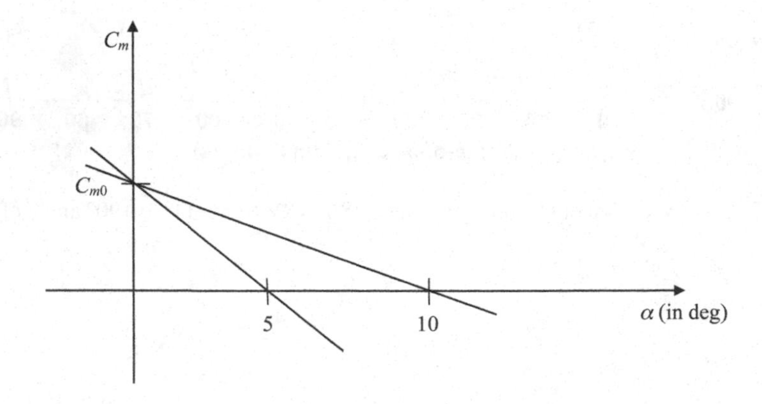

FIGURE 3.18 C_m versus α curve of a general aviation airplane.

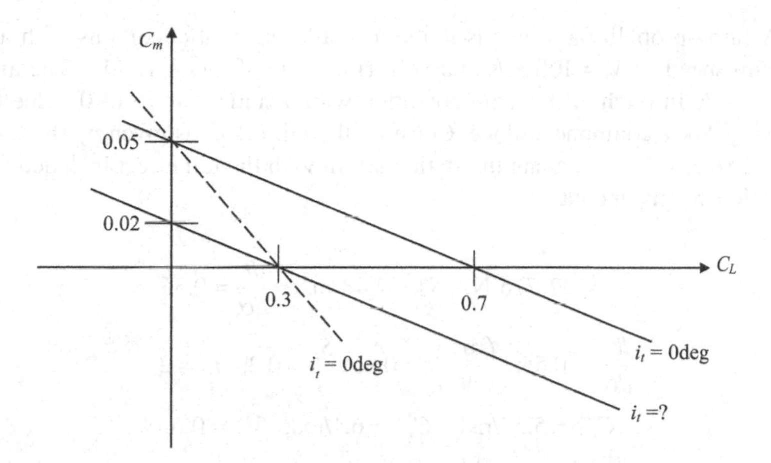

FIGURE 3.19 C_m–C_L curve for an airplane for different tail setting angles.

3.5 Carry out trim and stability analysis of a canard plus wing configuration (similar to the one with a rear horizontal tail presented in this chapter) and show that $C_{mCG} = C_{mNP} + C_L\left(h_{CG} - h_{NP}\right)$ in this case. [*Hint*: Instead of downwash, here, the upwash at the canard due to wing-bound vortex needs to be accounted for, which adds to the canard effective angle of attack. Neglect the downwash effect at the wing due to canard.]

3.6 A missile (body plus tail configuration) has a fuselage with circular cross section consisting of a cylinder with diameter D, cross-sectional area S and a forebody and a tail. The tail lift-curve slope is 0.1/°. Other data: $(S_t/S) = 1.33$, $(l_t/D) = 4.5$, $C_{m\alpha,fuselage} = 0.4/°$. Find the tail setting angle i_t required to trim at 5° angle of attack.

3.7 Airflow from the tip of the running propeller of a turbo-prop engine, known as propeller slipstream, indirectly affects the pitching moment characteristics and hence the trim and stability of an airplane. This effect is primarily due to the tail lying in the wake of the slipstream, which in turn affects the downwash. The tail pitch stability contribution due to propeller stream can be modelled using the formula (*Airplane Performance Stability and Control* by C.D. Perkins and R.E. Hage, Chapter 5, John Wiley and Sons Publications, 1949).

$$\left(\frac{dC_m}{dC_L}\right)_t = -\frac{C_{L\alpha}^t}{C_{L\alpha}^{wb}} V_H \left(1 - \frac{d\varepsilon}{d\alpha} - \frac{d\varepsilon_p}{d\alpha}\right)\left(\frac{\upsilon_s}{V}\right)^2 - V_H C_L^t \frac{d(\upsilon_s/V)^2}{dC_L}$$

where v_s is the slipstream velocity and $d\varepsilon_p/d\alpha$ is propeller downwash slope, which is always positive (a destabilizing effect) and $\left(d\left(v_s/V\right)^2\right)/dC_L$ is always positive (resulting in a stabilizing effect when C_L^t is positive and destabilizing when C_L^t is negative). The second term largely helps in controlling the destabilizing effect due to aft movement of CG. Can you reason out how?

3.8 A turbo-propeller airplane is flying at sea-level flight conditions with a level trim speed of $V = 100$ m/s and at a trim angle of attack $\alpha = 1°$. The aircraft is stable in pitch at this trim condition with a static margin of 0.3 due to the wing–body component alone. Consider the tail effect as given by the formula in Exercise 3.7. Calculate the static margin with the tail effect included for the following given data:

$$W = 27{,}776 \text{ N}; \quad S = 22.48 \text{ m}^2; \quad \frac{d\varepsilon}{d\alpha} = 0.35$$

$$\frac{d\varepsilon_p}{d\alpha} = 0.55; \quad \left(\frac{\upsilon_s}{V}\right)^2 = 0.9; \quad \frac{S_t}{S} = 0.2; \quad i_w = 2°$$

$$C_{L\alpha}^{wb} = 5.57/\text{rad}; \quad C_{L\alpha}^{t} = 6.1/\text{rad}; \quad V_H = 0.6$$

where, i_w is the wing incidence angle. The variation of $(v_s/V)^2$ with airplane lift coefficient C_L is given via the relation $(\upsilon_s/V)^2 = (1+(8T_c/\pi))$. $T_c = (T/\rho V^2 D^2)$ is the thrust coefficient depending on the thrust output T, ρ the density of air and D the propeller disk diameter. Further, T_c depends on C_L via $T_c = K\eta_p C_L^{3/2}$. K is a constant of proportionality that depends on engine power rating, wing loading and propeller disk diameter and η_p is the propeller efficiency (*Airplane Performance Stability and Control* by C.D. Perkins and R.E. Hage, Chapter 5, John Wiley and Sons Publications, 1949). Assume $K = 0.5$, and $\eta_p = 0.8$ in your calculations.

3.9 The contribution of a turbojet engine to the pitching moment coefficient can be arrived at as:

$$(\Delta C_{mCG})_{jet} = \left(\frac{Th}{\bar{q}Sc}\right)$$

h is the vertical distance between the thrust line and the CG of the airplane. Assuming an unaccelerated flight condition, discuss the contribution of a turbojet engine to the pitch stability of the airplane.

3.10 Consider the $C_m - C_L$ curves for wing–body and wing–body–tail (total airplane) in Figure 3.20. Use the following data to determine the horizontal tail volume ratio V_H and ε_0.

$$C_{L\alpha}^{t} = 6.1/\text{rad}; \quad C_{L\alpha}^{w} = 5.1/\text{rad}; \quad AR_w = 10; \quad i_w = 2°$$

3.11 Fuselage contribution to aircraft pitch stability is usually small and destabilizing. Max Munk (1920s) and later Multhopp developed analytical expressions for estimating fuselage contribution to $C_{m\alpha}$. Much later, simpler expressions based on experimental correlations were proposed by Hoak in 1960. The formula (from *Mechanics of Flight* by W.F. Phillips, Chapter 4, John Wiley and

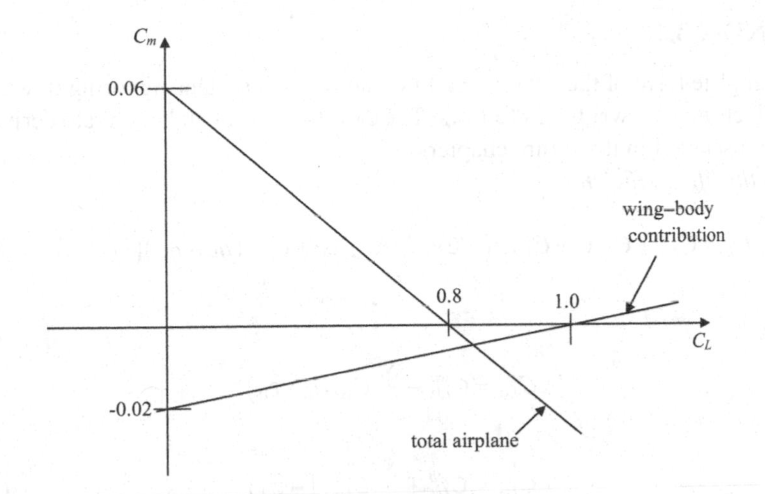

FIGURE 3.20 C_m–α curve for an airplane with contributions from individual components

Sons Publications, 2009) proposed by Hoak can be equally well applied to estimate contributions to $C_{m\alpha}$ from engine nacelles and external stores (bodies of revolution). These expressions are given by:

$$\Delta C_{m,f} = \Delta C_{m0,f} + \Delta C_{m\alpha,f}\alpha$$

Where,

$$\Delta C_{m0,f} = -2\frac{S_f l_f}{Sc}\left[1-1.76\left(\frac{2\sqrt{S_f/\pi}}{c_f}\right)^{3/2}\right]\alpha_{0f}$$

$$\Delta C_{m\alpha,f} = -2\frac{S_f l_f}{Sc}\left[1-1.76\left(\frac{2\sqrt{S_f/\pi}}{c_f}\right)^{3/2}\right]$$

S_f in the above expressions is the maximum cross-sectional area of the fuselage along its length, c_f is the length of the fuselage, l_f is the distance that the fuselage centre of pressure is aft of the CG of the airplane and α_{0f} is the angle that minimum drag axis of fuselage makes with the fuselage reference line. S and c are the wing planform area and the wing mean aerodynamic chord, respectively. l_f is usually negative, thus making the fuselage contribution destabilizing. Determine the fuselage contribution to pitch stability for an airplane with the following geometric properties:

Maximum fuselage diameter, d = 2 m; length of the fuselage, c_f = 8.5 m; location of CG from the nose of the fuselage, X_{CG} = 3 m; distance of fuselage centre of pressure from nose of the fuselage, $X_{CP,f}$ = 3.2 m; S = 16 m^2; c = 2 m.

APPENDIX 3.1

The complete form of the lift and pitching moment coefficients ignoring flow curvature effects may be written as follows. The elevator control and pitch rate derivatives will be discussed in the future chapters.

For the lift coefficient:

$$C_L = C_{L0} + C_{L\alpha}\alpha + C_{L\dot{\alpha}}\dot{\alpha}\left(c/2V^*\right) + C_{L\delta e}\delta e + C_{Lq1}\left(q_b - q_w\right)\left(c/2V^*\right) \quad (3A.1)$$

where,

$$C_{L0} = C_{L0}^{wb} - \frac{S_t}{S} C_{L\alpha}^t \left(i_t + \varepsilon_0\right) \quad (3A.2)$$

$$C_{L\alpha} = C_{L\alpha}^{wb} + \frac{S_t}{S} C_{L\alpha}^t \left(1 - \varepsilon_\alpha\right) \quad (3A.3)$$

$$C_{L\dot{\alpha}} = -\frac{S_t}{S} C_{L\alpha}^t \varepsilon_{\dot{\alpha}} \quad (3A.4)$$

$$C_{L\delta e} = \left(\frac{S_t}{S}\right) C_{L\delta e}^t \quad (3A.5)$$

$$C_{Lq1} = 2V_H C_{L\alpha}^t \quad (3A.6)$$

For the pitching moment coefficient:

$$C_{mCG} = C_{m0} + C_{m\alpha}\alpha + C_{m\dot{\alpha}}\dot{\alpha}\left(c/2V^*\right) + C_{m\delta e}\delta e + C_{mq1}\left(q_b - q_w\right)\left(c/2V^*\right) \quad (3A.7)$$

where,

$$C_{m0} = C_{mAC}^{wb} + \left(h_{CG} - h_{AC}^{wb}\right) C_{L0} + V_H^l C_{L\alpha}^t \left(i_t + \varepsilon_0\right) \quad (3A.8)$$

$$C_{m\alpha} = \left(h_{CG} - h_{AC}^{wb}\right) C_{L\alpha} - V_H^l C_{L\alpha}^t \left(1 - \varepsilon_\alpha\right) \quad (3A.9)$$

$$C_{m\dot{\alpha}} = \left(h_{CG} - h_{AC}^{wb}\right) C_{L\dot{\alpha}} + V_H^l C_{L\alpha}^t \varepsilon_{\dot{\alpha}} \quad (3A.10)$$

$$C_{m\delta e} = \left(h_{CG} - h_{AC}^{wb}\right) C_{L\delta e} - V_H^l C_{L\delta e}^t \quad (3A.11)$$

$$C_{mq1} = \left(h_{CG} - h_{AC}^{wb}\right) C_{Lq1} - 2V_H^l C_{L\alpha}^t \left(\frac{l_t}{c}\right) \quad (3A.12)$$

4 Longitudinal Control

We have seen that the horizontal tail plays a critical role in trimming and stabilizing the airplane pitching motion. By judiciously selecting the horizontal tail setting angle i_t and the horizontal tail volume ratio V_H, one can trim at a desired angle of attack, and hence velocity, as well as render the pitching motion (short-period dynamics) stable (in the sense discussed in Chapter 2). We have also seen that any change in the centre of gravity (CG) location induces a change in the trim and stability. The other way round, by deliberately moving the CG position, one can obtain a different trim velocity in flight, but there are a couple of issues here. Firstly, the new trim comes with a different degree of stability, and, secondly, moving the CG may not be such an easy task (except for hang gliders, perhaps). So, we need a better device that lets us move from one trim state to another, or even make more general manoeuvres in pitch, that does not share the drawbacks of CG movement.

4.1 ALL-MOVING TAIL

One obvious answer comes from studying the expression for the pitching moment coefficient that we derived in Chapter 3. Let us look back at those.

$$C_{mCG} = C_{m0} + C_{m\alpha}\alpha + C_{m\dot{\alpha}}\dot{\alpha}\left(c/2V^*\right) \tag{3.24}$$

$$C_{m0} = C_{mAC}^{wb} + C_{L0}^{wb}\left(h_{CG} - h_{AC}^{wb}\right) + V_H C_{L\alpha}^{t}\left(i_t + \varepsilon_0\right) \tag{3.24a}$$

$$C_{m\alpha} = C_{L\alpha}^{wb}\left(h_{CG} - h_{AC}^{wb}\right) - V_H C_{L\alpha}^{t}\left(1 - \varepsilon_\alpha\right) \tag{3.24b}$$

$$C_{m\dot{\alpha}} = V_H C_{L\alpha}^{t}\varepsilon_{\dot{\alpha}} \tag{3.24c}$$

From the sketch in Figure 3.4b, we know that the ideal way to shift from one trim α^* to another is to move the straight line with slope $C_{m\alpha}$ parallel to itself—to the right for a higher trim α^*, and to the left for a lower trim α^*—so that the slope itself remains unchanged. Thus, the stability is unaffected; only the trim point on the X axis changes. For this to be possible, the Y-intercept – C_{m0} must change. Examining Equation 3.24a, we discover that by changing i_t, we can indeed change C_{m0}, and it also leaves $C_{m\alpha}$ unchanged. A horizontal tail whose setting angle i_t can be changed in flight is called an *all-moving tail*.

A typical trim curve with different setting angles for an all-moving tail is shown in Figure 4.1. With changing tail settings, the $C_m - \alpha$ curve shifts (up for $i_t > 0$, leading edge down and downward for $i_t < 0$, leading edge up) without a change in slope. This also changes the intercept of the curve on the α – axis ($C_m = 0$), thus changing the trim angle of attack.

DOI: 10.1201/9781003096801-4

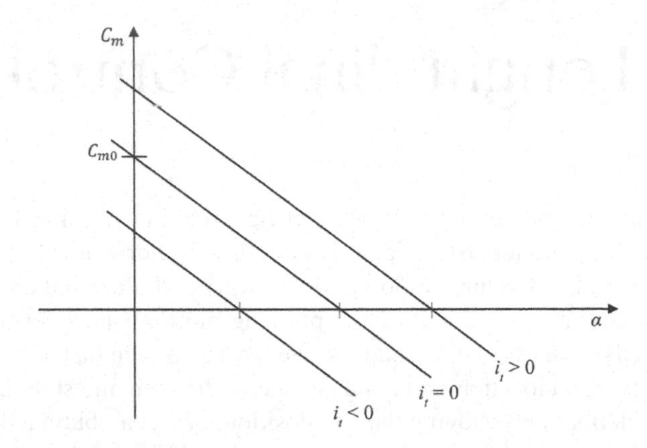

FIGURE 4.1 $C_m - \alpha$ curves for different tail settings (tail leading edge down is positive i_t) showing trimming at different angles of attack.

How does it work? When i_t is changed, it changes the tail lift, which on acting about the airplane CG, creates an unbalanced moment that swivels the airplane nose either up or down, changing its angle of attack. This leads to a change in lift, which manoeuvres the airplane into a pull-up (trajectory curves upward) or a push-down (trajectory curves down). Eventually, if i_t is held at a new fixed value, the airplane may be expected to settle into a new trim state at a different velocity V^* such that $L = W$ still holds.

Example 4.1

Many high-speed airplanes employ an all-moving tail. Can you guess why? An example is the F-111 shown in Figure 4.2, with its tail deflected at an extreme angle.

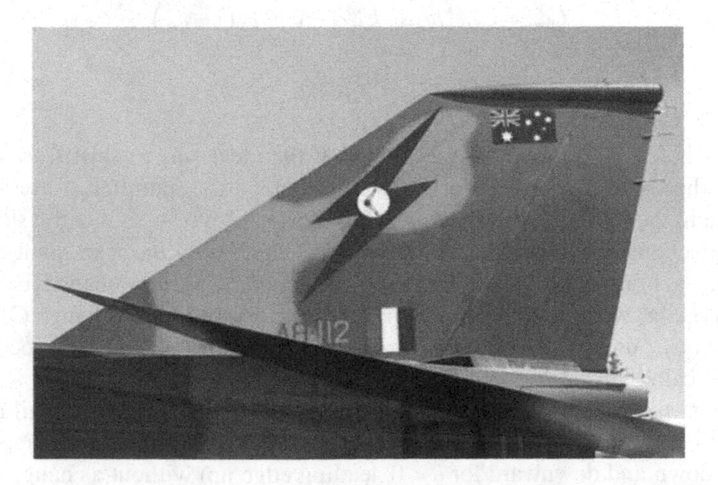

FIGURE 4.2 All-moving tail deflection on F-111 (http://lh6.ggpht.com/_kIWY2DV0KnE/TPFmFd9fwbI/AAAAAAAAIDE/WUnvwGvsbyg/F-111%20tail%20showing%20all-moving%20horizontal%20stabilizer.jpg).

However, deflecting an entire tail surface about its hinge line usually requires a powerful (and heavy) actuator, and the deflected surface can add to the tail drag. For most conventional airplanes with wing–body plus tail configuration, the same effect can be achieved by moving a small flap attached to the tail at its trailing edge, known as 'elevator' as we observed in Chapter 2. In the rest of this chapter, we shall study airplane trim and control using an elevator.

4.2 ELEVATOR

Figure 4.3 shows a sketch of an aircraft having a wing–body plus horizontal tail. At the trailing edge of the tail, you can notice a small flap attached; this is called the 'elevator'. As we shall see, an elevator does the same job as an all-moving tail, except that it needs far less effort to deflect and hold it in its deflected position.

When the trailing edge of the elevator is deflected down with respect to the tail axis (usually the same as the chord line), as in Figure 4.3, it is taken to be a positive elevator deflection. In the opposite sense, it is a negative elevator deflection.

The sign convention for control surfaces comes from the sense in which the corresponding angular rate is assigned. Since the elevator predominantly causes a pitching motion, the corresponding angular rate is the pitch rate. The positive sense of pitch rate is obtained by the right-hand rule—when pointing the right thumb along the positive Y^B axis (into the plane of the paper in Figure 4.3) the direction of the curling fingers gives the sense of positive pitch rate. Thus, a nose-up (and tail-down) pitch rate is taken as positive. At the trailing edge of the horizontal tail, where the elevator is located, the curling fingers indicate a downward deflection; hence, the positive sense of elevator deflection is (trailing edge) downwards. In other words, in Figure 4.3, the sense of the arrows for both the pitch rate q and the elevator deflection δe is clockwise. Likewise, for a canard, the positive clockwise sense would mean that a nose-up, tail-down deflection is positive.

Note: A positive elevator deflection (downward) results in a negative pitching moment and hence a negative pitch rate, q.

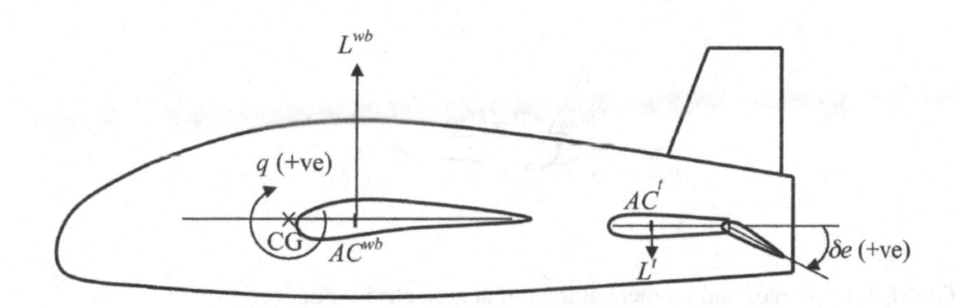

FIGURE 4.3 Wing–body plus horizontal tail configuration with positive elevator (down) deflection.

4.3 TAIL LIFT WITH ELEVATOR

From the sketch in Figure 4.3, clearly the purpose of the tail and an elevator on it conveniently placed at a significant distance behind the CG of the airplane is to make use of changing aerodynamic forces (predominantly lift) over the tail to achieve a pitching moment about the CG. Let us look at how the elevator deflection affects the tail lift.

A horizontal tail is usually made up of a symmetric airfoil section, as shown in the sketch in Figure 4.4. When this tail is kept at an angle of attack, α_t, with respect to the relative wind, it results in a lift L^t and a drag D^t acting at the aerodynamic centre of the tail. The moment $M^t_{AC} \approx 0$ in this case because the tail section is symmetric. This is so when the elevator is in its neutral (or undeflected) position.

We have seen that the tail lift coefficient is defined as:

$$C^t_L = \frac{L^t}{\overline{q}S_t} \tag{4.1}$$

where we do not quibble over a possible slight difference between the freestream dynamic pressure $\overline{q}$ and its value at the tail, $\overline{q}_t$. In terms of the tail angle of attack:

$$C^t_L = \frac{\partial C^t_L}{\partial \alpha_t} \cdot \alpha_t = C^t_{L\alpha}\alpha_t \tag{4.2}$$

Example 4.2

Calculation of Lift at the Tail

The following data are given for a large transport airplane (wing–body plus tail configuration) flying at a speed $V = 100$ m/s and angle of attack of 15 deg at sea level. Consider wing details as follows:

$$C^{wb}_{L0} = 0.5; \quad C^{wb}_{L\alpha} = 4.44/\text{rad}; \quad S = 300 \text{ m}^2; \quad b = 60 \text{ m}$$

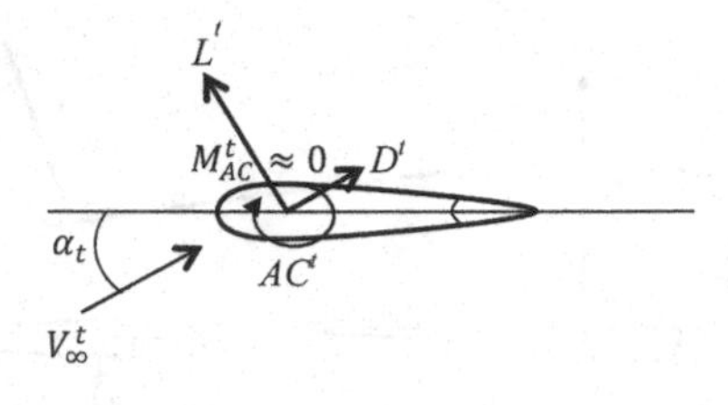

FIGURE 4.4 Forces and moment at the tail at zero elevator deflection.

The tail is unswept of planform area $S_t = 0.1S$ made up of a symmetric airfoil section and is set at an angle $i_t = 3°$ with respect to the reference axis of the aircraft. The tail lift curve slope can be approximated as:

$$C^t_{L\alpha} = 2\pi/\text{rad (from thin airfoil theory)}.$$

To determine the angle of attack of the tail, use $\alpha_t = \alpha - i_t - \varepsilon$, with an estimate for the downwash angle at the tail given by:

$$\varepsilon = \frac{2C^{wb}_L}{\pi AR_w} = \frac{2C^{wb}_{L0}}{\pi AR_w} + \alpha \frac{2C^{wb}_{L\alpha}}{\pi AR_w} = \varepsilon_0 + \varepsilon_\alpha \alpha$$

Using the given data, the wing aspect ratio AR_w can be found as:

$$AR_w = \frac{b^2}{S} = \frac{60^2}{300} = 12$$

Thus,

$$\varepsilon_0 = \frac{2C^{wb}_{L0}}{\pi AR_w} = \frac{2 \times 0.5}{\pi \times 12} = 0.0265 \text{ rad} = 1.52°$$

and

$$\varepsilon_\alpha = \frac{2C^{wb}_{L\alpha}}{\pi AR_w} = \frac{2 \times 4.44}{\pi \times 12} = 0.236/\text{rad} = 0.004/°$$

Therefore,

$$\begin{aligned}\alpha_t &= \alpha - \varepsilon_0 - \varepsilon_\alpha \alpha - i_t \\ &= 15° - 1.52° - 0.004 \times 15° - 3° \\ &= 10.41°\end{aligned}$$

$$C^t_L = C^t_{L\alpha} \times \alpha_t = 2\pi \times 10.41 \times \frac{\pi}{180} = 1.141$$

$$\begin{aligned}L^t &= \frac{1}{2}\rho_{\text{air}} V^2 S_t C^t_L = \frac{1}{2}\rho_{\text{air}} \times V^2 \times (0.1 \times S) \times C^t_L \\ &= \frac{1}{2} \times 1.225 \times 100^2 \times 0.1 \times 300 \times 1.141 = 209{,}658.75 \text{ N}\end{aligned}$$

$$\begin{aligned}L^{wb} &= \frac{1}{2}\rho_{\text{air}} V^2 S C^{wb}_L = \frac{1}{2} \times 1.225 \times 100^2 \times 300 \times \left(C^{wb}_{L0} + C^{wb}_{L\alpha}\alpha\right) \\ &= 0.5 \times 1.225 \times 100^2 \times 300 \times \left(0.5 + 4.44 \times 15 \times \frac{\pi}{180}\right) \\ &= 3{,}054{,}640.31 \text{ N}\end{aligned}$$

The ratio of tail lift to the lift produced at the wing is $(L^t/L^{wb}) = 0.0686$.

Note that the total lift produced at the tail for zero elevator deflection in this case is only 6.86% of that produced at the wing.

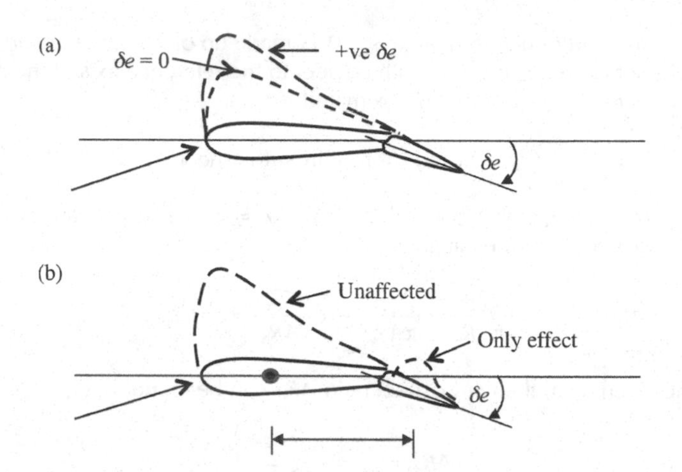

FIGURE 4.5 Tail plus elevator combination in: (a) subsonic and (b) supersonic flow.

How can we visualize and model the tail lift with the elevator in place? Elevator deflection, at least in subsonic flight, induces a change in the tail chordwise lift distribution, as depicted approximately in Figure 4.5a. A down (positive) elevator deflection is seen to create an increase in tail lift (and vice versa). Thus, the effect of deflecting the elevator is qualitatively similar to deflecting the entire horizontal tail as far as tail lift is concerned. In supersonic flight, however, the elevator is not very effective in changing the lift distribution on the horizontal tail, which is upstream to it; rather, its main contribution is from the lift it generates over itself when deflected, as shown approximately in Figure 4.5b. Understandably, therefore, the elevator is not such an effective mechanism for changing the tail lift in supersonic flight, and so many supersonic airplanes prefer to use an all-moving horizontal tail.

One way of understanding the elevator effect on horizontal tail lift is to imagine that the elevator deflection changes the effective camber of the tail airfoil section. You may remember from a course on *aerodynamics* that for an airfoil of given thickness, lift can be induced either by setting it at an angle of attack or by providing it with a camber. The symmetric tail section by itself has no camber. As sketched in Figure 4.6, a downward deflected elevator may be thought of as introducing an effective positive camber to the tail airfoil section. Thus, elevator deflection adds to tail lift, over and above the lift due to the tail being at an angle of attack to the relative wind.

To a first approximation, these two effects can be assumed to be independent, and the net tail lift can be written as a sum of lift due to each of these effects—tail angle of

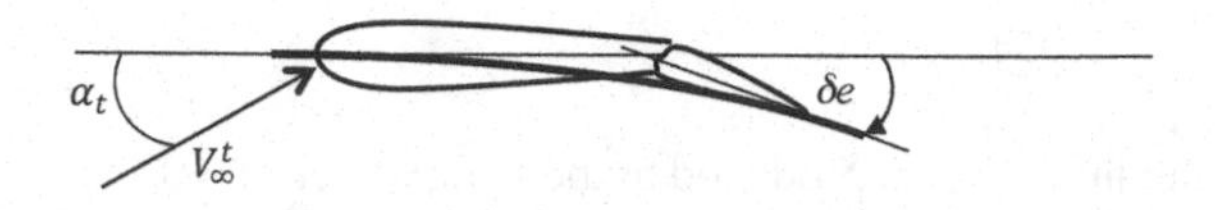

FIGURE 4.6 Tail at positive elevator deflection and equivalent camber line (shown thick).

attack (with no elevator deflection) and elevator angle (with zero tail angle of attack)—as below:

$$C_L^t = \frac{\partial C_L^t}{\partial \alpha_t}\alpha_t + \frac{\partial C_L^t}{\partial \delta e}\delta e$$
$$= C_{L\alpha}^t \alpha_t + C_{L\delta e}^t \delta e \tag{4.3}$$

The new derivative $C_{L\delta e}^t$ is the additional tail lift coefficient due to unit elevator deflection. Typically, $C_{L\alpha}^t \sim 0.1$/deg (similar to the wing) while $C_{L\delta e}^t \sim 0.05$/deg. Note that a degree change in tail setting angle i_t would change α_t by the same amount (1°) and hence change C_L^t by about 0.1. However, a degree change in δe would give a change in C_L^t of about 0.05—approximately half of that obtained from a degree of tail setting angle change. In that sense, elevator deflection is less effective than the all-moving tail. However, the elevator being a small surface with much lesser aerodynamic loads, it is far easier to deflect than moving the entire tail, so a smaller actuator with lesser weight will suffice. Also, the additional drag due to elevator deflection is far less than in the case of an all-moving tail. So, it actually makes sense to employ the elevator as the preferred pitch-control device on airplanes. Typically, elevators may undergo a full range of deflection around 30–40°.

4.4 AIRPLANE LIFT COEFFICIENT WITH ELEVATOR

Let us now rewrite the airplane lift coefficient for a wing–body plus tail configuration with an additional effect due to elevator deflection.

The inclusion of the elevator provides an additional source of lift at the tail, which is modelled in the tail lift coefficient as shown in Equation 4.3. With this addition, the total airplane lift coefficient (Equation 3.11) can now be written as:

$$C_L = C_L^{wb} + \left(\frac{S_t}{S}\right)C_L^t$$
$$= C_{L\alpha}^{wb}(\alpha_0 + \alpha) + \left(\frac{S_t}{S}\right)\left[C_{L\alpha}^t \alpha_t + C_{L\delta e}^t \delta e\right] \tag{4.4}$$
$$= C_{L\alpha}^{wb}(\alpha_0 + \alpha) + \left(\frac{S_t}{S}\right)C_{L\alpha}^t \alpha_t + \underbrace{\left(\frac{S_t}{S}\right)C_{L\delta e}^t \delta e}_{\text{Due to elevator}}$$

The only addition when compared with Equation 3.12 is the extra term due to elevator.

As before, we write the tail angle of attack as:

$$\alpha_t = \alpha - i_t - \varepsilon \tag{4.5}$$

Where the downwash angle ε is modelled as:

$$\varepsilon = \varepsilon_0 + \varepsilon_\alpha \alpha + \varepsilon_{\dot{\alpha}} \dot{\alpha}\left(\frac{c}{2V^*}\right)$$

Inserting Equation 4.5 in the expression for the lift coefficient in Equation 4.4, and rearranging the terms, yields the final expression for total airplane lift coefficient as shown below:

$$\begin{aligned}
C_L &= C_{L\alpha}^{wb}(\alpha_0+\alpha)+\left(\frac{S_t}{S}\right)C_{L\alpha}^{t}\alpha_t+\left(\frac{S_t}{S}\right)C_{L\delta e}^{t}\delta e \\
&= C_{L0}^{wb}+C_{L\alpha}^{wb}\alpha+\left(\frac{S_t}{S}\right)C_{L\alpha}^{t}\left(\alpha-i_t-\left\{\varepsilon_0+\varepsilon_\alpha\alpha+\varepsilon_{\dot\alpha}\dot\alpha\left(\frac{c}{2V^*}\right)\right\}\right)+\left(\frac{S_t}{S}\right)C_{L\delta e}^{t}\delta e \\
&= \left\{C_{L0}^{wb}+\left(\frac{S_t}{S}\right)C_{L\alpha}^{t}(-i_t-\varepsilon_0)+\left(\frac{S_t}{S}\right)C_{L\delta e}^{t}\delta e\right\} \\
&\quad +\alpha\left\{C_{L\alpha}^{wb}+\left(\frac{S_t}{S}\right)C_{L\alpha}^{t}(1-\varepsilon_\alpha)\right\}+\dot\alpha\left(\frac{c}{2V^*}\right)\left\{-\left(\frac{S_t}{S}\right)C_{L\alpha}^{t}\varepsilon_{\dot\alpha}\right\}
\end{aligned} \tag{4.6}$$

Using the aerodynamic derivative symbols we introduced in Chapter 3, this appears as:

$$C_L=\underbrace{\left(C_{L0}+C_{L\delta e}\delta e\right)}_{C_{LZA}}+C_{L\alpha}\alpha+C_{L\dot\alpha}\dot\alpha\left(c/2V^*\right) \tag{4.7}$$

where, by comparison with Equation 4.6, we can write:

$$C_{L0}=C_{L0}^{wb}-\left(\frac{S_t}{S}\right)C_{L\alpha}^{t}(i_t+\varepsilon_0) \tag{4.7a}$$

$$C_{L\alpha}=C_{L\alpha}^{wb}+\left(\frac{S_t}{S}\right)C_{L\alpha}^{t}(1-\varepsilon_\alpha) \tag{4.7b}$$

$$C_{L\dot\alpha}=-\frac{S_t}{S}C_{L\alpha}^{t}\varepsilon_{\dot\alpha} \tag{4.7c}$$

These three expressions are identical to those obtained in Equation 3.17 for the no-elevator case. Comparing Equations 4.6 and 4.7, we can also write:

$$C_{L\delta e}=\left(\frac{S_t}{S}\right)C_{L\delta e}^{t} \tag{4.7d}$$

Here, $C_{L\delta e}$ is the change in airplane lift coefficient due to a change in the elevator deflection angle, and it is defined as:

$$C_{L\delta e}=\left.\frac{\partial C_L}{\partial \delta e}\right|_* \tag{4.8}$$

where, * as always denotes that the partial derivative is to be evaluated at an equilibrium (or trim) point.

Examining Equations 4.6 and 4.7, we can see that C_{L0} is the lift coefficient at $\alpha = 0°$, $\delta e = 0°$ and is fixed once the tail setting angle is fixed, that is, i_t = constant. The effect of deflecting the elevator in Equation 4.7 is to add another component to C_{L0} while leaving the terms $C_{L\alpha}$ and $C_{L\dot{\alpha}}$ unaffected. The combination $(C_{L0} + C_{L\delta e}\delta e)$ in Equation 4.7 has been labelled as C_{LZA}, which stands for the zero-alpha lift coefficient.

Example 4.3

Let us evaluate the change in total airplane lift coefficient due to elevator deflection as given by Equation 4.7 for an example airplane. Using the data:

$$\left(\frac{S_t}{S}\right) = 0.1;\ C^t_{L\delta e} = 0.05/\text{deg}$$

for a change in elevator deflection by 1°

$$\Delta C_L = C_{L\delta e}\Delta\delta e = \left(\frac{S_t}{S}\right)C^t_{L\delta e}\Delta\delta e$$

$$\Delta C_L = 0.1 \times 0.05 \times 1.0 = 0.005$$

Usually, the airplane total lift coefficient at cruise, $C_L \sim 0.3$–0.5. Therefore, the elevator contribution to the total lift of the airplane is insignificantly small.

Notice from Equations 4.6 and 4.7 that both the tail setting angle i_t and the elevator deflection δe act to change the term C_{LZA}. On a plot of C_L versus angle of attack, α, C_{LZA} is the intercept on the Y axis, as sketched in Figure 4.7.

For different deflections of δe (or i_t for that matter), the Y-intercept shifts, but leaves the slope (that is, $C_{L\alpha}$) unchanged. In this respect, the use of both the all-moving tail and the elevator affects the airplane lift in a very similar manner. However, one big difference between the two is in the amount of moment that needs to be applied at

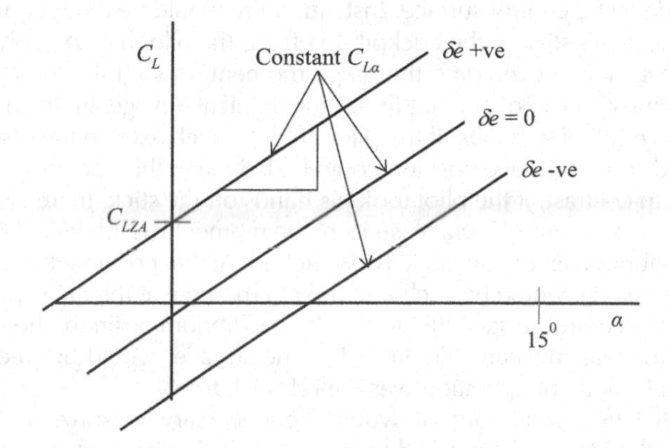

FIGURE 4.7 C_L versus α curves for different elevator deflections.

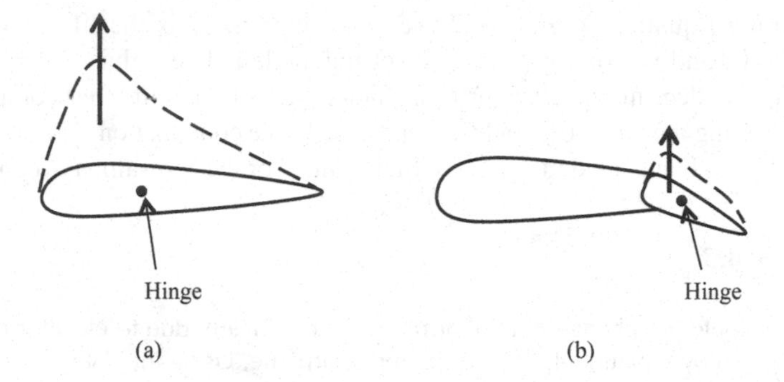

FIGURE 4.8 (a) Tail hinge moment and (b) elevator hinge moment.

their respective hinge points to effect and hold a deflection. Refer to Figure 4.8a that shows an all-moving tail and a typical lift distribution over it with the hinge point about which the tail is rotated and Figure 4.8b that shows the same for the elevator with its hinge point about which it is deflected with respect to the tail.

When you evaluate the net moment about the hinge, called the hinge moment, due to the lift distribution over the surface (tail or elevator, as the case may be), usually the hinge moment at the tail hinge point is much larger than that at the elevator hinge point. To hold the tail or the elevator at its deflected position, the actuator has to exert a moment equal (and opposite) to the hinge moment due to the lift distribution. In other words, left to itself without the restraining moment from an actuator, the surface would 'float' to a new deflection angle, one at which the net hinge moment would be zero. A larger hinge moment means that a more powerful (and hence heavier) actuator is needed for an all-moving tail than for the elevator.

Example 4.4

In the olden days, especially on smaller, slow-flying airplanes, there would be no actuator to deflect a control surface. Instead, there would be a direct mechanical linkage to the pilot's stick in the cockpit. In effect, the pilot had to apply and hold a force on the stick to overcome the hinge moment, though the force necessary was made lighter by use of gearing in the mechanical linkage and devices on the control surface called trim tabs. If the pilot held the stick fixed in the cockpit, then the deflected control surface position could be held, and this operation was called 'stick-fixed'. In contrast, if the pilot took his hands off the stick, there would be no force to restrain the control surface, so its hinge moment would deflect the surface until an equilibrium deflection angle was reached. At that position, the lift distribution over the surface would be such that the net moment at the hinge point would be zero. As the control surface 'floated' to its equilibrium position, the stick in the cockpit, which was mechanically linked to the surface, would be free to move, and hence this mode of operation was called 'stick-free'.

Under stick-free conditions, it would be necessary to solve an additional equation for the hinge moment and trim would require the control surface hinge moment to be also made zero. If the airplane was to be perturbed from a trim state, say in angle of attack, then the lift distribution over the control surface would also change, and hence the control surface would get a deflection, which would change the moment about the airplane CG. Thus, under 'stick-free' operation, both trim and stability would be affected by the free movement of the control surface. Usually, leaving the stick free has a destabilizing effect on the airplane.

In the past, both stick-fixed and stick-free stability would be topics for study; but, with stick-free operation becoming rare on modern airplanes, its significance has reduced. So, we will not spend more time on stick-free flying in this book.

4.5 AIRPLANE PITCHING MOMENT COEFFICIENT WITH ELEVATOR

Now, let us rewrite the expression for the airplane pitching moment with the elevator included. The net pitching moment of a wing–body plus tail configuration was derived in Chapter 3 (Equation 3.21) as:

$$C_{mCG} = C_{mAC}^{wb} + C_L^{wb}\left(h_{CG} - h_{AC}^{wb}\right) - V_H C_L^t \tag{3.21}$$

As seen in Section 4.4, the effect of introducing the elevator is to bring in an additional source of lift on the tail, and this was modelled in Equation 4.3 as:

$$C_L^t = C_{L\alpha}^t \alpha_t + C_{L\delta_e}^t \delta e$$

All we need to do is to introduce the form for C_L^t from Equation 4.3 into the pitching moment coefficient of Equation 3.21, as follows:

$$C_{mCG} = C_{mAC}^{wb} + C_L^{wb}\left(h_{CG} - h_{AC}^{wb}\right) - V_H\left(C_{L\alpha}^t \alpha_t + C_{L\delta_e}^t \delta e\right) \tag{4.9}$$

And rearrange the terms to give the expression for the net airplane pitching moment as:

$$C_{mCG} = \underbrace{\left(C_{m0} + C_{m\delta e}\delta e\right)}_{C_{mZA}} + C_{m\alpha}\alpha + C_{m\dot{\alpha}}\dot{\alpha}\left(c/2V^*\right) \tag{4.10}$$

where the terms:

$$C_{m0} = C_{mAC}^{wb} + C_{L0}^{wb}\left(h_{CG} - h_{AC}^{wb}\right) + V_H C_{L\alpha}^t\left(i_t + \varepsilon_0\right)$$

$$C_{m\alpha} = C_{L\alpha}^{wb}\left(h_{CG} - h_{AC}^{wb}\right) - V_H C_{L\alpha}^t\left(1 - \varepsilon_\alpha\right)$$

$$C_{m\dot{\alpha}} = V_H C_{L\alpha}^t \varepsilon_{\dot{\alpha}}$$

are exactly the same as in Equation 3.24 and

$$C_{m\delta e} = -V_H C_{L\delta e}^t \tag{4.11}$$

Since $V_H > 0$ and $C^t_{L\delta e} > 0, C_{m\delta e} < 0$ usually. $C_{m\delta e}$ is called the elevator control effectiveness and it measures the change in airplane pitching moment created by a unit deflection of the elevator. A down (positive) elevator deflection increases the tail lift, which creates a nose-down (negative) pitching moment at the airplane CG, and vice versa. This explains the negative sign of the derivative $C_{m\delta e}$.

Example 4.5

For the airplane data used in Example 3.4:

$$V_H = 0.34; \quad C^t_{L\delta e} = 0.05/\text{deg}$$

Hence, $C_{m\delta e} = -0.34 \times 0.05 = -0.017/\text{deg}$

The only difference between the wing–body–tail formula in Equation 3.24 and the wing–body–tail plus elevator formula in Equation 4.10 is the $(C_{m\delta e}\delta e)$ term appearing alongside C_{m0}. C_{m0} is the airplane pitching moment coefficient at $\alpha = 0$; $\delta e = 0$. The combination $(C_{m0} + C_{m\delta e}\delta e)$ in Equation 4.10 has been labelled as C_{mZA}, which stands for the zero-alpha pitching moment coefficient. On a plot of C_m versus α, such as that in Figure 4.9, C_{mZA} is the Y-intercept for the graph. The effect of the $(C_{m\delta e}\delta e)$ term in Equation 4.10 is to shift the Y-intercept and thus move the graph in Figure 4.9 parallel to itself. For a positive δe, the change in C_{mZA} is negative; the graph therefore shifts below and parallel to itself, intersecting the X axis at a smaller value of trim α. Likewise, a negative δe shifts the trim to a higher α.

Notice that the slope remains unaltered as the Y-intercept changes; hence, at each trim, $C_{m\alpha}$ is the same, and provided the damping derivatives are also unchanged (which is usually the case at low angles of attack), the short-period dynamics is essentially identical at each of the trim points. In other words, the stability and

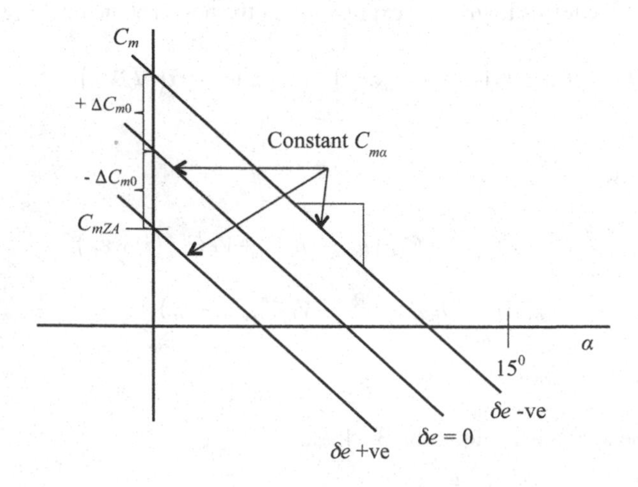

FIGURE 4.9 C_m versus α curves for different elevator deflections.

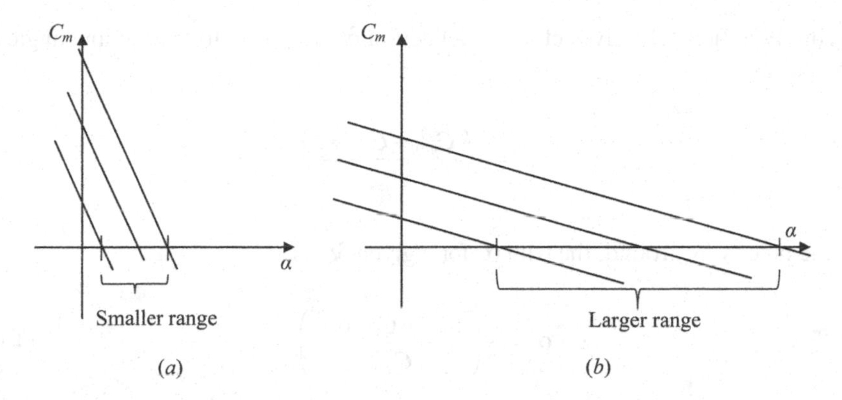

FIGURE 4.10 C_m versus α curves with different slopes, giving: (a) smaller range and (b) larger range of trim α^*.

response of the airplane to small disturbances are identical at each of the trim states. Thus, by changing the elevator deflection angle, the pilot can change the trim angle of the attack, and hence the trim velocity, with no noticeable variation in the airplane's dynamic behaviour.

Homework Exercise: Contrast the change in trim obtained by the use of an elevator with that obtained by changing the CG location. Which is preferable and why?

What is the range of trim α^* over which an airplane can be trimmed for a given range of elevator deflection and elevator control effectiveness, $C_{m\delta e}$? That depends on the slope of the curves in Figure 4.9. Figure 4.10 shows the pitching moment coefficient plots for two cases with different slopes, $C_{m\alpha}$.

In Figure 4.10a, the slope is more negative (steeper), characteristic of an airplane with greater pitch stiffness, whereas Figure 4.10b is a case with less stiffness and the slope is less negative (flatter). For the same elevator action (same range of change of the Y-intercept), the range of trim α^* in case (a) is smaller than that in case (b). In other words, for the more stable airplane in case (a), either the range of elevator deflection must be larger or the elevator action must be more powerful (larger absolute value of $C_{m\delta e}$), for it to have the same range of trim α^* as the airplane in case (b).

4.6 ELEVATOR INFLUENCE ON TRIM AND STABILITY

Let us formalize the discussion in Section 4.5 by evaluating the elevator deflection required to attain a certain trim state and the stability at that state.

As described earlier, at trim, the net pitching moment about the airplane CG must be zero. With $\dot{\alpha} = 0$ already at a trim, setting the left-hand side of Equation 4.10 to zero yields the following relation between the trim angle of attack α^* and the elevator deflection at that trim, δe^*:

$$C_{mCG} = 0 \Rightarrow C_{m0} + C_{m\alpha}\alpha^* + C_{m\delta e}\delta e^* = 0 \tag{4.12}$$

Solving Equation 4.12 gives elevator deflection δe^* required to trim at any angle of attack α^* as:

$$\delta e^* = -\left(\frac{C_{m0} + C_{m\alpha}\alpha^*}{C_{m\delta e}}\right) \tag{4.13}$$

Or, the other way around, the trim α^* for a given δe^* is:

$$\alpha^* = -\left(\frac{C_{m0} + C_{m\delta e}\delta e^*}{C_{m\alpha}}\right) \tag{4.14}$$

Differentiating both sides of Equation 4.14 with δe^* gives the change in α per unit change in elevator deflection δe between two trim states (the underlined words are critical):

$$\frac{d\alpha^*}{d\delta e^*} = -\frac{C_{m\delta e}}{C_{m\alpha}} \tag{4.15}$$

Since both the aerodynamic derivatives on the right-hand side are usually negative, the sign of the ratio in Equation 4.15 is normally negative. That is, a positive (downward) elevator deflection trims the airplane at a lower angle of attack, and vice versa. Also, the change in trim α^* for a given change in δe^* depends inversely on the aerodynamic stiffness derivative, $C_{m\alpha}$. The larger the stiffness, the smaller the range of trim α^* for a given change in δe^*. Thus, Equation 4.15 is a quantitative statement of the discussion in Figure 4.10.

4.6.1 Change in Trim Lift Coefficient

For any trim state defined by $(\alpha^*, \delta e^*)$ as related in Equation 4.13, the trim C_L can be found as follows, where * denotes the value at trim:

$$\begin{aligned} C_L^* &= C_{L0} + C_{L\alpha}\alpha^* + C_{L\delta e}\delta e^* \\ &= C_{L0} + C_{L\alpha}\alpha^* - \left(\frac{C_{m0} + C_{m\alpha}\alpha^*}{C_{m\delta e}}\right)C_{L\delta e} \\ &= \left(C_{L0} - \frac{C_{m0}}{C_{m\delta e}}C_{L\delta e}\right) + \left(C_{L\alpha} - \frac{C_{m\alpha}}{C_{m\delta e}}C_{L\delta e}\right)\alpha^* \end{aligned} \tag{4.16}$$

You will remember from Equations 4.7d and 4.11 that:

$$C_{L\delta e} = \left(\frac{S_t}{S}\right)C_{L\delta e}^t$$

and

$$C_{m\delta e} = -V_H C^t_{L\delta e} = -\left(\frac{S_t}{S}\right)\left(\frac{l_t}{c}\right)C^t_{L\delta e}$$

So, the ratio $C_{L\delta e}/C_{m\delta e} = -1/(l_t/c)$.

And we can write Equation 4.16 as:

$$C^*_L = \left(C_{L0} + \frac{C_{m0}}{(l_t/c)}\right) + \left(C_{L\alpha} + \frac{C_{m\alpha}}{(l_t/c)}\right)\alpha^* \tag{4.17}$$

It is important to understand the difference between the expression for trim C^*_L in Equation 4.17 and our earlier form of the airplane lift coefficient C_L in Equation 4.7:

$$C_L = \underbrace{\left(C_{L0} + C_{L\delta e}\delta e\right)}_{C_{L7A}} + C_{L\alpha}\alpha + C_{L\dot{\alpha}}\dot{\alpha}\left(c/2V^*\right) \tag{4.7}$$

In particular, Equation 4.7 says that the elevator deflection does not change the $C_{L\alpha}$ term, whereas in Equation 4.17, the coefficient of the term linear in α^* is clearly different from $C_{L\alpha}$ and that too by an additional term that comes from the trim elevator deflection δe^* in Equation 4.13.

The variation of C_L with α for different elevator deflection angles is represented in Figure 4.11. However, for each elevator deflection, there is only one trim angle of attack given by Equation 4.14 and the corresponding trim C_L is given by Equation 4.17. These are marked in Figure 4.11 by the triangle symbols. The triangle symbols for different elevator deflections are joined together by the dashed line—this is a line of trim states and represents Equation 4.17.

Thus, each full line representing Equation 4.7 for one value of δe may be obtained by artificially restraining the airplane, for example, in a wind tunnel, setting the elevator angle to a fixed value, varying the angle of attack and plotting the lift

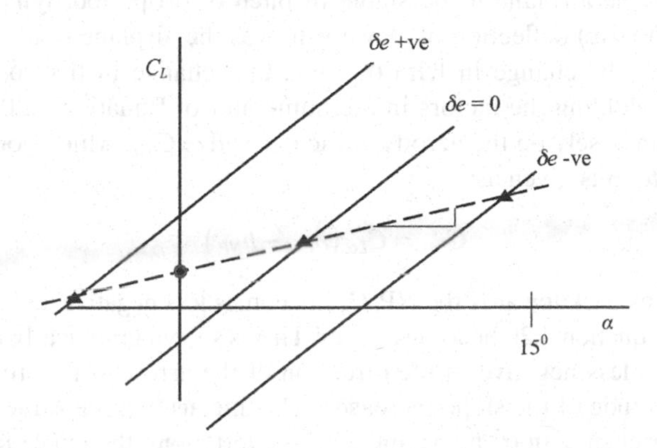

FIGURE 4.11 (Solid line) C_L versus α curves for different elevator deflections and (dashed line) C_L versus α-trim curve.

coefficient measured. This is regardless of the balance of the pitching moment. In flight, α and δe are not independent, but are related at a trim state by the pitching moment balance condition, Equation 4.12. To trim at a higher α, we need to use an up (negative) elevator deflection, which creates a little down-lift—this is precisely the term that subtracts from $C_{L\alpha}$ in Equation 4.17. Hence, the trim lift curve (dashed line in Figure 4.11) has a lesser slope than the full lines (slope equal to $C_{L\alpha}$), assuming $C_{m\alpha}$ to be negative, of course.

From Equation 4.17, the change in airplane trim lift coefficient per unit change in trim angle of attack can be obtained as:

$$\frac{dC_L^*}{d\alpha^*} = C_{L\alpha} - \frac{C_{m\alpha}}{C_{m\delta e}} C_{L\delta e} = C_{L\alpha} + \frac{C_{m\alpha}}{(l_t/c)} \tag{4.18}$$

which may be referred to as the airplane trim lift curve slope (the dashed line of Figure 4.11). Interestingly, from the rightmost expression in Equation 4.18, it turns out that the trim lift curve slope is independent of all elevator properties and depends on l_t—the location of the horizontal tail aerodynamic centre behind the CG!

Homework Exercise: Putting Equations 4.15 and 4.18 together, derive:

$$\frac{d\delta e^*}{dC_L^*} = -\frac{C_{m\alpha}}{C_{m\delta e} C_{L\alpha} - C_{m\alpha} C_{L\delta e}} \tag{4.19}$$

4.6.2 Another Viewpoint of Stability

Let us review Equation 4.15, which can be written as:

$$\frac{d\alpha^*}{d\delta e^*} = -\frac{C_{m\delta e}}{C_{m\alpha}} = -\frac{-V_H C_{L\delta e}^t}{C_{m\alpha}} = \frac{V_H C_{L\delta e}^t}{C_{m\alpha}} \tag{4.20}$$

The sign comes from the sign of $C_{m\alpha}$, which, as we have seen in Chapter 2, must be negative for the airplane to be stable in pitch (short-period dynamics). In that case, an up (negative) deflection of elevator moves the airplane to a higher α (positive) trim state. The change in trim α^* for a unit change in trim δe^* is given by Equation 4.20. Holding the factors in the numerator of Equation 4.20 constant, the slope depends inversely on the aerodynamic derivative $C_{m\alpha}$, which you will remember from Chapter 3 is given as:

$$C_{m\alpha} = C_{L\alpha}(h_{CG} - h_{NP}) \tag{3.45}$$

As the CG moves back towards the NP, $C_{m\alpha}$ becomes less negative and the magnitude of the slope in Equation 4.20 becomes larger. This is shown graphically in Figure 4.12.

$C_{m\alpha}$ becomes less negative in the direction of the arrow in Figure 4.12; accordingly, the magnitude of the slope increases. This means that the same change in δe^* fetches a larger change in α^* as we move to the left along the curve in Figure 4.12. In other words, the airplane appears to be more sensitive to elevator deflections—smaller deflections create the same change in trim α^*.

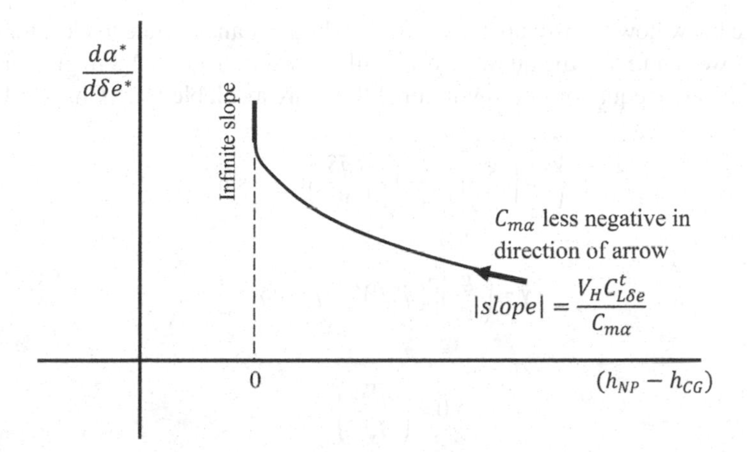

FIGURE 4.12 Trim alpha versus elevator deflection.

In the limit as CG approaches neutral point (NP) and $C_{m\alpha}$ tends to zero, the slope in Figure 4.12 tends to infinity. At the point where

$$\frac{d\alpha^*}{d\delta e^*} \to \infty \tag{4.21}$$

the slightest change in δe^* can (theoretically) cause an infinite change in α^*. Or, more practically speaking, the airplane becomes extremely sensitive, almost like an object placed on a knife's edge. The condition in Equation 4.20 gives an alternative view of stability with the condition in Equation 4.21 marking the onset of instability.

Unlike our approach to stability in Chapter 2, where we examined the system's response to a small perturbation from an equilibrium (trim) state and required that perturbation die down (in infinite time) for the system to be pronounced stable, the present viewpoint involves the transition between two equilibrium (trim) states. However, the two approaches are entirely equivalent, as we ask you to show below.

Homework Exercise: Think of arguments to establish that the alternative view of stability presented here is entirely equivalent to the more standard approach presented in Chapter 2.

4.7 LONGITUDINAL MANOEUVRES WITH THE ELEVATOR

Now, we can improve upon our analysis in Section 2.8 of the action of the elevator on aircraft flight. There we had considered elevator effect on pitching motion alone, assuming both V and γ to be held fixed. Thus, we had examined only the perturbed pitch equation in the form:

$$\Delta\ddot{\alpha} - \left(\frac{\bar{q}^* Sc}{I_{yy}}\right)\left(\frac{c}{2V^*}\right)\left(C_{mq1} + C_{m\dot{\alpha}}\right)\Delta\dot{\alpha} - \left(\frac{\bar{q}^* Sc}{I_{yy}}\right)C_{m\alpha}\Delta\alpha = \left(\frac{\bar{q}^* Sc}{I_{yy}}\right)C_{m\delta e}\Delta\delta e \tag{2.43}$$

Now, we know how to evaluate the changes to both C_L and C_m due to elevator deflection. So, we can revisit the question originally posed in Section 2.8 in greater detail.

The dynamic equations of longitudinal flight arc available from Chapter 1:

$$\frac{\dot{V}}{V} = \left(\frac{g}{V}\right)\left[\left(\frac{T}{W}\right) - \left(\frac{\bar{q}S}{W}\right)C_D - \sin\gamma\right] \tag{1.11a}$$

$$\dot{\gamma} = \left(\frac{g}{V}\right)\left[(\bar{q}S/W)C_L - \cos\gamma\right] \tag{1.11b}$$

$$\ddot{\theta} = \left(\frac{\bar{q}Sc}{I_{yy}}\right)C_m \tag{1.11c}$$

At trim, Equation 1.11 gives the following force/moment balance conditions:

$$C_D^* = \frac{W}{\bar{q}^* S}\left[\frac{T^*}{W} - \sin\gamma^*\right] \tag{1.12a}$$

$$C_L^* = \cos\gamma^* \frac{W}{\bar{q}^* S} \tag{1.12b}$$

$$C_m^* = 0 \tag{1.12c}$$

Consider a level flight trim condition, specified by the trim values:

$$V^*, \gamma^* = 0, \quad \theta^* = \alpha^*$$

We will continue to assume that the trim velocity is held fixed at V^*—that is, we consider the effect of elevator deflection on airplane drag to be negligible. So, ignoring Equation 1.11a, let us focus only on Equations 1.11b and 1.11c.

Note: The analysis below is for a specific purpose. In general, please be careful when using these equations with velocity held fixed at V^* as it impacts the nature of the phugoid mode (to be discussed in Chapter 5).

Under small perturbations from the trim state, we first define the perturbations as below:

$$V^* = \text{constant}, \ \gamma = \gamma^* + \Delta\gamma = \Delta\gamma \ \left(\text{since } \gamma^* = 0\right), \ \theta = \theta^* + \Delta\theta$$

And since $\alpha = \theta - \gamma$, it follows that:

$$\Delta\alpha = \Delta\theta - \Delta\gamma$$

Then insert them into Equations (1.11b and 1.11c) as shown below:

$$\dot{\gamma}^* + \Delta\dot{\gamma} = \left(\frac{g}{V^*}\right)\left[\left(\bar{q}^* S/W\right)\left(C_L^* + \Delta C_L\right) - \cos\left(\gamma^* + \Delta\gamma\right)\right] \tag{4.22}$$

$$\ddot{\theta}^* + \Delta\ddot{\theta} = \left(\frac{\bar{q}^* Sc}{I_{yy}}\right)\left(C_m^* + \Delta C_m\right) \tag{4.23}$$

And then apply the trim conditions in Equation 1.12 and drop the higher-order terms, since the perturbations are small, to get:

$$\Delta\dot{\gamma} = \left(\frac{g}{V^*}\right)\left[\left(\bar{q}^* S/W\right)\Delta C_L\right] \tag{4.24}$$

$$\Delta\ddot{\theta} = \left(\frac{\bar{q}^* Sc}{I_{yy}}\right)\Delta C_m \tag{4.25}$$

Equation 4.25 is the same as Equation 2.21, but we now have an additional Equation 4.24.

The incremental pitching moment coefficient ΔC_m has already been written out in Equation 2.42 as:

$$\Delta C_m = C_{m\alpha}\Delta\alpha + \left(C_{mq1} + C_{m\dot{\alpha}}\right)\Delta\dot{\alpha}\left(c/2V^*\right) + C_{m\delta e}\Delta\delta e \tag{2.42}$$

In an exactly similar manner, we can write the incremental lift coefficient ΔC_L as:

$$\Delta C_L = C_{L\alpha}\Delta\alpha + \left(C_{Lq1} + C_{L\dot{\alpha}}\right)\Delta\dot{\alpha}\left(c/2V^*\right) + C_{L\delta e}\Delta\delta e \tag{4.26}$$

Do check these incremental coefficients with Equation 4.7 for C_L and Equation 4.10 for C_m derived earlier in this chapter—they are the same, but for the inclusion of the dynamic derivatives $(C_{Lq1} + C_{L\dot{\alpha}})$ and $(C_{mq1} + C_{m\dot{\alpha}})$.

So, Equations 4.24 and 4.25 appear in full as:

$$\Delta\dot{\gamma} = \left(\frac{g}{V^*}\right)\left[\left(\frac{\bar{q}^* S}{W}\right)\left\{C_{L\alpha}\Delta\alpha + \left(C_{Lq1} + C_{L\dot{\alpha}}\right)\Delta\dot{\alpha}\left(\frac{c}{2V^*}\right) + C_{L\delta e}\Delta\delta e\right\}\right] \tag{4.27}$$

$$\Delta\ddot{\theta} = \left[\left(\frac{\bar{q}^* Sc}{I_{yy}}\right)\left\{C_{m\alpha}\Delta\alpha + \left(C_{mq1} + C_{m\dot{\alpha}}\right)\Delta\dot{\alpha}\left(\frac{c}{2V^*}\right) + C_{m\delta e}\Delta\delta e\right\}\right] \tag{4.28}$$

where,

$$\Delta\ddot{\theta} = \Delta\ddot{\alpha} + \Delta\ddot{\gamma}$$

Notice that we still have the timescale factors T_1 and T_2 that we introduced in Chapter 1, on the right-hand side of Equations 4.27 and 4.28, but these scale factors are relevant when the dynamic response is unforced (also called 'natural'). When a linear dynamical system, such as that in Equation 4.27, is forced by an 'external' source, such as the perturbed elevator deflection $\Delta\delta e$, then in the steady state (after initial transients have died down) the frequency of the excitation and that of the system response is the same. The same principle applies to Equation 4.28, or any other linear dynamical system.

Homework Exercise: You may want to review the discussion on 'forced response' in Section 2.8 at this point. The time response of a linear dynamical system, such as that given by Equations 4.27 and 4.28, consists of a free ('natural') part and a forced part. Assuming the system to be stable, the free response dies down (as per the natural timescales, e.g., T_1 and T_2 in this particular case), while the forced response persists as long as the forcing function is alive.

Note: Even though the free response, in theory, takes infinite time to die out, in practice, after a certain time period, it decays to a negligible fraction of the initial perturbation.

In effect, we can no longer separately analyse Equations 4.27 and 4.28 based on different timescales when they are responding to an externally imposed elevator perturbation $\Delta\delta e(t)$. So, we shall resort to a numerical simulation instead.

Example 4.6

Consider the data for the F-18/high-angle-of-attack research vehicle (HARV) airplane, first introduced in Section 2.6.1.

Mean aerodynamic chord length, $c = 3.511$ m
Wing planform area, $S = 37.16$ m^2
Speed of sound at sea level, $v_s = 340.0$ m/s
Pitch moment of inertia, $I_{yy} = 239720.76$ kg-m^2
Aircraft mass, $m = 15119.28$ kg
Density of air at sea level, $\rho = 1.225$ kg/m^3

We take a level flight condition at Mach number, $Ma = 0.43$, at sea level. The trim values of the angle of attack and the elevator deflection at this flight condition are:

$$\alpha^* = 3.5°, \quad \delta e^* = 0°.$$

At this trim point, the relevant aerodynamic data are:

$C_{m\dot{\alpha}} = 0$; $C_{mq1} = -0.084/\text{deg} = -4.813/\text{rad}$; $C_{m\alpha} = -0.003/\text{deg} = -0.172/\text{rad}$;
$C_{m\delta e} = -0.007/\text{deg} = -0.46/\text{rad}$;
$C_{L\alpha} = 0.102/\text{deg} = 5.844/\text{rad}$; $C_{L\delta e} = 0.007/\text{deg} = 0.401/\text{rad}$;
$C_{Lq1} = 0.074/\text{deg} = 4.24/\text{rad}$; $C_{L\dot{\alpha}} = 0$

The timescale factors, T_1 and T_2, and other parameters are evaluated as:

$$\frac{g}{V^*} = \frac{9.81}{0.43 \times 340.0} = 0.067/\text{s}$$

$$\frac{\bar{q}^* Sc}{I_{yy}} = \frac{0.5\rho V^{*2} Sc}{I_{yy}} = \frac{0.5 \times 1.225 \times (0.43 \times 340.0)^2 \times 37.16 \times 3.511}{239720.76} = 7.125/\text{s}^2$$

$$\frac{c}{2V^*} = \frac{3.511}{2 \times 0.43 \times 340.0} = 0.012\ \text{s}$$

$$\frac{\bar{q}^* S}{W} = \frac{0.5\rho V^{*2} S}{mg} = \frac{0.5 \times 1.225 \times (0.43 \times 340.0)^2 \times 37.16}{15119.28 \times 9.81} = 3.28$$

Equations 4.27 and 4.28 with the above data may be evaluated as below:

$$\Delta\dot{\gamma} = \left(\frac{g}{V^*}\right)\left[(\bar{q}^* S/W)\left\{C_{L\alpha}\Delta\alpha + (C_{Lq1} + C_{L\dot{\alpha}})\Delta\dot{\alpha}(c/2V^*) + C_{L\delta e}\Delta\delta e\right\}\right]$$

$$\Delta\ddot{\theta} - \frac{\bar{q}^* Sc}{I_{yy}}(C_{mq1} + C_{m\dot{\alpha}})\Delta\dot{\alpha}\frac{c}{2V^*} - \frac{\bar{q}^* Sc}{I_{yy}} C_{m\alpha}\Delta\alpha = \frac{\bar{q}^* Sc}{I_{yy}} C_{m\delta e}\Delta\delta e$$

Substituting values we get:

$$\Delta\dot{\gamma} = 0.067\left[3.28\left\{5.84\Delta\alpha + (4.24 + 0)\Delta\dot{\alpha} \times 0.012 + 0.401\Delta\delta e\right\}\right]$$

$$\Delta\ddot{\theta} - 7.125(-4.813 + 0)\Delta\dot{\alpha} \times 0.012 - 7.125 \times (-0.172)\Delta\alpha = 7.125 \times (-0.46)\Delta\delta e$$

which gives:

$$\Delta\dot{\gamma} = 1.28\Delta\alpha + 0.011\Delta\dot{\alpha} + 0.088\Delta\delta e \tag{4.29}$$

$$\Delta\ddot{\theta} = -0.411\Delta\dot{\alpha} - 1.225\Delta\alpha - 3.277\Delta\delta e \tag{4.30}$$

The simulation results obtained by integrating Equations 4.29 and 4.30 (and using $\Delta\theta = \Delta\alpha + \Delta\gamma$) for a step positive (downward) input in elevator have been presented in Figure 4.13, highlighting the first second of simulation time.

What is notable and quite interesting is that, looking at the full curve in Figure 4.13, after applying the down elevator, the first response of the flight path angle is to go positive (i.e., climb) before eventually going negative (i.e., descent). Figure 4.13 also shows that in case the $C_{L\delta e}$ term is shut off (dashed curve for $C_{L\delta e} = 0$), then this initial contrary behaviour does not occur.

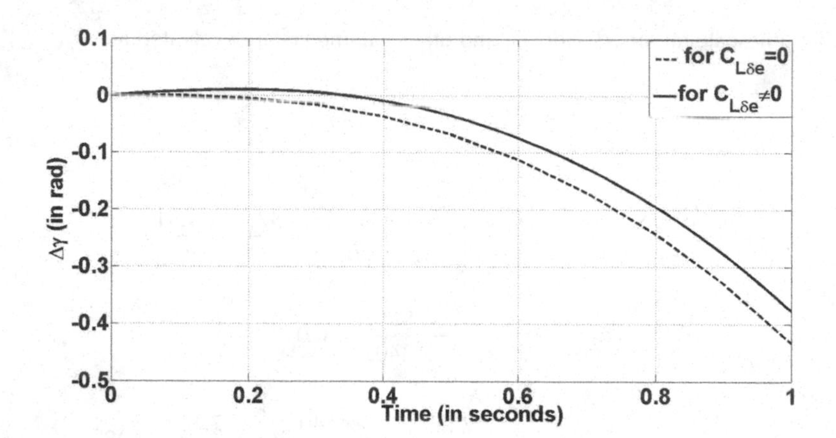

FIGURE 4.13 Flight path angle $\Delta\gamma$ response to a step elevator input (Equations 4.29 and 4.30).

Homework Exercise: Can you figure out why this happens? The answer is given below.

Example 4.7

The use of the elevator for inducing a pitching motion in airplanes has a peculiarity—this is referred to in the control literature as a *non-minimum phase* response.[1]

Imagine an airplane flying in level flight trim. The pilot decides to perform a pull-up manoeuvre (as we saw in Chapter 1, Figure 1.10b). For this, he/she needs to increase the lift so that the additional lift creates an upward acceleration that curves the flight path upwards into a pull-up. To increase the lift, he/she needs to increase the angle of attack and to do that he/she needs to deflect the elevators upwards from their trim position. So, he/she deflects the elevators upwards to produce an increase in lift. But, the first thing that happens when the elevators are deflected upwards is that they induce an additional down-lift, no matter how small, on the horizontal tail. So, in effect, the net airplane lift decreases slightly. The tail down-lift *then* creates a moment about the airplane CG that rotates the nose up, increasing the angle of attack. If the velocity is maintained, then the larger angle of attack creates more lift overcoming the initial loss in lift and the airplane does enter a pull-up. However, the first tendency of the airplane is to respond in exactly the opposite manner to the desired manoeuvre.

Figure 4.13 shows such a non-minimum phase response in pitch for the F-18/HARV airplane. Note that this is a common tendency for all airplanes that have a pitch control mechanism such as the elevator, or even all-moving tail, placed at the rear of the vehicle.

Homework Exercise: One way to avoid the non-minimum phase response of airplanes in pitch is to use the deflection of a control surface near the nose of the vehicle for pitch control—this surface is usually called a *canard*, as shown in Figure 4.14. Convince yourself about the pitch response of a canard-controlled vehicle. You may want to trawl the web to find some popular airplanes that use a canard design.

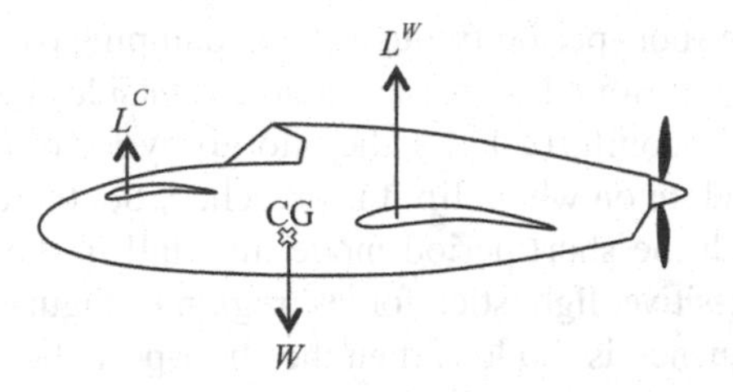

FIGURE 4.14 A pusher-type airplane with canard.

4.8 MOST FORWARD CG LIMIT

You will remember that the rearmost allowable location of CG was found to be the NP of the airplane, the point at which the aircraft becomes neutrally stable or just loses pitch stiffness ($C_{m\alpha} = 0$). As the CG moves forward, $C_{m\alpha}$ becomes more negative, which means an increased pitch stiffness or a higher short-period frequency. Is that always a good thing?

One answer comes from the field of handling qualities or flying qualities, which deals with how the airplane 'feels' to a pilot who is flying it. A 'feel' may be sluggish, oversensitive and so on. Based on this, there are different criteria which place limits on the frequency and damping of the various modes. One such criterion for the short-period mode is shown in Figure 4.15.

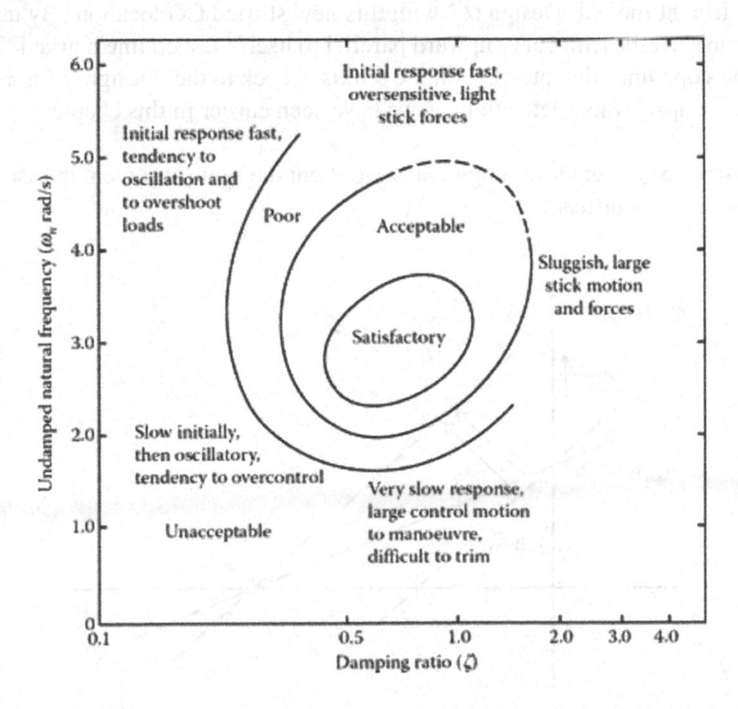

FIGURE 4.15 Handling qualities criterion for short-period mode. (F. O'Hara. Handling Criteria. *J. Royal Aero. Soc.*, Vol. 71, No. 676, pp. 271–291, 1967.)

There is a range of short-period frequency and damping parameters within which the airplane handling is found to be acceptable. Outside this range, the airplane becomes difficult or uncomfortable for the pilot to fly for one of the reasons listed in Figure 4.15, depending on which limit is breached. So, there is a forward limit of the CG beyond which the short-period mode may fall in the unacceptable 'initial response fast, oversensitive, light stick forces' region of Figure 4.15. Equally well, if the short-period frequency is too low, then the short-period characteristics are classified as unacceptable due to 'very slow response, large control forces to manoeuvre, difficult to trim'. Thus, the handling qualities requirement imposes a practical rear CG limit that is *ahead of the NP*, which is the theoretical rearmost CG location.

Now, we shall see another factor that limits the forward-most CG location.

4.8.1 Using Elevator to Compensate for CG Shift

Firstly, we must look at how the elevator can be used to recover trims that have been lost due to CG shift. Refer to Figure 4.16.

The full line labelled '0' shows trim at 'Design α^*' for zero elevator deflection and a particular CG location. Imagine the CG to shift forward for whatever reason. Then, $C_{m\alpha}$ becomes more negative, the slope in Figure 4.16 becomes steeper, but the Y-intercept does not change. So, the new trim curve, shown dashed and labelled '1' intersects the α axis at a lower value, called 'New α^*'. If the airplane maintains level flight, then the lower trim α^* implies a higher trim velocity, V^*. How can we recover trim at the old 'Design α^*' with this new shifted CG location? By using the elevator to shift the trim curve upward parallel to itself (dashed line marked '2') with the same slope until the intercept on the α axis is back to the 'Design α^*' point. This requires an up-elevator deflection, as we have seen earlier in this chapter.

Homework Exercise: Draw a figure and work out the same argument in case the CG shift is rearwards instead.

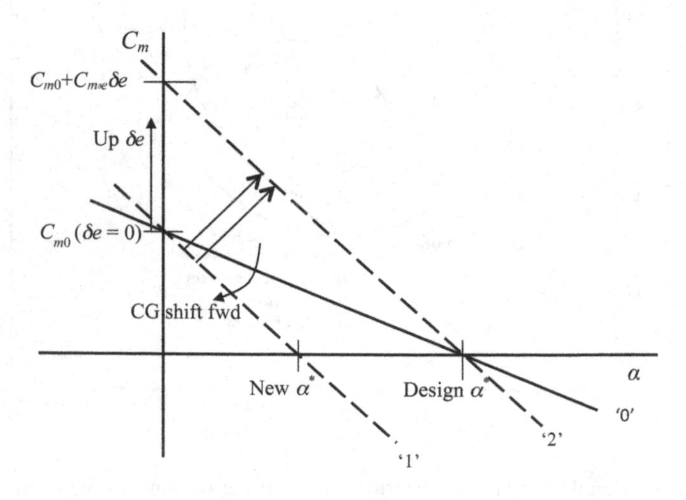

FIGURE 4.16 Effect of CG shift and elevator deflection to recover trims.

4.8.2 Typical Elevator Deflection Limits

On typical airplanes, such as commercial airliners, the up and down elevator deflection limits are not equal. There is a good reason for this. Let us explain this with reference to Figure 4.17.

Typically, the cruise trim angle of attack, which is usually selected as α^*_{design} marked in Figure 4.17, is around 4–5°. The minimum trim angle of attack, α^*_{min}, corresponds to the maximum level flight velocity—α^*_{min} is usually of the order of 1–2°. The maximum trim angle of attack is required during landing and corresponds to a trim velocity which is about 1.15–1.2 times the stall velocity V_{stall}. Thus, α^*_{max} is usually around 12–15°. As a result, almost always α^*_{design} is closer to α^*_{min} than α^*_{max}.

And, we would also like a null elevator deflection ($\delta e = 0$) at the cruise trim condition. This is simply because a deflected elevator adds to drag, and since much of the flight time is spent in cruise, it is ideal if the elevator is not deflected at the cruise condition.

With this information, we can interpret the solid lines in Figure 4.17. For the given CG location, the slopes of these lines are all the same. Their C_m-intercept is shifted by the use of elevator—positive (down) δe shifts the intercept down and negative (up) shifts the intercept up. At the down δe limit the airplane must be able to trim at α^*_{min}, and at the up δe limit the airplane must be able to trim at α^*_{max} plus a little more, as a safety measure. Clearly, the larger difference between α^*_{max} and α^*_{design} as against α^*_{min} and α^*_{design} on the α axis is reflected in the C_m axis. Hence, the C_m-intercept at the up-δe limit is farther away from the $C_{m0}(\delta e = 0)$ point than the C_m-intercept at the down-δe limit. In other words, a larger elevator deflection is needed to the up-δe limit than to the down-δe limit. Typically, $\delta e_{DN} \sim +10°$ and $\delta e_{UP} \sim -20°$.

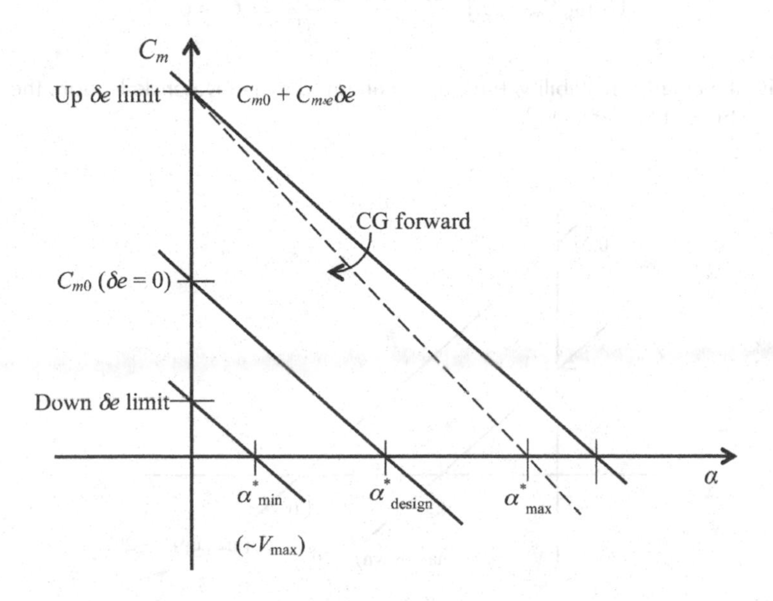

FIGURE 4.17 Forward CG limit and elevator deflection limits.

Example 4.8

Let us calculate the change in CG required to trim the aircraft at C_{Lmax} with max up elevator deflection for the data given in Figure 4.18 and $C_{Lmax} = 0.9$.

From Figure 4.18:

$$\frac{d\delta e^*}{dC_L^*} = \frac{-20-10}{1.0-0.5} \times \frac{\pi}{180} = -\frac{\pi}{3}\text{rad}$$

Also

$$\frac{dC_m}{dC_L} = (h_{CG} - h_{NP}) = \frac{0.2-0}{0.0-1.0} = -0.2$$

For the given $C_{Lmax} = 0.9$, the required pitching moment–lift curve slope is:

$$\frac{dC_m}{dC_L} = (h_{CGnew} - h_{NP}) = \frac{0.2-0}{0.0-0.9} = -0.22$$

The required change in CG thus can be found as:

$$\Delta h_{CG} = (h_{CGnew} - h_{NP}) - (h_{CG} - h_{NP}) = -0.22 - (-0.2) = -0.02$$

That is, a shift of $X = 0.02c$ forward.

With this new stability margin, the new value of C_{Ltrim} at $\delta e = +10°$ can be determined as follows:

$$\text{Using}, C_{m0} = 0.1; \quad \frac{0.1-0}{0-C_L} = \frac{0.2-0}{0-0.9} \Rightarrow C_L = 0.45$$

With the change in stability, thus, at maximum down elevator deflection, the airplane trims at higher speed.

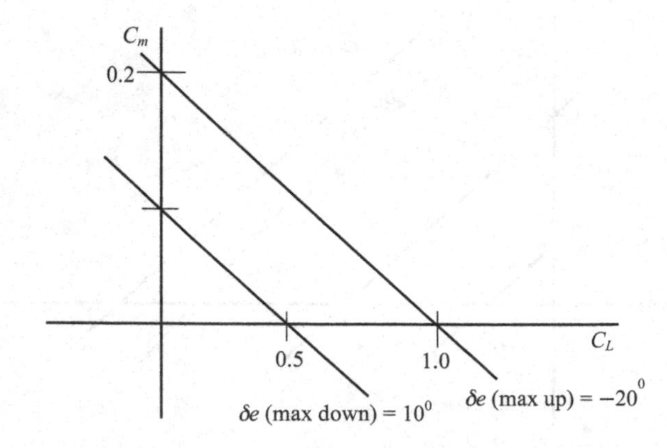

FIGURE 4.18 $C_m - C_L$ plot for maximum elevator deflections.

4.8.3 Forward-Most CG Limit due to Elevator Up-Deflection Limit

From Figure 4.17, it is easily seen that the elevator up-deflection limit imposes a limit on the forward CG travel of an airplane. Assume that the airplane has been trimmed at the nominal CG location with maximum up-elevator, as shown by the uppermost full line in Figure 4.17. As the CG moves forward, the trim curve remains hinged at the C_m axis and swings to the left, intersecting the α axis at lower values of α. After a certain amount of forward CG movement, the dashed trim curve in Figure 4.17 shows a trim at α^*_{max}. The airplane must be able to trim at α^*_{max} during landing.

If the CG moves any further forward, the maximum trim angle of attack will fall below α^*_{max}. Then, the only way to recover trim at α^*_{max} would be to further deflect the elevator up, as shown in Figure 4.16. But, the elevator deflection for the dashed line in Figure 4.17 has already reached its upper limit and there is no scope for further upward deflection. Hence, trim at α^*_{max} becomes impossible. This imposes a limit on the forward movement of CG.

The location of CG at which, with elevator at its upward deflection limit, the airplane trims at α^*_{max} is the forward-most allowable CG location.

Needless to say, if there are multiple criteria restricting the forward CG location, then the least (most conservative) forward CG location is the limiting case.

4.9 NP Determination from Flight Tests

One way of determining the NP of an airplane would be to evaluate one of the expressions we derived for the NP in Chapter 3:

$$h_{NP} = h_{AC}^{wb} + V_H \left\{ \frac{C_{L\alpha}^t}{C_{L\alpha}^{wb}} \right\} (1 - \varepsilon_\alpha) \tag{3.27}$$

$$h_{NP} = h_{AC}^{wb} + V_H^{/} \left(\frac{C_{L\alpha}^t}{C_{L\alpha}} \right) (1 - \varepsilon_\alpha) \tag{3.41}$$

But, there is always an uncertainty in estimating the various terms on the right-hand side of Equation 3.27 or 3.41. A more accurate way of estimating the NP is to actually flight test the airplane.

Essentially, we need to fly the airplane with different CG conditions and evaluate the pitch stiffness, or another measure of stability. We know that when the CG is at the NP, pitch stiffness is zero and the airplane has neutral stability. But then, it would be very dangerous to fly the airplane with the CG located at its NP. This is where the alternative viewpoint of stability discussed in Section 4.6.2 comes to our rescue.

Consider Equation 4.19, which relates the trim elevator deflection to the trim lift coefficient:

$$\frac{d\delta e^*}{dC_L^*} = -\frac{C_{m\alpha}}{C_{m\delta e} C_{L\alpha} - C_{m\alpha} C_{L\delta e}} \tag{4.19}$$

The slope on the left-hand side of Equation 4.19 is directly related to $C_{m\alpha}$ on the right-hand side, and we know that the NP corresponds to the location of the CG at which $C_{m\alpha} = 0$. Hence, at the NP, the slope $\left(d\delta e^* / dC_L^*\right) = 0$. So, the method to find the NP in a flight test is to evaluate the slope $\left(d\delta e^* / dC_L^*\right)$ for different CG locations. This requires a series of flights as follows:

- For one fixed CG location, set the airplane in level flight trim successively for different elevator deflections. We need to collect enough data to determine the trim C_L^* for each trim state after returning to the ground.
- Reset the CG at a different location and repeat the procedure on another flight. Do it at least one more time.
- After processing the data, we can create a plot like that shown in Figure 4.19.

From the data points, we can determine the slope on the left-hand side of Equation 4.19 for each case of CG location. The most negative slope is for the most forward CG location; as the CG moves aft, the slope becomes less negative. Ideally, if we could move the CG all the way back to the NP, the slope would become zero, and the zero slope will let us identify the NP location. But, that is not possible in flight. Hence, we extrapolate the slope from the available data all the way to the NP as follows.

Each dashed line in Figure 4.19 yields one value of slope each for a particular CG location. We can plot the value of the slopes thus obtained against the CG locations, as shown in Figure 4.20; join the points by a best-fit straight line and extend the line all the way to the h_{CG} axis (where $(d\delta e^* / dC_L^*)$ and hence $C_{m\alpha}$ is zero). The h_{CG}-intercept is the estimate of the NP location (marked h_{NP} in Figure 4.20).

Homework Exercise: Can you work out the data collection and analysis required for such a flight test procedure?

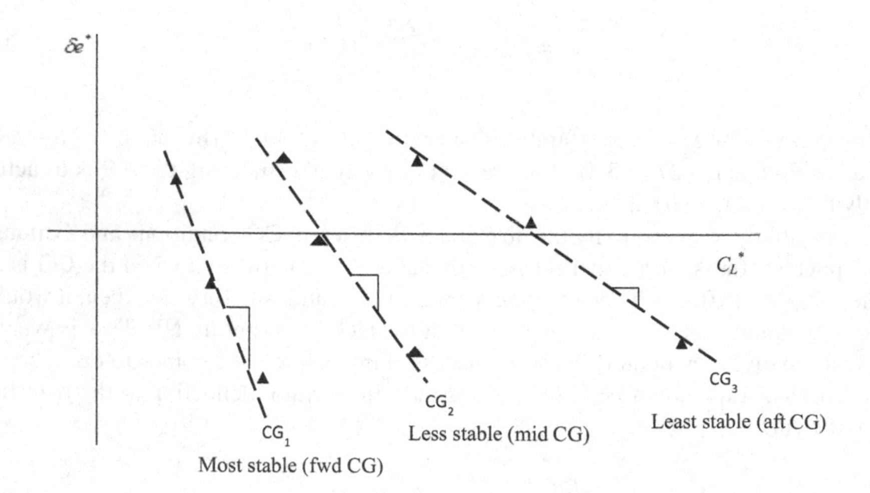

FIGURE 4.19 Elevator required to trim versus trim lift coefficient for different CG locations.

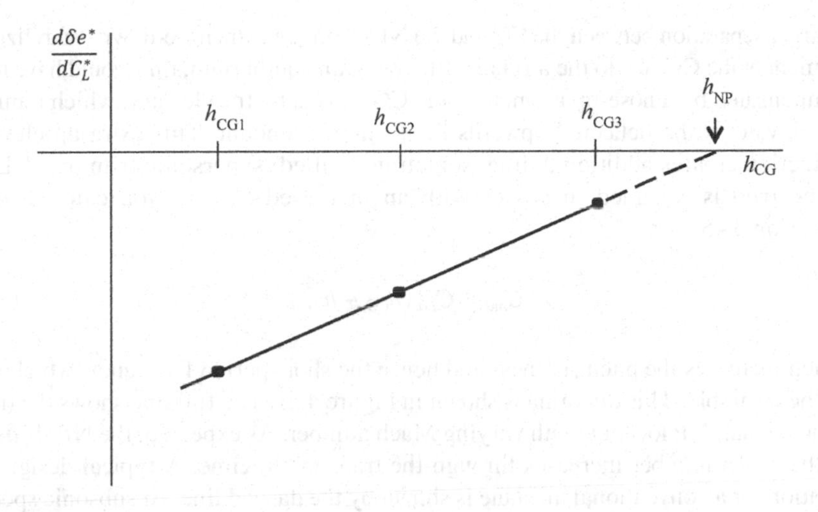

FIGURE 4.20 Neutral point estimation from flight test.

4.10 EFFECT OF NP SHIFT WITH MACH NUMBER

One more issue that we must address before we close this chapter is that of a shift in the NP location. From Equation 3.27 or 3.41, it is clear that the NP depends on the location of the wing–body aerodynamic centre, h_{AC}^{wb}. The aerodynamic centre is usually fairly invariant, but there are occasions when it may shift its location. One definite instance is during the transition between subsonic and supersonic flight. Aerodynamic theory says that ideally the subsonic AC location is at wing quarter-chord and the supersonic AC location is at wing half-chord. Thus, the AC moves aft (towards the tail) by the distance of about a quarter wing chord when the airplane accelerates from subsonic to supersonic flight. This is a common phenomenon on all supersonic airplanes.

Another instance is that of variable-sweep (also called swing-wing) airplanes that increase the wing sweep when going from low subsonic to high subsonic to supersonic flight. Though in that case there is a corresponding shift in CG location as well.

Examine Equation 3.27:

$$h_{NP} = h_{AC}^{wb} + V_H \left\{ \frac{C_{L\alpha}^t}{C_{L\alpha}^{wb}} \right\} (1 - \varepsilon_\alpha)$$

If the aerodynamic centre moves back, there is an equal corresponding shift in the NP as well. If the CG stays put, then the second term on the right-hand side of Equation 3.27 is largely unaltered in this process. Remember, we can place all the forces at the NP and write the moment balance at the CG as:

$$C_{mCG} = C_{mNP} + C_L \left(h_{CG} - h_{NP} \right) \tag{3.47}$$

A larger separation between the CG and the NP means an extra nose-down ('stabilizing') moment at the CG due to the airplane lift. To regain equilibrium, this would have to be compensated by a nose-up moment at the CG caused by the elevator, which requires the elevator to be deflected upwards by a suitable amount. This extra up-elevator deflection creates additional drag, sometimes called 'supersonic trim drag'. Even if the trim is regained, it comes with an increased $C_{m\alpha}$, as you can see from Equation 3.45:

$$C_{m\alpha} = C_{L\alpha}\left(h_{CG} - h_{NP}\right) \tag{3.45}$$

which increases the pitch stiffness and hence the short-period frequency, which may not be desirable. This dilemma is shown in Figure 4.21. The full line shows the (non-dimensional) NP location with varying Mach number. As expected, the NP shifts aft as the Mach number increases through the transonic regime. A typical design CG location for a conventional airplane is shown by the dashed line. At subsonic speeds, this gives a comfortable static margin (*SM*). But, the same CG location gives too large a static margin at supersonic Mach numbers. One solution to this problem is to locate the design CG at the dash-dot line in Figure 4.21. In subsonic flight, the CG is behind the NP; hence, the arrangement is clearly unstable—the short-period dynamics will be divergent. But, under supersonic conditions, the same CG location gives a reasonable static margin. Such an arrangement is called a 'relaxed static stability' (RSS) airplane, the static stability referring to the static margin, or the pitch stiffness, or the value of $C_{m\alpha}$. In subsonic flight, an RSS airplane needs an active flight control system to maintain stable flight, but it pays off by having better supersonic flight characteristics.

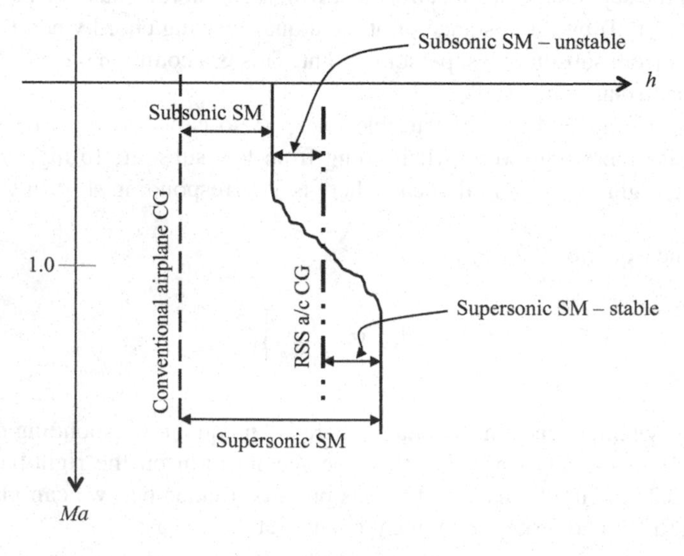

FIGURE 4.21 Shift in neutral point with Mach number and CG management in conventional and relaxed static stability (RSS) airplanes.

Example 4.9

To get a clearer picture of how trim and stability are affected during the transition from subsonic to supersonic flight, let us sketch the relevant trim curves on the usual plot of C_m versus α. See Figure 4.22.

Assume that with zero elevator deflection, the airplane is trimmed at an angle of attack called α^*_{design}, which corresponds to a low subsonic flight. This is marked by the full line in Figure 4.22. Going to trim at a higher subsonic flight speed, we need to reduce the trim angle of attack, so we now trim at α_1. Since neither the CG nor the NP has moved, the slope of the trim curve is unchanged. This is shown by the second parallel full line that intersects the C_m axis at a lower value of C_m, which is possible by applying down (positive) elevator. This is marked in Figure 4.22 as $\delta e_1 > 0$. So, to trim at a higher speed, we have to apply down elevator.

Now, we wish to go supersonic and trim at an increased speed corresponding to an angle of attack α_2. When we go supersonic, the slope of the trim curve ($C_{m\alpha}$) becomes more negative. So, if we do not change the elevator setting ($\delta e_1 > 0$), the trim curve remains hinged on the C_m axis, but swings to intersect the α axis at an angle of attack α_3, which is lower than the desired angle of attack α_2. To obtain trim at α_2, we need to shift the trim curve parallel to itself as shown by the dashed line in Figure 4.22. This requires a significant up-elevator deflection to $\delta e_2 < 0$. That is, to trim at a further higher but supersonic speed, we need to apply up elevator!

Note carefully: We are used to applying up elevator to trim at a higher angle of attack, but in this case, we need to use up elevator to trim at a *lower* angle of attack, $\alpha_2 < \alpha_1$.

Homework Exercise: Draw a similar plot to that in Figure 4.22 to show trim and elevator deflection with *Mach number* varying from subsonic to supersonic for the case of an RSS aircraft.

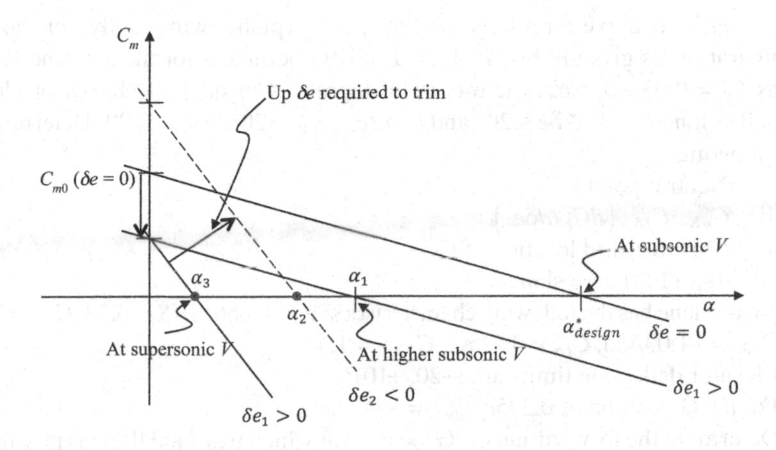

FIGURE 4.22 C_m versus α curve for different flight speed conditions.

EXERCISE PROBLEMS

4.1 Elevator hinge moment is given by the expression $H_e = \frac{1}{2}\rho V^2 S_e c_e C_{he}$, where S_e is the elevator planform area behind the hinge, c_e is the corresponding chord length and C_{he} is the hinge moment coefficient. This is the moment that a pilot needs to overcome using stick force F_s and the stick arm length l_s. The relation between the stick force and the hinge moment is defined by $F_s = GH_e$. The proportionality factor $G(= \delta e / l_s \times \delta_s)$, known as the gear ratio, is a function of the elevator deflection δe, stick arm length l_s and angular displacement of the stick about its own hinge point, δ_s. A pilot pulls a 0.75 m long stick towards oneself ($\delta_s = 5°$) to create an elevator up deflection of –15°. Determine the hinge moment if the stick force applied by the pilot is 2 N.

4.2 If the elevator is left free, that is, the pilot stick force is zero and consequently the elevator hinge moment is zero, the trim condition can still be defined by the expression for pitching moment coefficient about the CG. Discuss the stability of the trim condition in stick-free case. [*Hint:* The tail lift curve is modified due to elevator's free movement. Use the expression for hinge moment coefficient, $C_{he} = C_{h0} + C_{h\alpha_t}\alpha_t + C_{h\delta e}\delta e$.]

4.3 Use the data given in the table below to determine the neutral point location of the airplane:

h_{CG}	$C_L^*(@\delta e^* = 10°)$	$C_L^*(@\delta e^* = 5°)$	$C_L^*(@\delta e^* = -5°)$
0.10	0.06	0.11	0.15
0.20	0.13	0.28	0.48
0.30	0.20	0.45	0.87

4.4 Given the neutral point location for an airplane $h_{NP} = 0.9$, and other data as $\alpha^*_{\text{design}} = 3.5°$ ($\delta e = 0°$), $\alpha_{\min} = -1°$, $C_{m\delta e} = -0.43$/rad, $C_{L\alpha} = 5.815$/rad, δe (max up) = –15°, determine the forward-most location of CG, if $C_{m0} = 0.02$ and $\alpha_{\text{stall}} = 20°$.

4.5 C_m versus α curve for a large jet transport airplane (wing–body plus tail configuration) is given in Figure 4.23. The lift coefficient for the airplane is given as $C_L = 0.03 + 0.08\alpha$, where α is in degrees. Consider the limits of elevator deflection as $-24° \le \delta e \le 20°$ and $l_t = 2c$, $\alpha_{\text{stall}} = 20°$, $h_{CG} = 0.29$. Determine the following:
   i. Neutral point.
   ii. $C_{m\delta e}$, $C_{L\delta e}$, $\left(dC_L^*/d\delta e^*\right)$.
   iii. Most forward location of CG.
   iv. Trim lift curve slope.

4.6 An airplane has the following characteristics: $V_H = 0.66$, $S_t/S = 0.23$, $C_{m0} = 0.085$, $C_{m\delta e} = -1.03$/rad, $C_{L\alpha} = 4.8$/rad, $C_{Lmax} = 1.4$.
Elevator deflection limits are (–20, +10)°.
For a CG position of 0.295*c*, $C_{m\alpha} = -1.1$/rad.
Determine the forward-most CG location at which trimmed flight is possible.

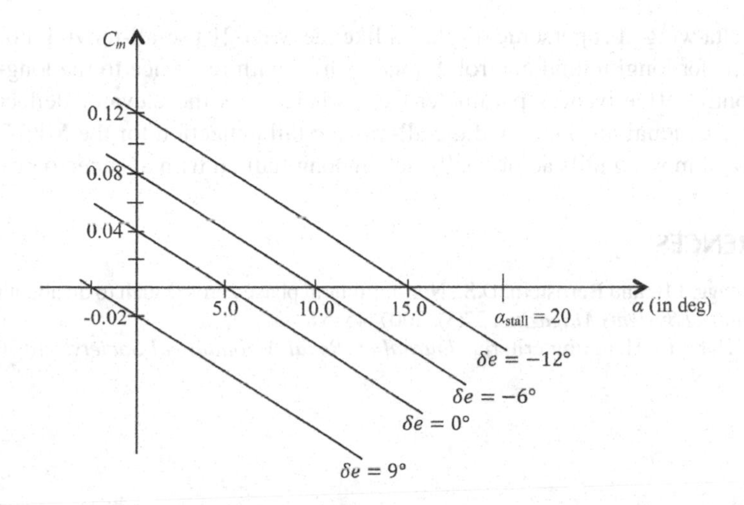

FIGURE 4.23 C_m versus α curves for different elevator deflections.

4.7 Derive expressions for elevator required per 'g' ($d\delta e/dn$) in a pull-up and level turn manoeuvres. Investigate the effect of derivative $C_{m\alpha}$ on elevator required per 'g'. [*Hint:* Consider the manoeuvres initiated from a level flight condition by use of elevator deflection.]

4.8 Consider an airplane with wing–body and canard configuration. Determine the elevator angle per 'g' using the following data ('c' superscript refers to canard):

$$S_w = 100\ \text{m}^2;\quad b_w = 25\ \text{m};\quad C_{L\alpha}^{wb} = 4.4/\text{rad};\quad (W/S) = 3500\ \text{N/m}^2;$$

$$C_{L0}^{wb} = -0.02;\quad S_c = 25\text{m}^2;\quad b_c = 12\ \text{m};\quad C_{L\alpha}^{c} = 5.3/\text{rad};$$

$$C_{m0}^{c} = 0;\quad C_{m\delta e}^{c} = 0.49/\text{rad};$$

$$h_{AC}^{c} - h_{AC}^{wb} = -10.7;\quad \varepsilon = 0.51°;\quad h_{CG} - h_{AC}^{wb} = -7.3$$

4.9 Airplanes that need to fly both subsonic and supersonic encounter problems due to excessive stability and trim drag at supersonic speeds. Besides relaxed static stability, suggest two possible remedies and describe how each would alleviate the two problems.

4.10 An airplane (wing–body–tail) is flying in a level flight trim state with unstable short-period dynamics. Describe the sequence of elevator deflections required to be given by the pilot to shift the airplane to a similar trim at a slightly higher angle of attack. Sketch this on a plot of C_m versus α.

4.11 Use the following airplane data: $C_{m0} = 0.085$, $C_{m\delta e} = -1.03/\text{rad}$, $C_{L\alpha} = 4.8/\text{rad}$. Elevator deflection limits are (+20, −10)°. The forward-most CG location $h_{CG} = 0.19$ corresponds to trim angle of attack $\alpha^* = 15°$. Find the location of the neutral point.

4.12 Delta-winged supersonic airplanes like the MiG-21 use all-moving horizontal tails for longitudinal control. Explain why—with reference to the longitudinal control effectiveness paramcter $C_{m\delta e}$, where δe is the elevator deflection for conventional airplanes and the all-moving tail deflection for the MiG-21. How do all-moving tails additionally help in longitudinal trim at supersonic speeds?

REFERENCES

1. Hoagg, J.B. and Bernstein, D.S., Non-minimum phase zeros: Much to do about nothing, *Control Systems Magazine*, 27(3), 2007, 45–57.
2. O'Hara, F., Handling criteria, *Journal of. Royal Aeronautical Society*, 71(676), 1967, 271–291.

5 Long-Period (Phugoid) Dynamics

5.1 PHUGOID MODE EQUATIONS

If you go back to Chapter 1 of this book, you will find that we had written the equations for the dynamics of the airplane in the longitudinal plane. These were written in terms of V and γ (Equations 1.5a and b) and θ (Equation 1.6). Then, we had identified a fast timescale T_1 (the pitch timescale) at which θ varies and a slower timescale T_2 (the heave timescale) at which the dynamics in the variables V and γ takes place.

From that point onwards, we had focussed on the fast timescale T_1 and the short period dynamics in θ (or equivalently in α), assuming that the variation of V and γ at the slower timescale T_2 is slow enough that V and γ are effectively held constant while θ (or equivalently α) changes. That led us to investigate the trim states obtained by solving the pitching moment equation, the stability of the short-period mode in terms of the derivatives $C_{m\alpha}$ and $C_{mq1} + C_{m\dot{\alpha}}$, and finally the effect of elevator control.

In this chapter, we investigate the slower dynamics in V and γ at the timescale T_2. This is called the long-period dynamics or the phugoid dynamics. For this, we assume that the short-period dynamics in α is much faster and also well damped. Thus, if an airplane flying at a certain trim is perturbed in all the variables, α, V and γ, then the change in α is immediate and also dies down quickly leaving behind a more leisurely variation in V and γ, which is typically less well damped, and hence persists for several cycles over tens of seconds. Over a typical long-period cycle, therefore, one may assume α to be held fixed at its trim value, α^*.

Then, neglecting the fast T_1 timescale dynamics in the $\ddot{\theta}$ Equation 1.6, the phugoid dynamics is contained in the slower T_2 timescale Equations 1.5a and 1.5b as below:

$$\frac{\dot{V}}{V} = \left(\frac{g}{V}\right)\left\{\frac{T}{W} - \frac{\bar{q}S}{W}C_D - \sin\gamma\right\} \tag{5.1}$$

$$\dot{\gamma} = \left(\frac{g}{V}\right)\left\{\frac{\bar{q}S}{W}C_L - \cos\gamma\right\} \tag{5.2}$$

5.2 ENERGY

Let us first analyse these equations in terms of the energy—sum of the airplane's kinetic and potential energies. This is how the phugoid mode was first analysed by Lanchester (see Box 5.1):

$$E = \frac{1}{2}mV^2 + mgh \tag{5.3}$$

DOI: 10.1201/9781003096801-5

BOX 5.1 LANCHESTER'S ANALYSIS OF THE PHUGOID MOTION (1908)

Lanchester considered the phugoid (long-period) motion of aircraft as an exchange of kinetic and potential energies, with the total energy being conserved. His formula for frequency of phugoid oscillation follows from the derivation below.

Assume the aircraft to be flying in level flight with the equilibrium condition stated as:

$$T - D = 0; \quad W - L = 0$$

Thus, $W = (1/2)\rho V^{*2} S C_L^*$; where superscript (*) represents the trim (steady state) value. Writing down the equation of vertical motion:

$$m\ddot{h} = L - W$$

Carrying out linearization of the above equation around the equilibrium state satisfying $W = L^*$ at fixed altitude $h = h^*$ as:

$$m(\ddot{h}^* + \Delta\ddot{h}) = \left(L^* + \Delta L\right) - W$$

From which, the equation for the perturbed dynamics in altitude is derived to be:

$$m\Delta\ddot{h} = \Delta L \quad \text{OR} \quad m\Delta\ddot{h} = \frac{1}{2}\rho\left(2V^*\Delta V\right)SC_L^*$$

The above equation can be rewritten as:

$$\Delta\ddot{h} = 2\frac{\Delta V}{V^*}\frac{L^*}{m} = 2\frac{\Delta V}{V^*}\frac{W}{m} = 2g\frac{\Delta V}{V^*} \tag{5.1a}$$

The perturbed motion thus modelled involves an exchange of kinetic and potential energies as the airplane exchanges altitude with velocity and vice versa. No change in the angle of attack is allowed, and thus C_L^* is assumed fixed at its trim value. It is also assumed that $T = D$ is satisfied at all time.

Conservation of energy yields (from Equation 5.3):

$$\Delta E = 0 = mV^*\Delta V + mg\Delta h$$

Which is $\dfrac{\Delta V}{V^*} = -\dfrac{g}{V^{*2}}\Delta h$

Substituting the above expression in Equation 5.1a results in:

$$\Delta\ddot{h} + 2\left(\frac{g}{V^*}\right)^2 \Delta h = 0 \quad (5.1b)$$

From Equation 5.1b, one can determine the frequency of the phugoid motion as $\omega_p = \sqrt{2}\left(g/V^*\right)$. Having assumed the energy to be conserved (that is, no damping), the Lanchester formulation could not provide an expression for damping ratio of the phugoid motion, although his expression for the phugoid frequency remained acceptable for a long time.

Differentiating Equation 5.3 with time gives:

$$\frac{dE}{dt} = mV\frac{dV}{dt} + mg\frac{dh}{dt} = mV\frac{dV}{dt} + mgV\sin\gamma$$

Dividing each term by mg:

$$\frac{1}{mg}\frac{dE}{dt} = \frac{V}{g}\frac{dV}{dt} + \frac{dh}{dt} = \underbrace{\frac{V}{g}\frac{dV}{dt}}_{\text{"forward acceleration"}} + \underbrace{V\sin\gamma}_{\text{climb rate}} \quad (5.4)$$

The first term of the right-hand side is marked as the forward acceleration in metre/second (with time made non-dimensional by $T_2 = V/g$) and the second term is the climb rate.

Replacing dV/dt in Equation 5.4 from Equation 5.1 we get:

$$\frac{1}{mg}\frac{dE}{dt} = V\left\{\frac{T}{W} - \frac{\bar{q}S}{W}C_D - \sin\gamma\right\} + V\sin\gamma = V\left\{\frac{T}{W} - \frac{\bar{q}S}{W}C_D\right\}$$

which results in:

$$\frac{1}{mg}\frac{dE}{dt} = \left\{\frac{T}{W} - \frac{\bar{q}S}{W}C_D\right\}V = \frac{(T-D)V}{W} \quad (5.5)$$

The right-hand side of Equation 5.5 is usually called the specific excess power (SEP). In other words, any excess power due to an excess of thrust over drag may be used to increase the airplane's energy, either kinetic or potential, resulting in a forward acceleration and/or climb.

What happens when the energy, E, is a constant, and hence $(dE/dt) = 0$? Then, there can be a mutual exchange between the kinetic and potential energies, that is, between the acceleration and climb, as per this relation:

$$\frac{V}{g}\frac{dV}{dt} + \frac{dh}{dt} = \frac{V}{g}\frac{dV}{dt} + V\sin\gamma = 0 \quad (5.6)$$

In other words, any change in the airplane's energy is only a redistribution between its kinetic and potential energies. As it loses altitude, it gains speed (and kinetic energy), and when it gains altitude, it bleeds speed and loses kinetic energy.

Before proceeding further, we need to consider one more factor.

5.2.1 Normal Acceleration

The normal acceleration of an airplane (in the vertical plane) while flying along a curved flight path is given by $V\dot{\gamma}$, which by use of Equation 5.2 can be written as:

$$V\dot{\gamma} = g\left\{\frac{\bar{q}S}{W}C_L - \cos\gamma\right\} \tag{5.7}$$

Or, dividing by g, gives the normal acceleration in units of g:

$$\frac{V\dot{\gamma}}{g} = \frac{\bar{q}S}{W}C_L - \cos\gamma \tag{5.8}$$

You may recall that the ratio L/W or $\bar{q}SC_L/W$ is called the *load factor* in aircraft performance, and is denoted by the symbol, n. Thus,

$$\frac{V\dot{\gamma}}{g} = n - \cos\gamma \tag{5.9}$$

In simple words, if the lift exceeds the weight, the airplane flight path tends to curve upwards, and when lift is less than the weight, the flight path curves downward.

Rewriting Equation 5.9:

$$n = \frac{V\dot{\gamma}}{g} + \cos\gamma(\text{in } g\text{'s}) \tag{5.10}$$

If, instantaneously, $\gamma = 0$, then:

$$n = \frac{V\dot{\gamma}}{g} + 1\ (\text{in } g\text{'s}) \tag{5.11}$$

Thus, $n = 1(1\ g)$ corresponds to level flight, n slightly greater than 1 is a shallow ascent along a curved flight path and n slightly less than 1 is a shallow descent along a curved flight path (see Figure 5.1).

Putting this together with the conclusions from Equation 5.6, we can describe the phugoid (long-period) motion next.

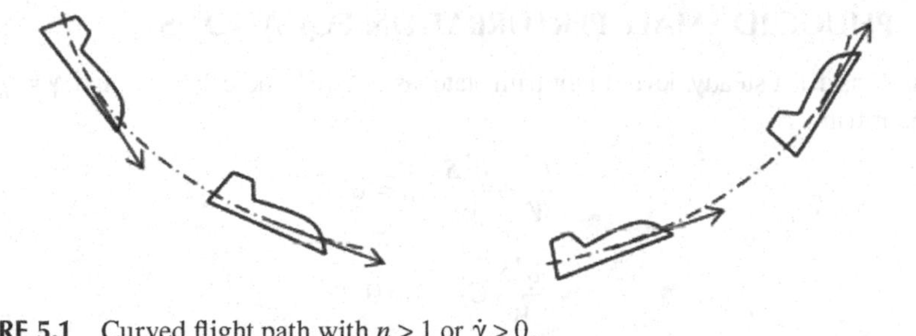

FIGURE 5.1 Curved flight path with $n > 1$ or $\dot{\gamma} > 0$.

5.3 PHUGOID MODE PHYSICS

The phugoid mode appears as a sinusoidal oscillation in altitude about the level flight trim state. A couple of cycles of the phugoid motion are shown in Figure 5.2.

From a peak to a trough—a descending segment—velocity is gained ($\dot{V} > 0$) in exchange for height lost ($h < 0, \gamma < 0$), and from that trough to the next peak—an ascending segment—velocity is lost ($\dot{V} < 0$) and height is gained ($h > 0, \gamma > 0$). This is the trade-off between kinetic and potential energy.

From a mid-point to the next mid-point, the flight path either curves up or down; where it curves up, $L > W$ and $\gamma > 0$, and on the downward curve segment, $L < W$ and $\dot{\gamma} < 0$.

Now, putting these two together, we can trace the path of the phugoid motion over one cycle. Let us start at a trough. Here, the velocity is greater than the trim velocity. Hence, the lift (proportional to V^2) is greater than the weight (remember, the lift at trim was just equal to the weight), so the airplane path curves upwards. The airplane is now gaining altitude at the expense of velocity. By the mid-point, L and W are again equal, so the flight path stops curving upwards, though $\gamma > 0$ still, and V is equal to the trim value V^* again.

Inertia now carries the airplane past this mid-point and the velocity continues to fall below V^* as height is gained. But now, the lower V implies a lift lesser than weight, so the flight path starts to curve downward. At the peak, $L < W$, so $\dot{\gamma} < 0$, and the airplane starts descending and gaining velocity at the expense of altitude.

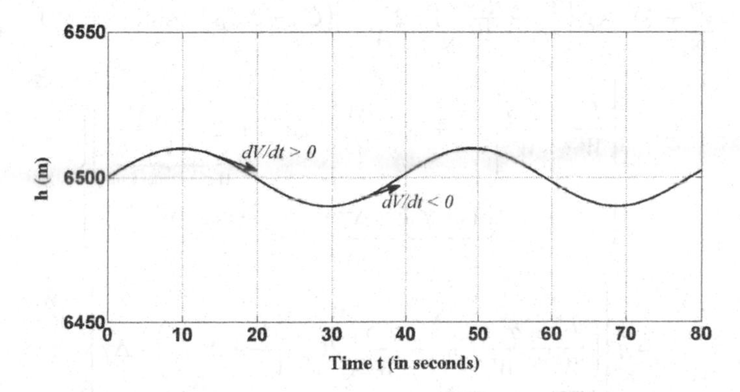

FIGURE 5.2 Typical phugoid motion.

5.4 PHUGOID SMALL-PERTURBATION EQUATIONS

Let us consider a steady, level flight trim state, as before, where $V = V^*$ and $\gamma = \gamma^* = 0$. Then, at trim:

$$\frac{T^*}{W} - \frac{\bar{q}^* S}{W} C_D^* = 0 \tag{5.12}$$

$$\frac{\bar{q}^* S}{W} C_L^* - 1 = 0 \tag{5.13}$$

which is simply a balance between thrust and drag in Equation 5.12, and lift and weight in Equation 5.13. T^* here is the thrust at trim, which is assumed to be unchanged with change in V (this is a fairly good approximation for turbojet engines).

Now, consider a small perturbation from this trim state. As discussed earlier, the perturbation in α is assumed to have died down at the faster short-period timescale T_1, leaving behind perturbations in V, γ given by:

$$V = V^* + \Delta V \quad \text{and} \quad \gamma = \gamma^* + \Delta\gamma \tag{5.14}$$

Note, $\gamma^* = 0$ and $\alpha = \alpha^*$.

First, let us evaluate the perturbation term in dynamic pressure:

$$\begin{aligned} \bar{q} &= \frac{1}{2}\rho V^2 = \frac{1}{2}\rho\left(V^* + \Delta V\right)^2 \\ &= \frac{1}{2}\rho V^{*2}\left(1 + \frac{\Delta V}{V^*}\right)^2 \\ &= \frac{1}{2}\rho V^{*2}\left\{1 + 2\frac{\Delta V}{V^*} + \left(\frac{\Delta V}{V^*}\right)^2\right\} \end{aligned} \tag{5.15}$$

$$\bar{q} = \bar{q}^*\left(1 + 2\frac{\Delta V}{V^*}\right) \text{(on neglecting the quadratic term as being of higher order)}$$

Now, writing Equations 5.1 and 5.2 in terms of the perturbation variables in Equation 5.14:

$$\begin{aligned} \underbrace{\dot{V}^*}_{0} + \Delta\dot{V} &= g\left\{\frac{T^*}{W} - \frac{\bar{q}^* S}{W}\left(1 + 2\frac{\Delta V}{V^*}\right)\left(C_D^* + \Delta C_D\right) - \sin\left(\gamma^* + \Delta\gamma\right)\right\} \\ &= g\left\{\begin{gathered} \frac{T^*}{W} - \frac{\bar{q}^* S}{W}\left(C_D^* + 2C_D^*\frac{\Delta V}{V^*} + \Delta C_D + \underbrace{2\frac{\Delta C_D \Delta V}{V^*}}_{\text{Higher-order term}}\right) \\ \underbrace{-(\sin\gamma^* \cos\Delta\gamma + \cos\gamma^* \sin\Delta\gamma)}_{\gamma^*=0} \end{gathered}\right\} \\ &= g\left\{\underbrace{\left(\frac{T^*}{W} - \frac{\bar{q}^* S}{W} C_D^*\right)}_{\text{Equilibrium}} - \frac{\bar{q}^* S}{W}\left(2C_D^*\frac{\Delta V}{V^*} + \Delta C_D\right) - \Delta\gamma\right\} \end{aligned} \tag{5.16}$$

$$\begin{aligned}\Delta\dot{\gamma}\left(V^{*}+\Delta V\right)&=g\left\{\frac{\bar{q}^{*}S}{W}\left(1+2\frac{\Delta V}{V^{*}}\right)\left(C_{L}^{*}+\Delta C_{L}\right)-\cos\left(\gamma^{*}+\Delta\gamma\right)\right\}\\&=g\left\{\frac{\bar{q}^{*}S}{W}\left(1+2\frac{\Delta V}{V^{*}}\right)\left(C_{L}^{*}+\Delta C_{L}\right)-\underbrace{\left(\cos\gamma^{*}\cos\Delta\gamma-\sin\gamma^{*}\sin\Delta\gamma\right)}_{1}\right\}\end{aligned}\tag{5.17}$$

Rearranging terms, and using the small angle assumption and the trim condition ($\cos\Delta\gamma\approx1$, $\sin\Delta\gamma\approx\Delta\gamma$; $\gamma^{*}=0$):

$$\Delta\dot{\gamma}V^{*}+\underbrace{\Delta\dot{\gamma}\Delta V}_{\text{Higher order}}=g\left\{\begin{array}{c}\underbrace{\left(\frac{\bar{q}^{*}SC_{L}^{*}}{W}-1\right)}_{\text{Equilibrium}}+\frac{\bar{q}^{*}S}{W}\left(\Delta C_{L}+2C_{L}^{*}\frac{\Delta V}{V^{*}}\right)\\+\underbrace{2\frac{\bar{q}^{*}S}{W}\left(\frac{\Delta V\Delta C_{L}}{V^{*}}\right)}_{\text{Higher-order term}}\end{array}\right\}\tag{5.18}$$

In Equations 5.16 and 5.18, the terms marked 'equilibrium' are zero as per Equations 5.12 and 5.13, and the terms marked 'higher-order' may be neglected as being the product of two small quantities, hence even smaller.

That yields the following perturbation equations for the phugoid dynamics:

$$\frac{\Delta\dot{V}}{V^{*}}=\left(\frac{g}{V^{*}}\right)\left\{-\frac{\bar{q}^{*}S}{W}\left(\Delta C_{D}+2C_{D}^{*}\frac{\Delta V}{V^{*}}\right)-\Delta\gamma\right\}\tag{5.19}$$

$$\Delta\dot{\gamma}=\left(\frac{g}{V^{*}}\right)\left\{\frac{\bar{q}^{*}S}{W}\left(\Delta C_{L}+2C_{L}^{*}\frac{\Delta V}{V^{*}}\right)\right\}\tag{5.20}$$

Clearly, both ΔV and $\Delta\gamma$ vary at the rate (g/V^{*}), which defines the slow use hyphen as time-scale, T_2.

Before we proceed further, we need to model the perturbed aerodynamic terms, ΔC_D and ΔC_L.

5.5 AERODYNAMIC MODELLING WITH MACH NUMBER

We have seen in Chapter 1 that the aerodynamic forces are a function of the angle of attack and the Mach number, in the static sense. Later, in Chapter 2, when modelling the aerodynamic moment for the short-period mode, we found that there is also a dynamic effect due to the difference between the perturbations in body-axis and wind-axis pitch rates $(\Delta q_b-\Delta q_w)$.

For the phugoid mode, we have assumed that the angle of attack is constant, so there is no static aerodynamic term due to $\Delta\alpha$. Also, as sketched in Figure 5.1, since the airplane nose rises and falls with the velocity vector in such a way as to keep the angle of attack constant at α^*, Δq_b the body-axis pitch rate is the same as the wind-axis pitch rate, Δq_w (equal to $\Delta\dot{\gamma}$). That is, $\Delta q_b - \Delta q_w = 0$. Hence, there is no effect of $(\Delta q_b - \Delta q_w)$ either (see Section 5.12 for a fuller explanation). Thus, only the static effect due to change in Mach number, ΔMa, needs to be modelled.

The expressions for the perturbed drag and lift coefficients can be written as:

$$\Delta C_D = C_{DMa}\Delta Ma \tag{5.21}$$

$$\Delta C_L = C_{LMa}\Delta Ma \tag{5.22}$$

where C_{DMa} and C_{LMa} are, respectively, defined as the partial derivatives of C_D and C_L with Ma (all other variables being held constant) at the equilibrium point (denoted by *).

$$C_{DMa} = \left.\frac{\partial C_D}{\partial Ma}\right|_* \tag{5.23}$$

$$C_{LMa} = \left.\frac{\partial C_L}{\partial Ma}\right|_* \tag{5.24}$$

with Mach number $Ma = (V/a)$, and Mach number at the trim state $Ma^* = \left(V^*/a^*\right)$, where a^* is the speed of sound at the level trim flight altitude.

Even though the airplane loses and gains altitude during the phugoid motion, the change in altitude is usually small enough that it is acceptable to assume a^* to be fixed. Then:

$$\Delta Ma = \frac{\Delta V}{a^*} \tag{5.25}$$

And so,

$$\Delta C_D = C_{DMa}\frac{\Delta V}{a^*} = \frac{V^*}{a^*}C_{DMa}\left(\frac{\Delta V}{V^*}\right) = Ma^* C_{DMa}\left(\frac{\Delta V}{V^*}\right) \tag{5.26}$$

And in similar fashion:

$$\Delta C_L = Ma^* C_{LMa}\left(\frac{\Delta V}{V^*}\right) \tag{5.27}$$

Example 5.1 Typical Variation of C_{LMa}

For airplanes where most of the lift arises due to the wing, and the wing has a reasonably large aspect ratio, the Prandtl–Glauert rule applies:

$$C_L(Ma) = \frac{C_{L(Ma=0)}}{\sqrt{1-Ma^2}} \tag{5.28}$$

in subsonic flow, where $C_{L(Ma=0)}$ is the lift coefficient for incompressible flow. In essence, Equation 5.28 captures the effect of compressibility that comes in with increasing Mach number on the lift coefficient.

Differentiating with Mach number, Ma, yields the derivative:

$$\begin{aligned} C_{LMa} = \left.\frac{\partial C_L}{\partial Ma}\right|_* &= \left.-\frac{1}{2}C_{L(Ma=0)}\left(1-Ma^2\right)^{-3/2}(-2Ma)\right|_* \\ &= \frac{C_{L(Ma=0)}}{\sqrt{1-Ma^{*2}}}\cdot\frac{Ma^*}{\left(1-Ma^{*2}\right)} \\ &= C_L\left(Ma^*\right)\cdot\frac{Ma^*}{\left(1-Ma^{*2}\right)} \end{aligned} \tag{5.29}$$

Usually, $C_{LMa} > 0$ and $C_{DMa} \sim 0$ for subsonic speeds.

Homework Exercise: Evaluate C_{LMa} and C_{DMa} for supersonic speeds.

5.6 PHUGOID DYNAMICS

Let us now insert these expressions for the perturbed aerodynamic force coefficients into the small-perturbation phugoid Equations 5.19 and 5.20:

$$\begin{aligned} \frac{\Delta\dot{V}}{V^*} &= \left(\frac{g}{V^*}\right)\left\{-\frac{\bar{q}^*S}{W}\left(Ma^*C_{DMa}\frac{\Delta V}{V^*}+2C_D^*\frac{\Delta V}{V^*}\right)-\Delta\gamma\right\} \\ &= \left(\frac{g}{V^*}\right)\left\{-\frac{\bar{q}^*S}{W}\left(Ma^*C_{DMa}+2C_D^*\right)\frac{\Delta V}{V^*}-\Delta\gamma\right\} \end{aligned} \tag{5.30}$$

$$\begin{aligned} \Delta\dot{\gamma} &= \left(\frac{g}{V^*}\right)\left\{\frac{\bar{q}^*S}{W}\left(Ma^*C_{LMa}\frac{\Delta V}{V^*}+2C_L^*\frac{\Delta V}{V^*}\right)\right\} \\ &= \left(\frac{g}{V^*}\right)\left\{\frac{\bar{q}^*S}{W}\left(Ma^*C_{LMa}+2C_L^*\right)\frac{\Delta V}{V^*}\right\} \end{aligned} \tag{5.31}$$

Next, we obtain a single equation from these two by differentiating Equation 5.30 and using Equation 5.31 in place of $\Delta\dot{\gamma}$.

$$\frac{\Delta\ddot{V}}{V^*}=\left(\frac{g}{V^*}\right)\left\{-\frac{\bar{q}^*S}{W}\left(Ma^*C_{DMa}+2C_D^*\right)\frac{\Delta\dot{V}}{V^*}-\Delta\dot{\gamma}\right\}$$

$$=\left(\frac{g}{V^*}\right)\left\{-\frac{\bar{q}^*S}{W}\left(Ma^*C_{DMa}+2C_D^*\right)\frac{\Delta\dot{V}}{V^*}-\left(\frac{g}{V^*}\right)\left[\frac{\bar{q}^*S}{W}\left(Ma^*C_{LMa}+2C_L^*\right)\frac{\Delta V}{V^*}\right]\right\} \tag{5.32}$$

Rearranging terms, we get:

$$\frac{\Delta\ddot{V}}{V^*}+\left(\frac{g}{V^*}\right)\left[\frac{\bar{q}^*S}{W}\left(Ma^*C_{DMa}+2C_D^*\right)\right]\frac{\Delta\dot{V}}{V^*}+\left(\frac{g}{V^*}\right)^2\left[\frac{\bar{q}^*S}{W}\left(Ma^*C_{LMa}+2C_L^*\right)\right]\frac{\Delta V}{V^*}=0 \tag{5.33}$$

which is a second-order differential equation in ΔV, where we have assumed $\Delta\alpha=0$. Note that we earlier had a second-order equation for the short-period mode in terms of the perturbed angle of attack $\Delta\alpha$, where we kept $\Delta V=0$. Now, we can analyse Equation 5.33 for the phugoid dynamics using the ideas in Chapter 2.

Example 5.2

A business jet aircraft is flying at sea-level condition with a speed $V^*=79.3$ m/s. Let us look at the phugoid dynamics of this aircraft at the given trim condition. Given data:

$$W=171136\text{ N};\ S=60.27\text{ m}^2;\ C_L^*=0.737;\ C_D^*=0.095.$$

$$\frac{\bar{q}^*S}{W}=\frac{0.5\times\rho\times V^{*2}\times S}{W}=\frac{0.5\times1.225\times79.3^2\times60.27}{171136}=1.35$$

Neglecting the effects due to compressibility at this speed, $C_{DMa}\sim0$ and $C_{LMa}\sim0$, and Equation 5.33 may be evaluated to arrive at the following equation in ΔV:

$$\Delta\ddot{V}+0.0294\Delta\dot{V}+0.032\Delta V=0 \tag{5.34}$$

The time response by numerically integrating Equation 5.34 is shown in Figure 5.3 for an initial perturbation in velocity, $\Delta V=0.2$ m/s. The time between two consecutive peaks or troughs (approximately 36 sec in Figure 5.3) is the damped time period, T_d, and the damping ratio may be calculated from the successive peaks/troughs as $\zeta=0.093$.

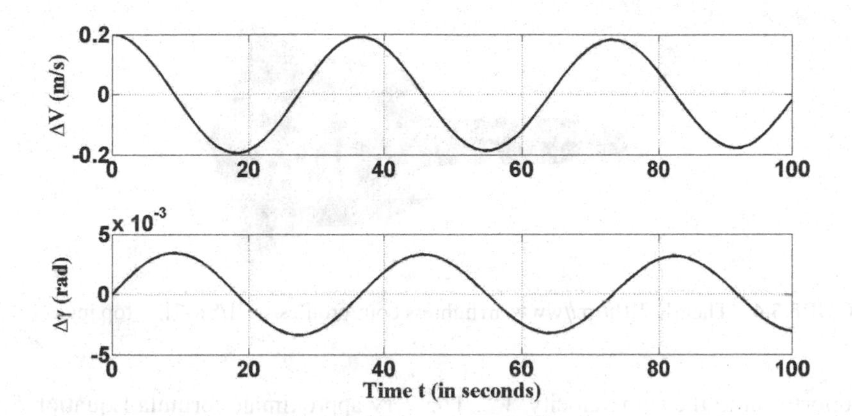

FIGURE 5.3 Simulation results showing aircraft response in phugoid.

5.7 PHUGOID MODE FREQUENCY AND DAMPING

By examining the phugoid dynamics Equation 5.33 and comparing it with the standard form of a second-order dynamical system as in Equation 2.4, we can extract the phugoid frequency and damping as:

$$\omega_{np}^2 = \left(\frac{g}{V^*}\right)^2 \left[\frac{\bar{q}^* S}{W}\left(Ma^* C_{LMa} + 2C_L^*\right)\right] \tag{5.35}$$

$$2\zeta_p \omega_{np} = \left(\frac{g}{V^*}\right)\left[\frac{\bar{q}^* S}{W}\left(Ma^* C_{DMa} + 2C_D^*\right)\right] \tag{5.36}$$

Of course, this assumes that the expression on the right-hand side of Equation 5.35 is always positive, which is generally the case.

At low speeds, $Ma^* C_{LMa}$ is usually quite small, and if we neglect it, we can get an approximate formula for the phugoid frequency as:

$$\omega_{np}^2 \approx \left(\frac{g}{V^*}\right)^2 \left[\frac{\bar{q}^* S}{W}\left(2C_L^*\right)\right] = 2\left(\frac{g}{V^*}\right)^2 \tag{5.37}$$

Since at level flight, $\bar{q}^* S C_L = L = W$. So, we conclude that:

$$\omega_{np} \approx \sqrt{2}\left(\frac{g}{V^*}\right) \tag{5.38}$$

which is the expression derived by Lanchester in 1908. This is quite an amazing result as the frequency (hence, time period) of the phugoid oscillations appears to be independent of the airplane or the flight altitude. The time period ($T = 2\pi/\omega$) is just

FIGURE 5.4 The SR-71(http://www.airfighters.com/profiles/sr-71/SR-71A_top.jpg).

proportional to the trim velocity, V^*. The very approximate formula Equation 5.38 for the phugoid frequency turns out to be fairly adequate for some airplanes, but not so good for others.[1,4]

Example 5.3

At high speeds, Ma^*C_{LMa} can be negative and comparable in magnitude to $2C_L^*$. Then, the term on the right-hand side of Equation 5.35 can be nearly zero or even negative, which implies that the phugoid mode is unstable. This is possible for high-supersonic or hypersonic airplanes such as the SR-71 (shown in Figure 5.4).

It is not enough for the right-hand side of Equation 5.35 to be positive to ensure phugoid stability. We also need the damping term in Equation 5.36 to be positive. That is:

$$\left(\frac{g}{V^*}\right)\left[\frac{\bar{q}^*S}{W}\left(Ma^*C_{DMa}+2C_D^*\right)\right]>0 \tag{5.39}$$

This also happens to be true in most cases, at least at low speeds.

If we ignore Ma^*C_{DMa} as compared to $2C_D^*$, then we get a simpler approximation to the phugoid damping as:

$$2\zeta_p\omega_{np}\approx\left(\frac{g}{V^*}\right)\left[\frac{\bar{q}^*S}{W}\left(2C_D^*\right)\right]=\left(\frac{g}{V^*}\right)\left(\frac{2C_D^*}{C_L^*}\right) \tag{5.40}$$

Thus, from Equations 5.38 and 5.40, we have:

$$\zeta_p\approx\frac{1}{\sqrt{2}}\frac{C_D^*}{C_L^*} \tag{5.41}$$

For typical subsonic airplanes, this works out to roughly $\zeta_p\approx 0.05$, which is fairly low. Interestingly, the phugoid damping turns out to be inversely proportional to the aerodynamic efficiency, C_L/C_D. So as designers strive to improve the aerodynamic efficiency of modern airplanes, they unwittingly make the phugoid mode less stable!

In up-and-away flight, a poorly damped phugoid mode is usually not a bother, and the airplane's flight controller usually has a 'Mach hold' mode that suppresses phugoid oscillations. But, closer to the ground, especially during the landing approach, this can be potentially dangerous. Remember, any change in velocity excites the phugoid mode and appears as an oscillation in altitude (and velocity). Oscillations in altitude cannot be accepted when flying close to the ground!

Already, C_L^* is fairly high at the landing approach, the increased C_L being achieved by partial use of flaps; so, in the process, C_D^* is also enhanced. But, a further deflection of flaps to full (maximum) increases C_D^* even further, even though there is no corresponding increase in C_L^*. This is done mainly to improve the phugoid damping (as per Equation 5.41) and also to provide extra drag for braking after touchdown. During take-off, the extra C_L is desirable, but not the additional C_D (some of it is inevitable), so the flaps are only partially deflected at take-off.

Example 5.4 Phugoid Mode Handling Qualities

Unlike short-period mode, there is no stringent flying or handling qualities requirement on the frequency of the phugoid mode oscillation. This is primarily because the phugoid mode instability takes a long time to develop (time period being of the order of 20–30 seconds) and a pilot can very well control it. However, the damping of the phugoid mode is usually of concern and depending on the pilot workload, it is required that the damping ratio of the phugoid mode be in the following ranges, irrespective of type or class of airplanes and/or flight phases:

Level 1 (adequate flying and handling qualities): $\zeta_p > 0.04$
Level 2 (with little pilot workload): $\zeta_p > 0.0$
Level 3 (excessive pilot workload): Time to double the amplitude
$T_2 > 55$ seconds

5.8 ACCURATE SHORT-PERIOD AND PHUGOID APPROXIMATIONS

The short-period frequency and damping calculated from expressions that we derived in Chapter 2 (Equations 2.36 and 2.37), usually match reasonably well with those observed in flight. However, the phugoid frequency and damping formulas of Equations 5.35 and 5.36, while good enough for a first-cut approximation, do not always match well with the observations in flight.

The reason for this discrepancy was revealed by Ananthkrishnan and Unnikrishnan.[2] In deriving the phugoid approximations in this chapter, we assumed that the α dynamics was much faster and well damped, hence the (short-period) oscillations in α would die down quickly, leaving behind only V and γ to change with the phugoid mode. This, it turns out, is not entirely true. The perturbed α dynamics does damp out at the faster timescale T_1, but not all the way to zero. There is a residual α perturbation, which then proceeds to change as per the slower, phugoid timescale, T_2.

In other words, the division between variables like α changing at the fast, short-period timescale, and those like V and γ changing at the slower phugoid timescale, totally independent of each other, is not entirely appropriate. Instead, the angle of attack must itself be split into two components—a faster one that varies as per the short-period mode and damps out relatively quickly, and a slower one that persists for longer and varies with V and γ at the slower phugoid scale.

The short-period and phugoid mode approximations based on this revised arrangement were derived by Raghavan and Ananthkrishnan.[3] We present here an abbreviated version of that derivation.

The complete set of small-perturbation longitudinal equations assuming a level flight trim state may be gathered from Equations 5.19, 5.20 and 2.21 as below:

$$\frac{\Delta \dot{V}}{V^*} = \left(\frac{g}{V^*}\right)\left\{-\frac{\bar{q}^* S}{W}\left(\Delta C_D + 2C_D^* \frac{\Delta V}{V^*}\right) - \Delta\gamma\right\} \tag{5.42}$$

$$\Delta\dot{\gamma} = \left(\frac{g}{V^*}\right)\left\{\frac{\bar{q}^* S}{W}\left(\Delta C_L + 2C_L^* \frac{\Delta V}{V^*}\right)\right\} \tag{5.43}$$

$$\Delta\ddot{\theta} = \left(\frac{\bar{q}^* Sc}{I_{yy}}\right)\Delta C_m \tag{5.44}$$

The most general model for the perturbed aerodynamic coefficients may be written as in Table 5.1 where the elevator control derivatives have not been included for convenience. For the moment we ignore the '$q2$' derivatives which will be discussed in Section 5.12.

5.8.1 Short-Period Mode Dynamics

First, we study the fast-pitch dynamics at timescale T_1 in Equation 5.44. At this time scale, we can assume the slower variables ΔV, $\Delta\gamma$ to be effectively constant; hence, their time rates of change are zero. That allows us to write the perturbed pitch dynamics equation: $\Delta\theta = \Delta\alpha$ as:

$$\Delta\ddot{\alpha} = \left(\frac{\bar{q}^* Sc}{I_{yy}}\right)\Delta C_m = \left(\frac{\bar{q}^* Sc}{I_{yy}}\right)\left\{Ma^* C_{mMa}\left(\frac{\Delta V}{V^*}\right) + C_{m\alpha}\Delta\alpha + \left(C_{mq1} + C_{m\dot{\alpha}}\right)\Delta\dot{\alpha}\left(\frac{c}{2V^*}\right)\right\} \tag{5.45}$$

Rearranging terms, we obtain the short-period dynamic model as:

$$\begin{aligned}\Delta\ddot{\alpha} - \left(\frac{\bar{q}^* Sc}{I_{yy}}\right)(c/2V^*)\left(C_{mq1} + C_{m\dot{\alpha}}\right)\Delta\dot{\alpha} - \left(\frac{\bar{q}^* Sc}{I_{yy}}\right)C_{m\alpha}\Delta\alpha \\ = \left(\frac{\bar{q}^* Sc}{I_{yy}}\right)Ma^* C_{mMa}\left(\frac{\Delta V}{V^*}\right)\end{aligned} \tag{5.46}$$

The left-hand side of Equation 5.46 is identical to Equation 2.32 and hence the short-period frequency and damping expressions in Equations 2.33 and 2.34 are retained unchanged. However, the presence of the 'slow' term on the right-hand side of Equation 5.46 implies that the perturbation in the angle of attack does not die down to zero; instead, it leaves behind a static residual value $\Delta\alpha_s$ given by:

$$-\left(\frac{\bar{q}^*Sc}{I_{yy}}\right)C_{m\alpha}\Delta\alpha_s = \left(\frac{\bar{q}^*Sc}{I_{yy}}\right)Ma^*C_{mMa}\left(\frac{\Delta V}{V^*}\right)$$
$$\Delta\alpha_s = -Ma^*\left(\frac{C_{mMa}}{C_{m\alpha}}\right)\left(\frac{\Delta V}{V^*}\right) \tag{5.47}$$

which now varies at the slower timescale T_2.

5.8.2 Phugoid Mode Dynamics

Combining Equations 5.42 and 5.43 at the slower timescale with the aerodynamic model in Table 5.1 yields a single second-order equation for the phugoid dynamics, as in Equation 5.33, as follows:

$$\frac{\Delta\ddot{V}}{V^*} + \left(\frac{g}{V^*}\right)\left[\frac{\bar{q}^*S}{W}\left(Ma^*C_{DMa} + 2C_D^*\right)\right]\frac{\Delta\dot{V}}{V^*} + \left(\frac{g}{V^*}\right)\frac{\bar{q}^*S}{W}C_{D\alpha}\Delta\dot{\alpha}$$
$$+\left(\frac{g}{V^*}\right)^2\left[\frac{\bar{q}^*S}{W}\left(Ma^*C_{LMa} + 2C_L^*\right)\right]\frac{\Delta V}{V^*} + \left(\frac{g}{V^*}\right)^2\frac{\bar{q}^*S}{W}C_{L\alpha}\Delta\alpha = 0 \tag{5.48}$$

TABLE 5.1
Summary of Small-Perturbation Longitudinal Aerodynamic Coefficients

$$\Delta C_D = C_{DMa}\Delta Ma + C_{D\alpha}\Delta\alpha + C_{Dq1}\left(\Delta q_b - \Delta q_w\right)\left(c/2V^*\right) + C_{Dq2}\Delta q_w\left(c/2V^*\right) + C_{D\dot{\alpha}}\Delta\dot{\alpha}\left(c/2V^*\right)$$
$$= Ma^*C_{DMa}\left(\frac{\Delta V}{V^*}\right) + C_{D\alpha}\Delta\alpha + \left(C_{D\dot{\alpha}} + C_{Dq1}\right)\Delta\dot{\alpha}\left(c/2V^*\right) + C_{Dq2}\Delta\dot{\gamma}\left(c/2V^*\right)$$

$$\Delta C_L = C_{LMa}\Delta Ma + C_{L\alpha}\Delta\alpha + C_{Lq1}\left(\Delta q_b - \Delta q_w\right)\left(c/2V^*\right) + C_{Lq2}\Delta q_w\left(c/2V^*\right) + C_{L\dot{\alpha}}\Delta\dot{\alpha}\left(c/2V^*\right)$$
$$= Ma^*C_{LMa}\left(\frac{\Delta V}{V^*}\right) + C_{L\alpha}\Delta\alpha + \left(C_{L\dot{\alpha}} + C_{Lq1}\right)\Delta\dot{\alpha}\left(c/2V^*\right) + C_{Lq2}\Delta\dot{\gamma}\left(c/2V^*\right)$$

$$\Delta C_m = C_{mMa}\Delta Ma + C_{m\alpha}\Delta\alpha + C_{mq1}\left(\Delta q_b - \Delta q_w\right)\left(c/2V^*\right) + C_{mq2}\Delta q_w\left(c/2V^*\right) + C_{m\dot{\alpha}}\Delta\dot{\alpha}\left(c/2V^*\right)$$
$$= Ma^*C_{mMa}\left(\frac{\Delta V}{V^*}\right) + C_{m\alpha}\Delta\alpha + \left(C_{m\dot{\alpha}} + C_{mq1}\right)\Delta\dot{\alpha}\left(c/2V^*\right) + C_{mq2}\Delta\dot{\gamma}\left(c/2V^*\right)$$

where we drop the $q2$ derivative terms as before, as well as the $q1$ derivatives—C_{Dq1} is usually not so important and C_{Lq1} gives a correction to the phugoid damping that is of higher order in g/V^* and hence may be ignored. The $\Delta\alpha$ in Equation 5.48 is the component that varies at the slow timescale and which we have determined as the static residual in Equation 5.47. $\Delta\dot{\alpha}$ may be obtained by differentiating Equation 5.47. The resulting model for the phugoid mode dynamics appears as:

$$\frac{\Delta\ddot{V}}{V^*} + \left(\frac{g}{V^*}\right)\left[\frac{\bar{q}^* S}{W}\left\{Ma^* C_{DMa} + 2C_D^* - Ma^*\left(C_{mMa}/C_{m\alpha}\right)C_{D\alpha}\right\}\right]\frac{\Delta\dot{V}}{V^*}$$
$$+\left(\frac{g}{V^*}\right)^2\left[\frac{\bar{q}^* S}{W}\left\{Ma^* C_{LMa} + 2C_L^* - Ma^*\left(C_{mMa}/C_{m\alpha}\right)C_{L\alpha}\right\}\right]\frac{\Delta V}{V^*} = 0 \quad (5.49)$$

The expressions for the phugoid frequency and damping may be read off from Equation 5.49 as follows:

$$\omega_{np}^2 = \left(\frac{g}{V^*}\right)^2\left[\frac{\bar{q}^* S}{W}\left\{Ma^* C_{LMa} + 2C_L^* - Ma^*\left(C_{mMa}/C_{m\alpha}\right)C_{L\alpha}\right\}\right] \quad (5.50)$$

$$2\zeta_p\omega_{np} = \left(\frac{g}{V^*}\right)\left[\frac{\bar{q}^* S}{W}\left\{Ma^* C_{DMa} + 2C_D^* - Ma^*\left(C_{mMa}/C_{m\alpha}\right)C_{D\alpha}\right\}\right] \quad (5.51)$$

The phugoid frequency and damping formulas in Equations 5.50 and 5.51 are different and more complex than the ones we obtained in Equations 5.35 and 5.36. You will notice that the first two terms on the right-hand side of each of Equations 5.50 and 5.51 are the same as the terms in Equations 5.35 and 5.36. However, in each of the revised expressions, there is an additional term involving the derivative C_{mMa}—the change in pitching moment due to a change in Mach number. This term did not appear in our derivation earlier in this chapter since we set out with only the dynamical equations for V and γ; we had ignored the pitch dynamics equation in θ (or α).

Some of the pitch derivatives are discussed next.

5.9 DERIVATIVE C_{mMa}

The derivative C_{mMa} arises from two effects. The first is due to the change in lift coefficient (the significant effect is from the horizontal tail) with Mach number. This follows the Prandtl–Glauert rule that we saw in Example 5.1. The variation of C_m with Ma appears as:

$$C_m(Ma) = \frac{C_{m(Ma=0)}}{\sqrt{1 - Ma^2}} \quad (5.52)$$

in subsonic flow, where $C_{m(Ma=0)}$ is the pitching moment coefficient for incompressible flow. In essence, Equation 5.52 captures the effect of compressibility that comes in with increasing Mach number on the lift coefficient.

In an exactly similar manner as in Equation 5.29, differentiating with Mach number, Ma, yields the derivative:

$$C_{mMa} = \left.\frac{\partial C_m}{\partial Ma}\right|_* = -\frac{1}{2} C_{m(Ma=0)} \left(1 - Ma^2\right)^{-3/2} \cdot \left.(-2Ma)\right|_*$$
$$= \frac{C_{m(Ma=0)}}{\sqrt{1 - Ma^{*2}}} \cdot \frac{Ma^*}{\left(1 - Ma^{*2}\right)} = C_m\left(Ma^*\right) \cdot \frac{Ma^*}{\left(1 - Ma^{*2}\right)} \tag{5.53}$$

There is another reason for the change of C_m with Ma and that is due to the shift in the aerodynamic centre and hence the neutral point (NP) as the airplane traverses the transonic flight regime. We discussed this in Chapter 4 (Section 4.10). The NP shifts further aft in supersonic flight; hence, the net lift acting at the airplane NP exerts a larger down-pitching moment at the airplane centre of gravity (CG). As the airplane traverses up the transonic Mach numbers, this down-pitching moment gradually comes into effect pitching the airplane nose down. This is often called the *tuck under* effect.

5.10 DERIVATIVE C_{mq1} IN PITCHING MOTION

We defined the derivative C_{mq1} in Chapter 2 as:

$$C_{mq1} = \left.\frac{\partial C_m}{\partial\left[(q_b - q_w)(c/2V)\right]}\right|_* \tag{2.25}$$

which captures the pitching moment induced due to a difference between the body-axis and wind-axis pitch rates. We have then seen that the derivative C_{mq1} is partly responsible for the damping of the short-period mode.

Let us first try to evaluate the derivative C_{mq1} for a wing–body–tail configuration. For simplicity, we assume that the airplane is flying along a straight line at constant velocity, that is, $V^* =$ constant and $\gamma^* = 0$, just as we assumed when deriving the short-period dynamics in Chapter 2. In that case, the wind axis is held fixed and does not pitch, so $q_w = 0$ all through.

Consider a small body-axis pitch rate perturbation Δq_b as shown in Figure 5.5 about the airplane CG. The distance from the airplane CG to the tail AC is l_t. The pitch rate Δq_b causes a displacement rate at the tail AC of magnitude $\Delta q_b \times l_t$ in the downward direction. From the point of view of the tail AC, this is equivalent to saying that the tail AC sees an induced flow velocity $\Delta q_b \times l_t$ in the upward sense as indicated in Figure 5.5. Since the wind axis is held fixed, the direction of the freestream velocity vector V^* is unchanged; that is, $\Delta q_w = 0$. Hence, the induced velocity at the tail may be written as $\left(\Delta q_b - \Delta q_w\right) \times l_t$.

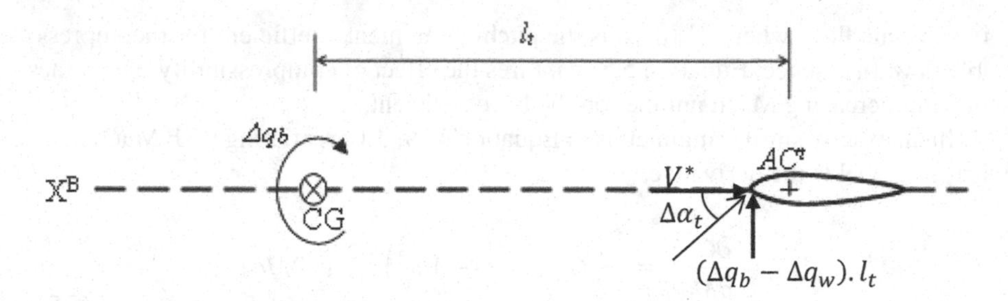

FIGURE 5.5 Body-axis pitch rate perturbation about airplane CG.

Composing the vectors V^* and $(\Delta q_b - \Delta q_w) \times l_t$ as shown in Figure 5.5, it is clear that there is an additional angle of attack:

$$\Delta\alpha_t = (\Delta q_b - \Delta q_w) \cdot \frac{l_t}{V^*} \tag{5.54}$$

induced at the tail AC. Sure enough, the additional angle of attack will induce an extra lift at the tail, which we can evaluate as below:

$$\Delta C_L^t = C_{L\alpha}^t \cdot \Delta\alpha_t = C_{L\alpha}^t (\Delta q_b - \Delta q_w) \cdot \frac{l_t}{V^*}$$

$$\Delta L_t = \bar{q} S_t \Delta C_L^t = \bar{q} S_t \cdot C_{L\alpha}^t (\Delta q_b - \Delta q_w) \cdot \frac{l_t}{V^*}$$

The additional airplane lift is the same as the extra tail lift:

$$\Delta L = \Delta L_t = \bar{q} S_t \cdot C_{L\alpha}^t (\Delta q_b - \Delta q_w) \cdot \frac{l_t}{V^*}$$

The additional airplane lift coefficient is defined as:

$$\Delta C_L = \frac{\Delta L}{\bar{q} S} = \left(\frac{S_t}{S}\right) \cdot C_{L\alpha}^t (\Delta q_b - \Delta q_w) \cdot \frac{l_t}{V^*} \tag{5.55}$$

Carrying this further, there will also be created an additional pitching moment at the airplane CG as follows:

$$\begin{aligned} \Delta M_{CG} &= -\Delta L_t \cdot l_t = -\bar{q} S_t \cdot C_{L\alpha}^t (\Delta q_b - \Delta q_w) \cdot \frac{l_t}{V^*} \cdot l_t \\ \Delta C_{mCG} &= \frac{\Delta M_{CG}}{\bar{q} S c} = -\left(\frac{S_t}{S}\right) \cdot C_{L\alpha}^t (\Delta q_b - \Delta q_w) \cdot \frac{l_t}{V^*} \cdot \frac{l_t}{c} \end{aligned} \tag{5.56}$$

From Equations 5.55 and 5.56, we can evaluate the two derivatives C_{Lq1} and C_{mq1}.

$$C_{Lq1} = \left.\frac{\partial C_L}{\partial\left[(q_b - q_w)(c/2V)\right]}\right|_*$$

From Equation 5.55, we get:

$$C_{Lq1} = 2\left[\left(\frac{S_t}{S}\right)\cdot\frac{l_t}{c}\right]\cdot C_{L\alpha}^t = 2V_H C_{L\alpha}^t \tag{5.57}$$

Likewise, the derivative:

$$C_{mq1} = \left.\frac{\partial C_m}{\partial\left[(q_b - q_w)(c/2V)\right]}\right|_*$$

From Equation 5.56, this works out to be:

$$C_{mq1} = -2\left[\left(\frac{S_t}{S}\right)\cdot\frac{l_t}{c}\right]\cdot C_{L\alpha}^t\cdot\frac{l_t}{c} = -2V_H C_{L\alpha}^t\left(\frac{l_t}{c}\right) \tag{5.58}$$

The horizontal tail volume ratio (HTVR) or V_H, which was one of the parameters that could be used to adjust the pitch stiffness term, $C_{m\alpha}$, is also seen to contribute to C_{mq1}. Thus, in general, a larger V_H will improve both pitch stiffness and pitch damping. In terms of $V_H^{/}$, C_{mq1} appears as given in Appendix 3.1, Equation 3A.12.

You can observe that for a wing–body–tail configuration, the sign of C_{mq1} is naturally negative. That is, it stabilizes or contributes to damping the short-period dynamics.

Homework Exercise: Derive the expressions for C_{Lq1} and C_{mq1} in Equations 5.57 and 5.58 for a general case with arbitrary $(\Delta q_b - \Delta q_w)$. You will find that these expressions hold true regardless of how the $(\Delta q_b - \Delta q_w)$ was created, whether by the special kind of pitching motion considered in this section or in any other manner. In essence, the derivatives C_{Lq1} and C_{mq1} in Equations 5.57 and 5.58 capture the effects induced by an additional vertical velocity component at the horizontal tail due to any combination of Δq_b and Δq_w in the form $(\Delta q_b - \Delta q_w)$.

5.11 DERIVATIVE C_{mq1} IN PHUGOID MOTION

It is equally interesting to observe that the derivative C_{mq1} is absent in the phugoid expressions of Equations 5.50 and 5.51. Why is this so?

Consider a short segment of the phugoid trajectory near a trough, as sketched in Figure 5.6. The airplane attitude at three instants of time is also shown alongside.

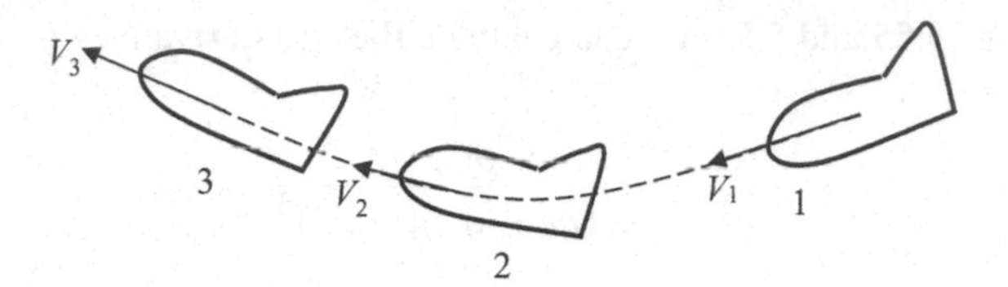

FIGURE 5.6 Phugoid trajectory near trough.

The velocity vector at each instant is tangent to the flight path. The airplane angle of attack is the same at all the three instants, and in fact at every point along the flight path. Clearly, the body axis attached to the airplane and the wind axis with X^W along the velocity vector at every instant move in tandem such that the angle between the X^B and X^W axes (namely, the angle of attack) is the same at all times. This implies that the body-axis and wind-axis pitch rates are identical. Hence, $(\Delta q_b - \Delta q_w) = 0$ for the phugoid motion, and the additional ΔC_L and ΔC_m as in Equations 5.55 and 5.56 are simply zero in this case.

Homework Exercise: Actually draw the velocity vectors at the tail arising from the body-axis pitch rate and from the wind-axis pitch rate (velocity vector tilting) and prove that these two precisely cancel each other if $\Delta q_b = \Delta q_w$. Hence, the tail does not see a resultant 'vertical' velocity component, and therefore the mechanism that caused the additional ΔC_L and ΔC_m in case of pitching motion is just not present here.

What would happen if the derivatives C_{Lq1} and C_{mq1} were not defined in terms of $(q_b - q_w)$ as:

$$C_{Lq1} = \left.\frac{\partial C_L}{\partial\left[(q_b - q_w)(c/2V)\right]}\right|_* \quad \text{and} \quad C_{mq1} = \left.\frac{\partial C_m}{\partial\left[(q_b - q_w)(c/2V)\right]}\right|_* \tag{5.59}$$

but in terms of q_b alone? Then one would have evaluated $\Delta C_L = C_{Lq1} \cdot \Delta q_b \cdot (c/2V^*)$ and $\Delta C_m = C_{mq1} \cdot \Delta q_b \cdot (c/2V^*)$, and since Δq_b is non-zero during the phugoid motion, both of them would have come out to be non-zero, implying a lift and pitching moment due to the airplane motion that is in reality physically non-existent.

From the cases of the pitching motion in Section 5.10 and the phugoid motion in Section 5.11, it is clearly brought out that the derivatives C_{Lq1} and C_{mq1} ought to be defined in the manner shown in Equation 5.59 if the physics is to be correctly captured.

5.12 FLOW CURVATURE EFFECTS

Since we have defined the derivatives C_{Lq1} and C_{mq1} with a subscript '1', is there perhaps another set of derivatives labelled C_{Lq2} and C_{mq2}, and if so what do they represent? Why have we not used them so far?

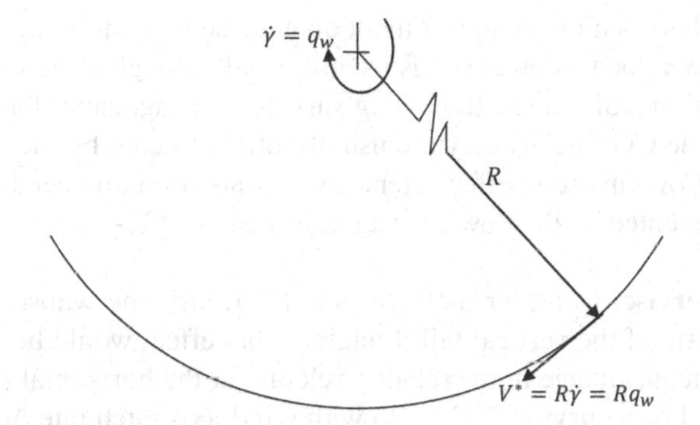

FIGURE 5.7 A curved path with constant $\dot{\gamma} = q_w$.

The answer is 'Yes'. The derivatives C_{Lq2} and C_{mq2} represent what may be called the flow curvature effect.

Consider again an airplane moving along a curved flight path as in Figure 5.6. For convenience, one may imagine this curved path as a segment of a circular vertical loop, as sketched in Figure 5.7. The angular velocity of the airplane about the centre of that loop is the same as the rate of change of the flight path angle ($\dot{\gamma}$), which is the same as the pitch rate q_w of the wind axis attached to the velocity vector. The point is that this $\dot{\gamma}$ (or q_w if you prefer) is the same at all points on the airplane, and so points that are closer to the centre of the loop must have a smaller relative wind velocity than points that are farther away. The velocity varies with the distance R from the centre of the loop and is equal to V^* only at the CG line whose trajectory is sketched in Figure 5.7.

Since the aerodynamic forces and moments depend on the local relative wind velocity, this difference in velocity that arises as a result of the curvature of the airplane's flight path can have an impact on the flight dynamics. The resulting aerodynamic effects are captured in terms of two derivatives C_{Lq2} and C_{mq2} defined below:

$$C_{Lq2} = \left.\frac{\partial C_L}{\partial\left[(q_w)(c/2V)\right]}\right|_* \quad \text{and} \quad C_{mq2} = \left.\frac{\partial C_m}{\partial\left[(q_w)(c/2V)\right]}\right|_* \tag{5.60}$$

and are included in the aerodynamic model for C_L and C_m through the terms:

$$\Delta C_L = C_{Lq2} \cdot (\Delta q_w) \cdot (c/2V^*) \tag{5.61}$$

$$\Delta C_m = C_{mq2} \cdot (\Delta q_w) \cdot (c/2V^*) \tag{5.62}$$

Thankfully, for most conventional airplanes, the vertical distance between the CG line and the farthest point, usually the tip of the vertical tail, is quite small (relative to the distance between the centre of the loop and the CG line—the radius), and so the difference in velocities observed is usually small enough to be neglected. Also, in most conventional airplanes, the lifting surfaces—wing, canard and tail—all lie pretty close to the CG line, so they are usually little affected by the velocity difference due to the flow curvature effect. Hence, we do not normally need to worry about the effects represented by the flow curvature derivatives C_{Lq2} and C_{mq2}.

Homework Exercise: Consider the case of a T-tail airplane whose horizontal tail is placed at the tip of the vertical tail. Imagine what effect would be caused in case there is a significant change in the relative velocity at the horizontal tail due to flow curvature caused by a curving flight path with wind-axis pitch rate Δq_w.

This completes the discussion on longitudinal dynamic modes for conventional airplanes and the aerodynamic derivatives that go into it.

EXERCISE PROBLEMS

5.1 An airplane is flying at sea level at a minimum drag speed $V^* = 60$ m/s in cruise conditions. Given the zero lift drag coefficient $C_{D0} = 0.015$, aspect ratio $AR = 7$ and Oswald efficiency factor $e = 0.95$, estimate the phugoid mode frequency and damping.

5.2 For an airplane in level flight trim, the phugoid response is shown in Figure 5.8. Determine the frequency and damping ratio of the airplane and the trim speed at which it is flying. Further, determine the trim drag coefficient using the following airplane data at sea level. Wing loading $W/S = 3000$ N/m^2.

5.3 Our analysis in this chapter has focused on level flying condition ($\gamma = 0$) to linearize the aircraft equations in the longitudinal plane and to derive the phugoid mode approximations. Carry out this analysis for $\gamma \neq 0$ conditions. Study the phugoid mode dynamics in take-off and landing approach.

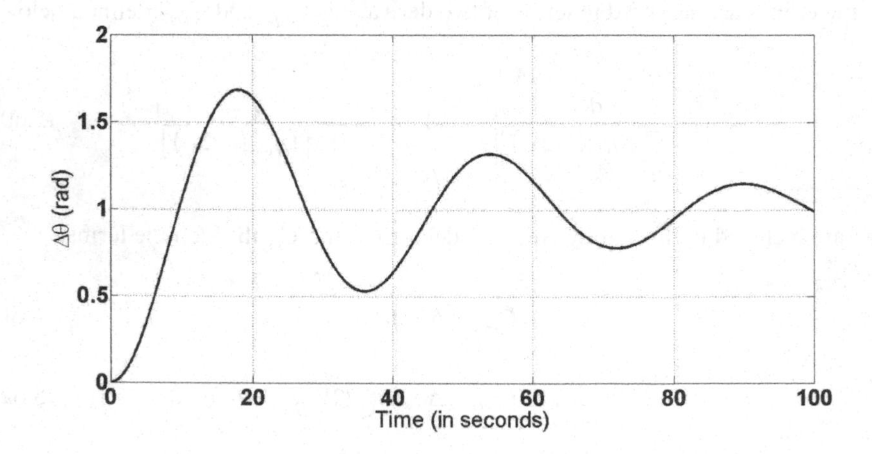

FIGURE 5.8 Phugoid response of an airplane.

5.4 An airplane accelerates from a cruise condition at a low subsonic speed to another cruise condition at high subsonic speed. Including the compressibility effect due to this transition in speed, study the effect of speed on the phugoid motion of airplane.

5.5 Find the short-period frequency for an airplane with the following data:

Wing–body data: $S = 94.5\ \text{m}^2$, $b = 32$ m, $c = 3.4$ m, AR = 9.75, $C_{L\alpha}^{wb} = 5.2/\text{rad}$, $C_L^* = 0.77$, $W = 177928.86$ N, $I_{yy} = 291500.86$ kg-m², $e = 0.75$, l_f (fuselage length) = 23.16 m, $h_{CG} = 40\%$ of root chord from the leading edge, location of ¼c_{root} = 31.6% of l_f, $C_{m\alpha}^f$ (contribution from fuselage, engine) = 0.93/rad.

Tail data: $S_t = 21.64\ \text{m}^2$, $b_t = 9.75$ m, $c_t = 2.13$ m, $AR_t = 4.4$, $l_t = 14.02$ m (from CG), $C_{L\alpha}^{wb} = 3.5/\text{rad}$.

5.6 Starting from (X^E, Z^E) = (0, –H), an airplane follows a sinusoidal flight path (as in Figure 5.2) of amplitude δh and wavelength X in the $X^E - Z^E$ plane given by the following mathematical relations:

$$Z^E = -H - \delta h \sin\left(2\pi X^E / X\right)$$

$$\gamma = \gamma_0 \cos\left(2\pi X^E / X\right)$$

Find the variation of C_L as a function of X^E during this manoeuvre.

5.7 The variation of the thrust coefficient with Mach number is represented by the derivative C_{TMa} (refer Box 5.2). Discuss the possible effects of this derivative on the phugoid motion.

BOX 5.2 DERIVATIVE C_{TMa}

This derivative accounts for the change in thrust with change in Mach number. The thrust of a typical aero-engine is primarily a function of three factors: (i) the flight altitude h, (ii) the flight Mach number Ma and (iii) the throttle setting η. It is standard to assume the variation of thrust with altitude to be proportional to the ratio of atmospheric densities at the respective altitudes. The thrust at a given altitude, Mach number and throttle setting can therefore be written in terms of a reference thrust as follows:

$$T(h, Ma, \eta) = \sigma C_T(Ma) T_{ref}(h_{SL}, Ma \to 0, \eta) \tag{5.2a}$$

where the reference thrust T_{ref} is evaluated at sea level altitude, for incompressible Mach number, and the same throttle setting η. $\sigma = \rho/\rho_{SL}$ is the ratio of atmospheric density at the operating altitude to the atmospheric density at sea level ('SL'). $C_T(Ma)$ captures the effect of flight Mach number Ma on the engine thrust and is called the thrust coefficient.

The derivative C_{TMa} is then defined as:

$$C_{TMa} = \left.\frac{\partial C_T}{\partial Ma}\right|_* \tag{5.2b}$$

where, * as always refers to the trim state about which the partial derivative is being defined. Usually, for turbojet engines, the value of this derivative is close to zero.

REFERENCES

1. Pradeep, S., A century of phugoid approximations, *Aircraft Design*, 1(2), 1998, 89–104.
2. Ananthkrishnan, N. and Unnikrishnan, S., Literal approximations to aircraft dynamic modes, *Journal of Guidance, Control, and Dynamics*, 24(6), 2001, 1196–1203.
3. Raghavan, B. and Ananthkrishnan, N., Small-perturbation analysis of airplane dynamics with dynamic stability derivatives redefined, *Journal of Aerospace Sciences and Technologies*, 61(3), 2009, 365–380.
4. Ananthkrishnan, N. and Sinha, N.K., Compact and accurate phugoid mode approximation with residualization, *Journal of Aerospace Sciences and Technologies*, 68(4), 2016, 935–949.

6 Lateral-Directional Motion

6.1 REVIEW

So far, we have been talking about the motion in the airplane's longitudinal plane. You will remember from Chapter 1 that this is the motion in the plane defined by the X^B–Z^B axes and, as we remarked earlier, for most airplanes this is a plane of symmetry, that is, it separates the airplane geometry into a left (port) side and a right (starboard) side, which are usually mirror images of each other. Most times, the airplane flies in its plane of symmetry, that is, its velocity vector lies in the X^B–Z^B plane. We have seen that this flight may be along a straight line trajectory and level (same altitude, $\gamma = 0$), ascending ($\gamma > 0$) or descending ($\gamma < 0$). We have also seen that the airplane flight path may curve upward ($\dot{\gamma} > 0$) or curve downward ($\dot{\gamma} < 0$). Any of these motions is possible depending on the combination of thrust and elevator deflection—the two controls which the pilot can use to manoeuvre the airplane in the longitudinal plane.

Even as it flies along a straight line path, the airplane's flight can be deflected due to disturbances, no matter what their source. We have seen that these may usually be resolved into two distinct modes—a faster short-period mode involving change in pitch angle and angle of attack that appears as a nose-bobbing motion, and a slower phugoid mode involving velocity and flight path angle, an exchange between the airplane's kinetic and potential energies, that appears as a lazy wave-like up-down motion. Of course, these motions are superposed over the mean forward motion of the airplane. As long as the short-period and phugoid modes are stable, the disturbances die out with time and the airplane recovers its original flight condition. How quickly the disturbances die out depends on the damping of the respective modes. We have derived expressions for the frequency and damping of these modes in terms of the flight condition, the aerodynamic coefficients and derivatives of the aerodynamic coefficients.

At the same time, the airplane may also be disturbed by perturbations out of the longitudinal plane. Our aim in this chapter is to study these disturbances and the airplane's response to them. Later, we shall examine deliberate manoeuvres out of the longitudinal plane, that is, the response of the airplane to aileron and rudder deflections.

6.2 DIRECTIONAL DISTURBANCE ANGLES

Let us first define the angles that take the airplane out of the longitudinal plane—these are similar to the angles θ, α, and γ that we defined in the case of longitudinal flight.

Consider an airplane flying due north with a velocity V as shown in Figure 6.1, with the body-fixed axes X^B–Y^B as indicated there. At a later instant, Figure 6.1

DOI: 10.1201/9781003096801-6

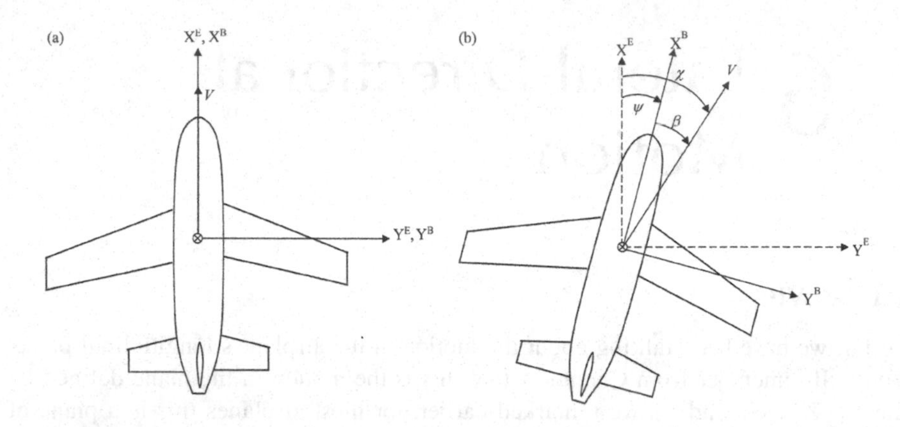

FIGURE 6.1 Angles in directional plane.

shows the airplane having yawed to the right by an angle ψ. That is, the body-fixed axes X^B–Y^B have swung to the right by an angle ψ, called the (body) yaw angle. The magnitude of the velocity vector is unchanged, but it is now reoriented by an angle χ in Figure 6.1. This is the direction in which the airplane is heading right now, and χ is called the heading angle. The difference between these two, $\chi - \psi$ in Figure 6.1, is called the sideslip angle β. That is, instead of flying in the same direction as its nose is pointing, the airplane is slipping to one side, a sort of crabbing motion.

The sense of the sideslip angle β shown in Figure 6.1 is positive. From the point of view of the airplane, the relative wind velocity V can be split into two components, $V\cos\beta$ along axis X^B and $V\sin\beta$ along axis Y^B. When the component $V\sin\beta$ shows a relative airflow coming over the right wing (i.e., from the direction of the positive axis Y^B), then the angle β is positive.

Comparing between the longitudinal and directional variables, we can make the following associations:

- ψ and θ are body-axis orientation angles—they show which way the body axis, and hence the airplane nose, is pointing with respect to the Earth axis.
- χ and γ are wind orientation angles—they show which way the airplane is flying, or equivalently what is the direction of the relative wind, with respect to the Earth axis.
- β and α are the orientation angles of the airplane relative to the wind, so they decide the static aerodynamic forces/moments on the airplane. The rate of change of β and α creates the dynamic forces/moments on the airplane.

We can imagine three special kinds of directional disturbances as follows with reference to Figure 6.2:

- The first one in Figure 6.2a shows the airplane yawed to the right by angle ψ, but the direction of the velocity vector V is unchanged. In this case, the airplane is side-slipping to its left and the sideslip angle β is marked out to be negative, and $\beta = -\psi$.

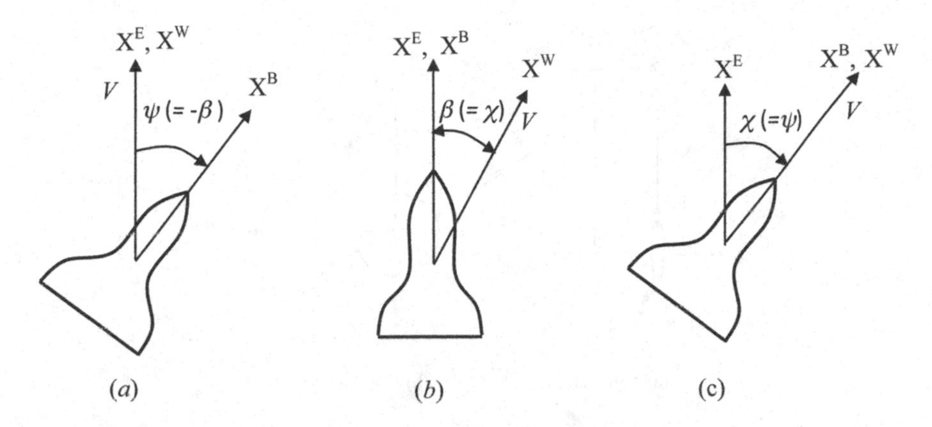

FIGURE 6.2 (a–c) Different ways in which an aircraft can be disturbed in the horizontal plane.

- In the second case, shown in Figure 6.2b, the airplane has changed its direction of flight by angle χ, but its nose has not rotated, that is, the X^B axis still points in the same direction as before. Once again, the airplane is side-slipping, but this time to the right; hence, β is now positive, and $\beta = \chi$.
- The third case in Figure 6.2c shows the airplane axis rotated by an angle ψ. Simultaneously, the direction of flight has also changed by an angle χ, in the same sense, and with $\chi = \psi$. Thus, the velocity vector and the airplane nose continue to be aligned, and there is no sideslip, that is, $\beta = 0$.

Example 6.1

An airplane is flying at 10 km altitude at a level trim speed $V^* = 200\ m/s$. The aircraft is heading northeast at an angle 45° with respect to an Earth-fixed inertial NED (North-East-Down) frame of reference. The aircraft starts side-slipping in the local horizontal plane due to a sidewind of magnitude $v = 40\ m/s$. Determine the wind orientation angle χ.

$$\text{Sideslip angle, } \beta = \tan^{-1}\left(\frac{v}{V^*}\right) = \tan^{-1}\left(\frac{40}{200}\right) = 11.31^\circ$$

The aircraft orientation with respect to Earth in the plane X^E–Y^E, $\psi = 45°$.
Wind orientation angle, $\chi = \beta + \psi = 11.31° + 45° = 56.3°$ (measured clockwise from North).

6.3 DIRECTIONAL VERSUS LONGITUDINAL FLIGHT

At first glance, it might appear that the directional dynamics ought to be very similar to what we have already seen for longitudinal flight, with a one-to-one correspondence between the angles we just stated. But it is not so for most conventional airplanes. Why?

Conventional airplanes have a powerful aerodynamic surface—the wings, and they bring with them a powerful aerodynamic force—the lift. The lift that operates in the

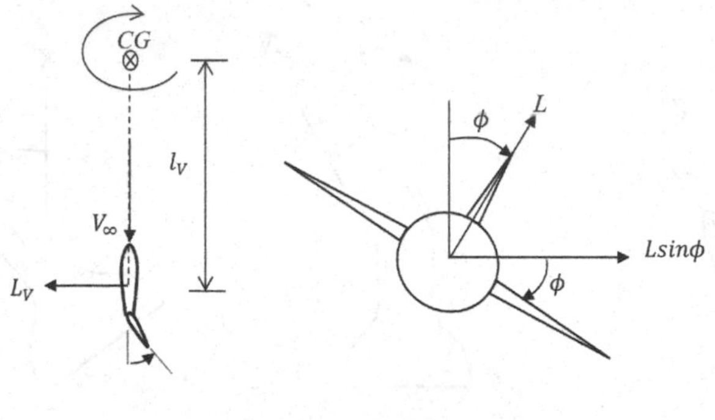

FIGURE 6.3 Two ways of changing the flight path in horizontal plane.

longitudinal plane dominates the airplane flight dynamics. Airplanes manoeuvre best by manipulating the lift. At the same time, the lift can create a moment imbalance that may require another lifting surface to provide trim—we have seen how to do this with the horizontal tail. To perform manoeuvres out of the longitudinal plane, airplanes simply do not have an equivalent source of creating a large side force. Instead, the preferred method for manoeuvring out of the longitudinal plane, such as executing a turn, is to first bank to that side and let the lift provide a component of force that creates a sideways acceleration. Figure 6.3 shows two possible ways of creating a side force to perform a turn manoeuvre—the first is by creating a side force by deflecting the rudder, and the second by banking the airplane into the turn and letting (a component of) the lift do the job. The second way of turning with the lift is the preferred method. What this means is that the directional dynamics is a little less complicated due to the absence of a significant source of side force like the lift, as long as the airplane is not banked.

In contrast, there is another source of complication in the directional dynamics due to the coupling with the lateral motion. This is because any directional motion usually creates a lift asymmetry between the right and left wings (remember that the left and right wings always had symmetric (equal) lift in the case of longitudinal flight). As a result, one of the wings will drop and the other will rise almost always in any directional flight. This means that the airplane will also roll along with a directional motion. The rolling motion is usually called the lateral dynamics. Later, we will observe that a rolling motion will also cause a directional motion. Hence, the directional and lateral motions are inextricably linked and must be taken together. Studying one without the other, which is often done in textbooks, is only of academic interest—most conventional airplanes just do not fly that way. They fly with their lateral and directional motions coupled together.

Example 6.2

An airplane is flying in a straight and level flight trim condition with speed $V^* = 100\ m/s$. With this speed, it enters into a coordinated level turn of radius $R = 500\ m$. Determine the rates of change of the wind and airplane orientation angles.

Since the velocity always points towards the aircraft nose along the circular path in a coordinated turn, the rate of change of the sideslip angle $\dot{\beta} = 0$.

In the X–Y plane, thus:

$$\dot{\chi} = \dot{\beta} + \dot{\psi} = 0 + \dot{\psi} = \frac{V^*}{R} = \frac{100}{500} = 0.2\ rad/s.$$

6.4 LATERAL DISTURBANCE ANGLES

Imagine an airplane flying along a straight, level trajectory at an angle of attack α^* as depicted in Figure 6.4a. There are two axes of interest—one is the body axis X^B and the other the wind axis along the velocity vector V^*. Figure 6.4b shows the airplane banked by an angle ϕ to the right about the axis X^B and Figure 6.4c shows another instance where the airplane has been banked by an angle μ to the right about the axis V^*.

A lateral disturbance by the angle ϕ is called a body-axis roll or bank—usually 'roll' is used when there is a continuous rolling as in having a roll rate, and 'bank' is used when there is no continuous rolling, but only a change in the angle. A bank to the right is the positive sense of bank angle, as shown in Figure 6.4. Likewise, a rotation by the angle μ about the axis V^* to the right as shown in Figure 6.4 is a positive wind-axis roll/bank.

When the angle of rotation or bank is large, then there is a difference between a body-axis roll and a wind-axis roll. Consider the airplane in Figure 6.4 banked

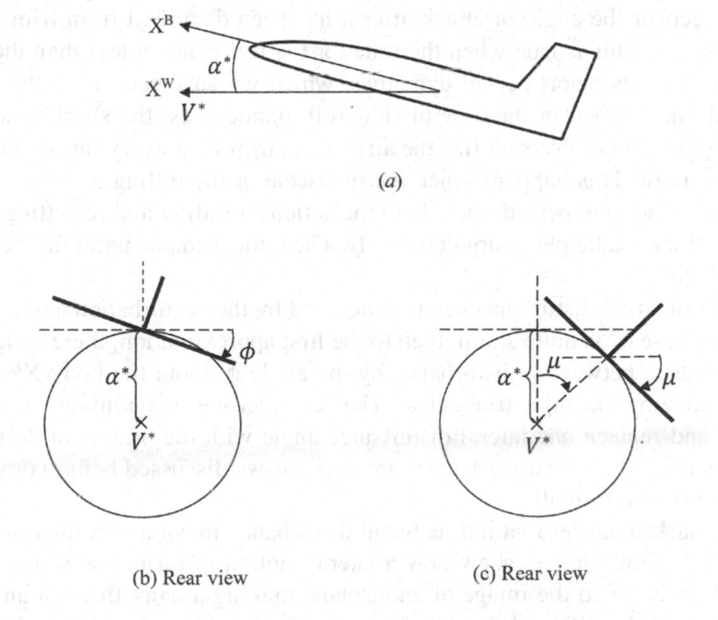

FIGURE 6.4 (a) An airplane in level flight at a positive angle of attack, (b) rear view of the airplane at a positive bank angle ϕ, and (c) rear view of the airplane rolled around the velocity vector by angle μ.

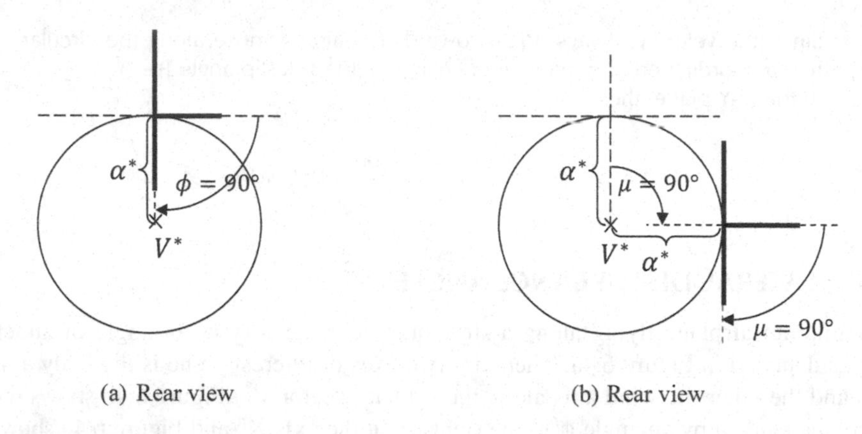

FIGURE 6.5 (a) Rear view of an airplane banked at an angle, $\phi = 90°$, and (b) rear view of the airplane rolled around the velocity vector by an angle, $\mu = 90°$.

through an angle $\phi = 90°$ as seen in Figure 6.5a—this is a rotation about the body axis X^B. In this case, the initial angle of attack α^* has been converted into an equivalent sideslip angle β after the 90° bank. In contrast, look at the wind-axis or velocity-vector roll sketched in Figure 6.5b. Here, the airplane nose executes a coning motion such that the angle of attack α^* is maintained even after banking by $\mu = 90°$ and no sideslip angle is created.

Typically, rapid roll manoeuvres happen to be body-axis rolls for then the airplane's inherent stability mechanism—remember how the short-period dynamics tends to recover the angle of attack after it has been disturbed from trim α^*—has no time to act. This is true when the time for the roll is much less than the typical time period of the short-period dynamics, which we have seen is of the order of 1 second. In contrast, in the case of slow roll manoeuvres, the short-period mode has the opportunity to re-stabilize the airplane at trim α^* at every instant during the roll manoeuvre. This happens when the timescale of the rolling motion is close to the short-period time period; then, both the actions—rolling and re-setting the trim angle of attack—take place simultaneously. Then, the airplane naturally performs a wind-axis roll.

In case of small disturbances in roll, denoted by the perturbation angles $\Delta\phi$ and $\Delta\mu$, and in case α^* is quite small, then to the first approximation, there is no significant difference between a disturbance by the angle ϕ about the body X^B axis and a disturbance by the angle μ about V^*. Hence, under these conditions, we can take $\Delta\phi \approx \Delta\mu$ and replace one lateral disturbance angle with the other, but do note that this is not true when the disturbances are large (as we discussed before) or when the trim angle α^* is not small.

We remarked earlier that a directional disturbance inevitably results in a lateral one as well. Now, we can show how a lateral motion in turn creates a directional disturbance. Refer to the image of an airplane making a bank through an angle ϕ in Figure 6.6. The lift and the weight are marked in Figure 6.6. By taking components of the weight, we can write a $W\cos\phi$ component along the axis of the lift, and a $W\sin\phi$ component normal to it directed along the wing that has banked down.

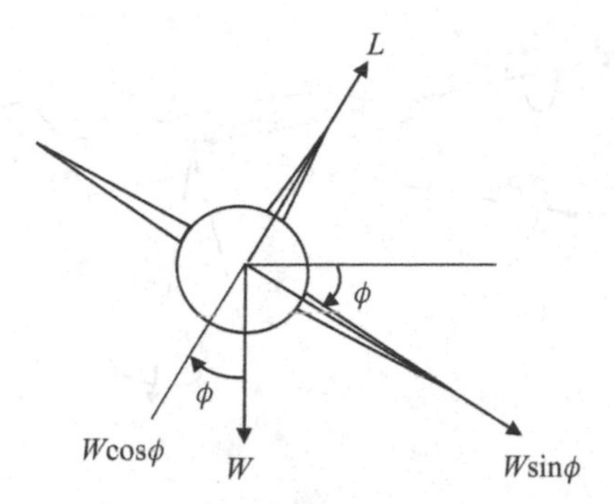

FIGURE 6.6 Rear view of an airplane banked right at an angle ϕ with lift and weight components.

If ϕ is small, then $W\cos\phi \sim W$, so it still balances out the lift (approximately), but the $W\sin\phi \sim W\phi$ component does not have a balancing force. So, the airplane begins slipping along the $W\sin\phi$ direction, but this is precisely what we defined earlier as a sideslip. In fact, as shown in Figure 6.6, the sideslip is positive as the relative wind due to this slipping motion approaches the airplane over its right wing. Similarly, for an airplane banked to the left (negative ϕ), the sideslip angle that develops is negative.

It is important to note that the connection between the lateral disturbance (bank) and the resulting directional disturbance (sideslip) is due to the effect of gravity (weight component). In the absence of gravity, this effect does not exist. When we write the equations, do look out for this connection.

6.5 LATERAL-DIRECTIONAL RATE VARIABLES

Just as we defined a body-axis pitch rate, q_b, and a wind-axis pitch rate, q_w, we shall define the corresponding rate variables in the lateral and directional axes. The angular velocity about the Z axis is called the yaw rate, r_b is the body-axis yaw rate about Z^B, and r_w, is the wind-axis yaw rate about Z^W. Likewise, the angular velocity about the body X^B axis is called the body-axis roll rate p_b, and that about the X^W axis is the wind-axis roll rate p_w. The axes along with their corresponding rate variables are shown in the side view of the airplane sketched in Figure 6.7. A roll rate to the right (starboard) is positive, and a yaw rate to the right (starboard) is positive.

Each pair of body-axis and wind-axis angular rates can be related to each other in terms of the aerodynamic angles, α and β, and their rates. We reproduce the relations below without deriving them (they are derived later in Chapter 8). We have seen this already in case of the pitch rates, q_b and q_w, in Chapter 2.

$$\begin{aligned} q_b - q_w &= \dot{\alpha} \\ p_b - p_w &= \dot{\beta}\sin\alpha \\ r_b - r_w &= \dot{\beta}\cos\alpha \end{aligned} \tag{6.1}$$

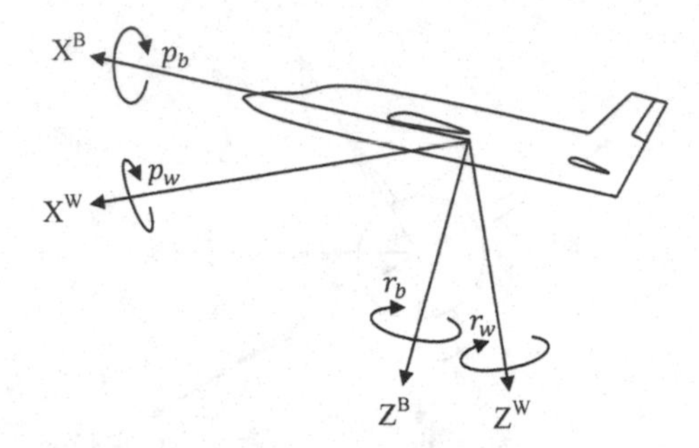

FIGURE 6.7 Angular rate variables in lateral-directional motion about the airplane body- and wind-fixed axes.

6.6 SMALL-PERTURBATION LATERAL-DIRECTIONAL EQUATIONS

Unlike the longitudinal dynamic equations, it is not easy to write down the lateral-directional equations without first deriving the complete six-degree-of-freedom equations of motion of an airplane, which we shall do in a later chapter (Chapter 8). However, for the present, we can write down the lateral-directional dynamic equations under the condition of small perturbations far more quickly and use them to carry forward our analysis.

Let us first start with a convenient trim state, which will be a level, straight line flight as before. Refer Figure 6.8 that shows a top view and a rear view of the airplane in trim with the various forces marked on it. $\mathcal{L}$ is the rolling moment, N is the yawing moment and Y is the side force—all of them are shown in Figure 6.8 in their positive sense.

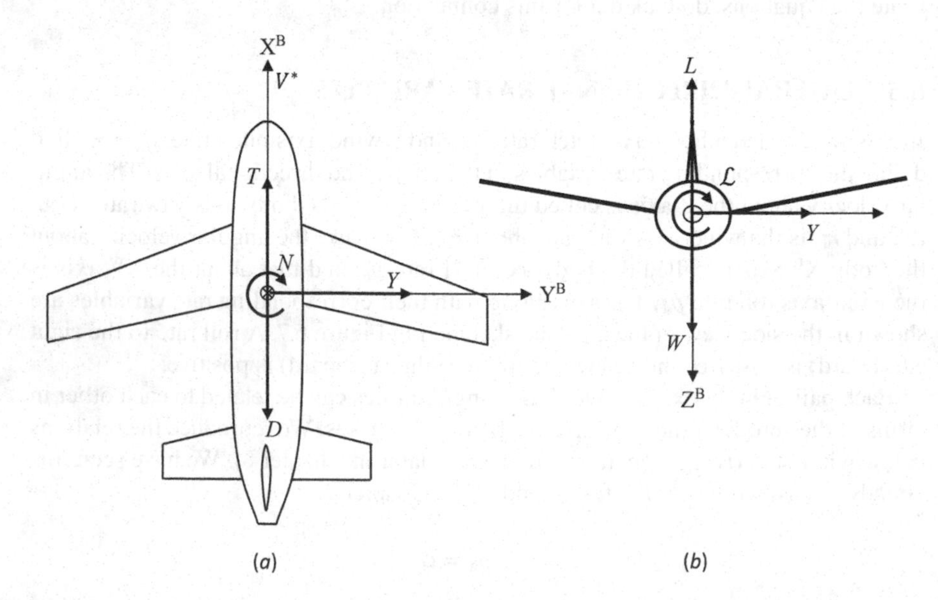

FIGURE 6.8 (a) Top view: X^B–Y^B, V^*, forces: T, D, Y (side force), moment: N. (b) Rear view: Y^B–Z^B, forces: L, W, Y, moment: $\mathcal{L}$.

The trim condition is specified by the trim velocity and angle of attack, V^* and α^*, respectively. The trim values of the other variables are:

$$\gamma^* = 0, \quad \theta^* = \alpha^*, \quad \beta^* = 0, \quad \mu^* = \phi^* = 0$$

The forces and moments balance at trim as follows:

$$\begin{aligned} T &= D; & \mathcal{L} &= 0 \\ Y &= 0; & M &= 0 \\ L &= W; & N &= 0 \end{aligned}$$

Now, we shall *not* perturb the longitudinal variables from their trim values. So, the following are maintained at their trim values:

$$V^*, \alpha^*, \gamma^* = 0, \quad \theta^* = \alpha^*$$

And there is no pitch rate about either body Y^B or wind Y^W axis, that is:

$$q_b = 0; \quad q_w = 0$$

There is a small issue with these assumptions—a lateral-directional manoeuvre typically involves loss of altitude, which means that the plane descends. So, it is not truly possible to hold $\gamma^* = 0$ fixed. However, since we shall assume small perturbations in the lateral-directional variables, we may justifiably consider $\gamma^* = 0$ to be held.

The perturbations in the lateral-directional variables are shown in Figure 6.9—the directional variables are seen in the top view and the lateral ones in the rear view.

The top view in Figure 6.9 shows the airplane yawed by a small angle $\Delta\psi$. The velocity vector V^* has yawed by a small angle $\Delta\chi$. From Figure 6.9,

$$\Delta\chi = \Delta\beta + \Delta\psi \tag{6.2}$$

But, this does not account for the lateral disturbance $\Delta\phi \approx \Delta\mu$. Equation 6.2 must be modified in case $\Delta\phi \neq 0$, but as long as α^* is small and the perturbations are small as well, the relation in Equation 6.2 is good to a first approximation. There is also a perturbation in yaw rate, Δr_b and Δr_w about the body and wind axes, respectively. The wind-axis rate Δr_w corresponds to the curvature in the flight path approximately in the horizontal plane. The thrust and drag are unaltered from their trim values, but the perturbation creates a side force ΔY and a yaw moment ΔN, also marked.

The rear view in Figure 6.9 shows a perturbed bank angle $\Delta\phi \approx \Delta\mu$. There is also a perturbation in roll rate, Δp_b and Δp_w about the body and wind axes, respectively. The lift L is unchanged from its trim value, but it is tilted now by the bank angle $\Delta\phi$. Taking components of W along and normal to the lift direction, there is an unbalanced component of W along Y^B, which is $W\sin\Delta\phi \approx mg\Delta\phi \approx mg\Delta\mu$. This is in addition to the aerodynamic side force ΔY along Y^B marked in the top view. There is also a perturbed roll moment $\Delta\mathcal{L}$ at centre of gravity (CG).

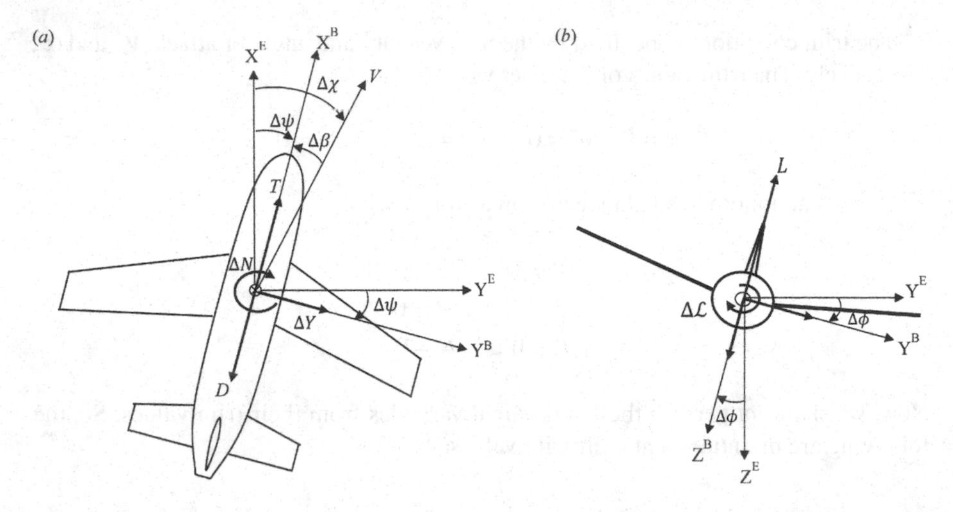

FIGURE 6.9 (a) Top view: X^B–Y^B axis yawed by $\Delta\psi$, V^* yawed by $\Delta\chi$, difference between them $\Delta\beta$ (positive as shown), forces T, D along X^B, ΔY along Y^B, ΔN at CG, and (b) rear view: Y^B–Z^B axis banked by $\Delta\phi \approx \Delta\mu$, forces L along Z^B, W along Z^E, $W\sin\Delta\phi \approx W\Delta\phi$ along Y^B, $\Delta\mathcal{L}$ at CG.

Note that the angle of attack is constrained to be constant; hence, the rolling motion must occur about the wind axis X^W, also called a velocity-vector roll. Also remember that as long as perturbations are small and α^* is small as well, we can assume $\Delta\phi \approx \Delta\mu$.

A note on the resolution of the aerodynamic forces: The net aerodynamic force is first resolved into a component ΔY along the Y^B axis and a resultant in the X^B–Z^B plane. The net velocity vector V is also resolved into a component v along the Y^B axis and a resultant V_{XZ} in the X^B–Z^B plane. The resultant force in the X^B–Z^B plane is then resolved into two components—a lift perpendicular to V_{XZ} and a drag parallel to V_{XZ}.

The lateral-directional equations under small perturbations can now be written by comparison with the small-perturbation longitudinal dynamics equations in Chapter 2.

The centripetal acceleration is equal to the net unbalanced force along the Y^B direction:

$$mV^*\Delta\dot{\chi} = \Delta Y + mg\Delta\mu \tag{6.3}$$

The moment equations about the X^B and Z^B axes are similar to the pitch dynamics equation about the Y^B axis in Equation 2.21:

$$I_{xx}\Delta\ddot{\phi} = \Delta\mathcal{L} \tag{6.4}$$

$$I_{zz}\Delta\ddot{\psi} = \Delta N \tag{6.5}$$

Thus, Equations 6.3 through 6.5 form the small-perturbation lateral-directional dynamics equations. To this, we may add the equations for the range and cross-range covered by the airplane:

$$\dot{x}_E = V^* \cos\chi \tag{6.6}$$

$$\dot{y}_E = V^* \sin\chi \tag{6.7}$$

which when integrated gives the distance covered by the airplane over the integration time along the X^E and Y^E directions.

6.7 LATERAL-DIRECTIONAL TIMESCALES

Firstly, let us write the side force and the roll and yaw moments in terms of their respective coefficients, again along exactly similar lines as we did for the longitudinal forces and moment.

$$\begin{aligned} Y &= \bar{q}SC_Y \\ \mathcal{L} &= \bar{q}SbC_l \\ N &= \bar{q}SbC_n \end{aligned} \tag{6.8}$$

The only difference from the longitudinal case is the use of the wing span b as the length factor in the definitions of the moment coefficients instead of the mean aerodynamic chord c that was used for defining the pitching moment coefficient.

The perturbed force/moment coefficients may be obtained from Equations 6.8 as:

$$\begin{aligned} \Delta Y &= \bar{q}^* S\Delta C_Y \\ \Delta\mathcal{L} &= \bar{q}^* Sb\Delta C_l \\ \Delta N &= \bar{q}^* Sb\Delta C_n \end{aligned} \tag{6.9}$$

Using the forms in Equation 6.9, we can rewrite Equations 6.3 through 6.5 as below:

$$\begin{aligned} \Delta\dot{\chi} &= \left(\frac{1}{mV^*}\right)\bar{q}^* S\Delta C_Y + \left(\frac{g}{V^*}\right)\Delta\mu \\ &= \left(\frac{g}{V^*}\right)\left\{\left(\frac{\bar{q}^* S}{W}\right)\Delta C_Y + \Delta\mu\right\} \end{aligned} \tag{6.10}$$

$$\Delta\ddot{\mu} = \left(\frac{\bar{q}^* Sb}{I_{xx}}\right)\Delta C_l = \left(\frac{I_{zz}}{I_{xx}}\right)\left(\frac{\bar{q}^* Sb}{I_{zz}}\right)\Delta C_l \tag{6.11}$$

$$\Delta\ddot{\psi} = \left(\frac{\bar{q}^* Sb}{I_{zz}}\right)\Delta C_n \tag{6.12}$$

where we have replaced $\Delta\ddot{\phi}$ with $\Delta\ddot{\mu}$. And, we have the velocity equation along the Earth axes from Equations 6.6 and 6.7 as:

$$\frac{\dot{x}_E}{H} = \left(\frac{V^*}{H}\right)\cos\chi \tag{6.6'}$$

$$\frac{\dot{y}_E}{H} = \left(\frac{V^*}{H}\right)\sin\chi \tag{6.7'}$$

where, H is a measure of airplane altitude (such as the ceiling, for instance) as in Chapter 1.

From Equations 6.10 through 6.12, we can readily identify two timescales that are similar to those in the longitudinal case in Chapter 1:

- A slow timescale:

$$T_s = \left(\frac{V^*}{g}\right) \sim 10 \text{ s} \tag{6.13}$$

- A fast timescale:

$$T_f = \sqrt{\frac{I_{zz}}{\bar{q}^* Sb}} \sim 1 \text{ s} \tag{6.14}$$

And from Equations 6.6′ and 6.7′:

- A third slower timescale:

$$T_3 = \left(\frac{H}{V^*}\right) \sim 100 \text{ s} \tag{6.15}$$

However, it turns out that the lateral-directional modes are not as easily classified according to these timescales as was the case of the longitudinal modes. We need to introduce an even faster timescale:

$$T_r = \left(\frac{b}{2V^*}\right) \sim 0.1 \text{ s} \tag{6.16}$$

The third and the slowest timescale T_3 can be kept aside as it is concerned with aircraft performance—issues such as the time taken to travel from one point to another, the time taken to complete a full (360°) turn in the horizontal plane, and so on.

For a conventional airplane, there are three lateral-directional modes corresponding to the other three timescales, as follows:

- A fast roll (rate) mode, which goes with the timescale T_r.
- An intermediate Dutch roll mode, which corresponds to the timescale T_f.
- A slow spiral mode, which occurs at the slow timescale, T_s.

Before we work on these modes, we need to write out the perturbed aerodynamic force and moment coefficients in terms of the aerodynamic derivatives.

6.8 LATERAL-DIRECTIONAL AERODYNAMIC DERIVATIVES

As in the case of the longitudinal modes, the perturbed lateral-directional force/moment coefficients—ΔC_Y, ΔC_l, ΔC_n—are expressed in terms of the perturbed aerodynamic variables. We have seen earlier that there are four different aerodynamic effects to model:

- *Static aerodynamics*—due to Mach number and the relative orientation of the aircraft (body-axis) to the wind (wind-axis) given by the aerodynamic angles, α, β.
- *Dynamic aerodynamics*—due to the angular velocity of the airplane (body-axis) with respect to the angular velocity of the relative wind (wind-axis) given by the difference between the two vectors, $\omega_b - \omega_w$.
- *Flow curvature effects*—due to the angular velocity of the wind axis, ω_w arising from the airplane flying along a curved flight path.
- *Downwash lag effect*—due to the wing-tip trailing vortices impacting the aft lifting surfaces with a time delay of approximately, l_t/V^*.

Now, remembering that we have held the velocity and all the longitudinal variables fixed for our present analysis, the lateral-directional perturbed force/moment coefficients can only be functions of the lateral variables. Hence, we can ignore effects due to the Mach number, angle of attack α and the body- and wind-axis pitch rates, respectively denoted as q_b and q_w. Also, the downwash lag effect in case of the lateral-directional dynamics, which happens because of the interaction between the wing-tip trailing vortices and the vertical tail, is usually not such a dominant effect. We shall include it later as necessary. Thus, we are left with the following variables whose effect is to be modelled:

- *Static*—$\Delta\beta$
- *Dynamic*—$\Delta p_b - \Delta p_w$, $\Delta r_b - \Delta r_w$
- *Flow curvature*—Δp_w, Δr_w

Besides these, there is the effect due to control deflection—usually, there are two lateral-directional controls of interest, the aileron and the rudder. We shall introduce those in Chapter 7.

The perturbed aerodynamic force/moment coefficients may now be modelled as:

$$\begin{aligned}\Delta C_Y = C_{Y\beta}\Delta\beta + C_{Yp1}\left(\Delta p_b - \Delta p_w\right)\left(b/2V^*\right) + C_{Yp2}\Delta p_w\left(b/2V^*\right)\\ + C_{Yr1}\left(\Delta r_b - \Delta r_w\right)\left(b/2V^*\right) + C_{Yr2}\Delta r_w\left(b/2V^*\right)\end{aligned} \tag{6.17}$$

$$\begin{aligned}\Delta C_l = C_{l\beta}\Delta\beta + C_{lp1}\left(\Delta p_b - \Delta p_w\right)\left(b/2V^*\right) + C_{lp2}\Delta p_w\left(b/2V^*\right)\\ + C_{lr1}\left(\Delta r_b - \Delta r_w\right)\left(b/2V^*\right) + C_{lr2}\Delta r_w\left(b/2V^*\right)\end{aligned} \tag{6.18}$$

$$\Delta C_n = C_{n\beta}\Delta\beta + C_{np1}(\Delta p_b - \Delta p_w)(b/2V^*) + C_{np2}\Delta p_w(b/2V^*) + C_{nr1}(\Delta r_b - \Delta r_w)(b/2V^*) + C_{nr2}\Delta r_w(b/2V^*) \quad (6.19)$$

where the timescale $(b/2V^*)$ from Equation 6.16 has now appeared.

Each of the aerodynamic derivatives in Equations 6.17 through 6.19 is defined as follows:

$$C_{Y\beta} = \left.\frac{\partial C_Y}{\partial \beta}\right|_*; \quad C_{l\beta} = \left.\frac{\partial C_l}{\partial \beta}\right|_*; \quad C_{n\beta} = \left.\frac{\partial C_n}{\partial \beta}\right|_*$$

$$C_{Yp1} = \left.\frac{\partial C_Y}{\partial (p_b - p_w)(b/2V)}\right|_*; \quad C_{lp1} = \left.\frac{\partial C_l}{\partial (p_b - p_w)(b/2V)}\right|_*;$$

$$C_{np1} = \left.\frac{\partial C_n}{\partial (p_b - p_w)(b/2V)}\right|_*$$

$$C_{Yp2} = \left.\frac{\partial C_Y}{\partial p_w(b/2V)}\right|_*; \quad C_{lp2} = \left.\frac{\partial C_l}{\partial p_w(b/2V)}\right|_*; \quad C_{np2} = \left.\frac{\partial C_n}{\partial p_w(b/2V)}\right|_* \quad (6.20)$$

$$C_{Yr1} = \left.\frac{\partial C_Y}{\partial (r_b - r_w)(b/2V)}\right|_*; \quad C_{lr1} = \left.\frac{\partial C_l}{\partial (r_b - r_w)(b/2V)}\right|_*;$$

$$C_{nr1} = \left.\frac{\partial C_n}{\partial (r_b - r_w)(b/2V)}\right|_*$$

$$C_{Yr2} = \left.\frac{\partial C_Y}{\partial r_w(b/2V)}\right|_*; \quad C_{lr2} = \left.\frac{\partial C_l}{\partial r_w(b/2V)}\right|_*; \quad C_{nr2} = \left.\frac{\partial C_n}{\partial r_w(b/2V)}\right|_*$$

where, * refers to the trim state.

We shall examine the physical mechanism and the significance of some of these derivatives a little later in Chapter 7.

6.9 LATERAL-DIRECTIONAL SMALL-PERTURBATION EQUATIONS (CONTD.)

From Equation 6.1, we can write:

$$\Delta p_b - \Delta p_w = \Delta\dot{\beta}\sin\alpha^* \quad \text{and} \quad \Delta r_b - \Delta r_w = -\Delta\dot{\beta}\cos\alpha^* \quad (6.21)$$

When α^* is small, the right-hand side of the $\Delta p_b - \Delta p_w$ equation can be taken to be the product of two small terms, namely $\Delta\dot{\beta}$ and $sin\alpha^*$, and hence dropped. Also, we may assume $cos\alpha^* \approx 1$. Thus,

$$\Delta p_b - \Delta p_w \approx 0 \quad \text{and} \quad \Delta r_b - \Delta r_w = -\Delta\dot{\beta} \quad (6.22)$$

That is, we do not need to distinguish between a body-axis roll rate and a wind-axis roll rate when dealing with small perturbations.

The wind-axis angular rates can themselves be written in terms of the rate of change of the wind-axis Euler angles as follows (refer to Chapter 8 for details):

$$\begin{aligned} p_w &= \dot{\mu} - \dot{\chi}\sin\gamma \\ r_w &= -\dot{\gamma}\sin\mu + \dot{\chi}\cos\gamma\cos\mu \end{aligned} \tag{6.23}$$

Equation 6.23 can be linearized (**Homework Exercise**!) to obtain:

$$\Delta p_w = \Delta\dot{\mu} - \Delta\dot{\chi}\sin\gamma^* \quad \text{and} \quad \Delta r_w = \Delta\dot{\chi}\cos\gamma^*\cos\Delta\mu \tag{6.24}$$

Since we have assumed level flight trim ($\gamma^* = 0$), and taking $cos\Delta\mu \approx 1$ for small $\Delta\mu$, this results in:

$$\Delta p_w \approx \Delta\dot{\mu} \quad \text{and} \quad \Delta r_w \approx \Delta\dot{\chi} \tag{6.25}$$

With the relations in Equations 6.22 and 6.25, the perturbed aerodynamic force/moment coefficients in Equations 6.17 through 6.19 may be updated as below:

$$\begin{aligned} \Delta C_Y &= C_{Y\beta}\Delta\beta + C_{Yp2}\Delta\dot{\mu}\left(b/2V^*\right) + C_{Yr1}\left(-\Delta\dot{\beta}\right)\left(b/2V^*\right) + C_{Yr2}\Delta\dot{\chi}\left(b/2V^*\right) \\ \Delta C_l &= C_{l\beta}\Delta\beta + C_{lp2}\Delta\dot{\mu}\left(b/2V^*\right) + C_{lr1}\left(-\Delta\dot{\beta}\right)\left(b/2V^*\right) + C_{lr2}\Delta\dot{\chi}\left(b/2V^*\right) \\ \Delta C_n &= C_{n\beta}\Delta\beta + C_{np2}\Delta\dot{\mu}\left(b/2V^*\right) + C_{nr1}\left(-\Delta\dot{\beta}\right)\left(b/2V^*\right) + C_{nr2}\Delta\dot{\chi}\left(b/2V^*\right) \end{aligned}$$

where, 'p_1' derivative terms have dropped out since $\Delta p_b - \Delta p_w \approx 0$. Now, we can insert this small-perturbation aerodynamic model in the lateral-directional Equations 6.10 through 6.12 to give the complete set of equations as below:

$$\Delta\dot{\chi} = \left(\frac{g}{V^*}\right)\left\{\left(\frac{\bar{q}^* S}{W}\right)\left[\begin{array}{c} C_{Y\beta}\Delta\beta + C_{Yp2}\Delta\dot{\mu}\left(b/2V^*\right) + C_{Yr1}\left(-\Delta\dot{\beta}\right) \\ \times\left(b/2V^*\right) + C_{Yr2}\Delta\dot{\chi}\left(b/2V^*\right) \end{array}\right] + \Delta\mu\right\} \tag{6.26}$$

$$\Delta\ddot{\mu} = \left(\frac{\bar{q}^* Sb}{I_{xx}}\right)\left[\begin{array}{c} C_{l\beta}\Delta\beta + C_{lp2}\Delta\dot{\mu}\left(b/2V^*\right) + C_{lr1}\left(-\Delta\dot{\beta}\right)\left(b/2V^*\right) \\ + C_{lr2}\Delta\dot{\chi}\left(b/2V^*\right) \end{array}\right] \tag{6.27}$$

$$\Delta\ddot{\psi} = \left(\frac{\bar{q}^* Sb}{I_{zz}}\right)\left[\begin{array}{c} C_{n\beta}\Delta\beta + C_{np2}\Delta\dot{\mu}\left(b/2V^*\right) + C_{nr1}\left(-\Delta\dot{\beta}\right)\left(b/2V^*\right) \\ + C_{nr2}\Delta\dot{\chi}\left(b/2V^*\right) \end{array}\right] \tag{6.28}$$

While this set of equations looks quite formidable, several reasonable assumptions can be made to reduce its complexity and arrive at a first-cut approximation to the lateral-directional dynamic modes.

Step 1:
First of all, the side force coefficient derivatives with the rates, C_{Yp2}, C_{Yr1}, C_{Yr2}, are usually of lesser importance and may be ignored. Then, Equation 6.26 reduces to:

$$\Delta\dot{\chi} = \left(\frac{g}{V^*}\right)\left\{\left(\frac{\bar{q}^* S}{W}\right) C_{Y\beta}\Delta\beta + \Delta\mu\right\} \tag{6.29}$$

Before proceeding further, for ease of algebraic manipulation, let us define some short symbols:

$$\left(\frac{\bar{q}^* S}{W}\right) C_{Y\beta} \equiv Y_\beta$$

So, Equation 6.29 can be compactly written as:

$$\Delta\dot{\chi} = \left(\frac{g}{V^*}\right)\left\{Y_\beta\Delta\beta + \Delta\mu\right\} \tag{6.30}$$

Step 2:
Likewise, with the following short symbols:

$$\left(\frac{\bar{q}Sb}{I_{zz}}\right) C_{n\beta} \equiv N_\beta; \quad \left(\frac{\bar{q}Sb}{I_{zz}}\right) C_{np2}\left(\frac{b}{2V^*}\right) \equiv N_{p2}$$

$$\left(\frac{\bar{q}Sb}{I_{zz}}\right) C_{nr1}\left(\frac{b}{2V^*}\right) \equiv N_{r1}; \quad \left(\frac{\bar{q}Sb}{I_{zz}}\right) C_{nr2}\left(\frac{b}{2V^*}\right) \equiv N_{r2}$$

Equation 6.28 can be transcribed as:

$$\Delta\ddot{\psi} = N_\beta\Delta\beta + N_{p2}\Delta\dot{\mu} + N_{r1}\left(-\Delta\dot{\beta}\right) + N_{r2}\Delta\dot{\chi} \tag{6.31}$$

From Equation 6.2, by differentiating all terms twice, we have:

$$\Delta\ddot{\psi} = \Delta\ddot{\chi} - \Delta\ddot{\beta} \tag{6.32}$$

And by differentiating Equation 6.30 once, we get:

$$\Delta\ddot{\chi} = \left(\frac{g}{V^*}\right)\left\{Y_\beta\Delta\dot{\beta} + \Delta\dot{\mu}\right\} \tag{6.33}$$

which we may insert in Equation 6.32 along with Equation 6.31 and rearrange terms to yield the following:

$$\Delta\ddot{\beta}+\left[-N_{r1}-\left(\frac{g}{V^*}\right)Y_\beta\right]\Delta\dot{\beta}+N_\beta\Delta\beta+\left[N_{p2}-\left(\frac{g}{V^*}\right)\right]\Delta\dot{\mu}+N_{r2}\Delta\dot{\chi}=0 \quad (6.34)$$

Then, substituting for $\Delta\dot{\chi}$ from Equation 6.30, we finally get:

$$\begin{aligned}\Delta\ddot{\beta}+\left[-N_{r1}-\left(\frac{g}{V^*}\right)Y_\beta\right]\Delta\dot{\beta}+\left[N_\beta+\left(\frac{g}{V^*}\right)Y_\beta N_{r2}\right]\Delta\beta \\ +\left[N_{p2}-\left(\frac{g}{V^*}\right)\right]\Delta\dot{\mu}+\left(\frac{g}{V^*}\right)N_{r2}\Delta\mu=0\end{aligned} \quad (6.35)$$

Usually, the derivative N_{p2} is not significant enough and may be dropped. Then, the yawing moment equation appears as:

$$\begin{aligned}\Delta\ddot{\beta}+\left[-N_{r1}-\left(\frac{g}{V^*}\right)Y_\beta\right]\Delta\dot{\beta}+\left[N_\beta+\left(\frac{g}{V^*}\right)Y_\beta N_{r2}\right]\Delta\beta \\ +\left[-\left(\frac{g}{V^*}\right)\right]\Delta\dot{\mu}+\left(\frac{g}{V^*}\right)N_{r2}\Delta\mu=0\end{aligned} \quad (6.36)$$

Step 3:

Turning our attention next to Equation 6.27, we first define the following short symbols:

$$\left(\frac{\bar{q}Sb}{I_{xx}}\right)C_{l\beta}\equiv\mathcal{L}_\beta;\quad \left(\frac{\bar{q}Sb}{I_{xx}}\right)C_{lp2}\left(\frac{b}{2V^*}\right)\equiv\mathcal{L}_{p2}$$

$$\left(\frac{\bar{q}Sb}{I_{xx}}\right)C_{lr1}\left(\frac{b}{2V^*}\right)\equiv\mathcal{L}_{r1};\quad \left(\frac{\bar{q}Sb}{I_{xx}}\right)C_{lr2}\left(\frac{b}{2V^*}\right)\equiv\mathcal{L}_{r2}$$

And then write Equation 6.27 as:

$$\Delta\ddot{\mu}=\mathcal{L}_\beta\Delta\beta+\mathcal{L}_{p2}\Delta\dot{\mu}+\mathcal{L}_{r1}\left(-\Delta\dot{\beta}\right)+\mathcal{L}_{r2}\Delta\dot{\chi} \quad (6.37)$$

Yet again, we have $\Delta\dot{\chi}$ from Equation 6.30, and using that we can rewrite Equation 6.37 as:

$$\Delta\ddot{\mu}=\left[\mathcal{L}_\beta+\left(\frac{g}{V^*}\right)Y_\beta\mathcal{L}_{r2}\right]\Delta\beta+\mathcal{L}_{p2}\Delta\dot{\mu}+\mathcal{L}_{r1}\left(-\Delta\dot{\beta}\right)+\left(\frac{g}{V^*}\right)\mathcal{L}_{r2}\Delta\mu \quad (6.38)$$

TABLE 6.1
Lateral-Directional Small-Perturbation Equations

Source	Equation	Modes and Timescales
Rolling moment and side force	$\Delta\ddot{\mu} = -\mathcal{L}_{r1}\Delta\dot{\beta} + \left[\mathcal{L}_{\beta} + \left(\frac{g}{V^*}\right)Y_{\beta}\mathcal{L}_{r2}\right]\Delta\beta + \mathcal{L}_{p2}\Delta\dot{\mu} + \left(\frac{g}{V^*}\right)\mathcal{L}_{r2}\Delta\mu$	Roll (T_r) and Spiral (T_s)
Yawing moment and side force	$\Delta\ddot{\beta} + \left[-N_{r1} - \left(\frac{g}{V^*}\right)Y_{\beta}\right]\Delta\dot{\beta} + \left[N_{\beta} + \left(\frac{g}{V^*}\right)Y_{\beta}N_{r2}\right]\Delta\beta + \left[-\left(\frac{g}{V^*}\right)\right]\Delta\dot{\mu} + \left(\frac{g}{V^*}\right)N_{r2}\Delta\mu = 0$	Dutch roll (T_f)

To summarize, we collect the final set of lateral-directional small-perturbation equations in Table 6.1. The table contains two equations of second order in the variables $\Delta\mu$ and $\Delta\beta$. These equations are coupled because of the $\Delta\dot{\beta}$, $\Delta\beta$ terms in the $\Delta\mu$ (rolling moment) equation and the $\Delta\dot{\mu}$, $\Delta\mu$ terms in the $\Delta\beta$ (yawing moment) equation.

Note that the $\Delta\chi$ (side force) equation has been absorbed into these two equations; hence, there is no separate equation for $\Delta\chi$. In fact, the variable $\Delta\chi$ has itself been eliminated. To that extent, the lateral-directional equations in Table 6.1 appear similar to the two equations in ΔV and $\Delta\alpha$ for the longitudinal case.

However, you will remember that the two longitudinal equations were easily separated based on their timescales—the ΔV equation (for the phugoid mode) operates at the slow timescale T_2, whereas the $\Delta\alpha$ dynamics (short-period mode) occurred at the faster T_1 timescale. Unfortunately, the division of the lateral-directional equations based on timescales is not so simple for most conventional airplanes. As noted in the last column of Table 6.1, the $\Delta\mu$ (rolling moment) equation contains dynamics both at the slowest (T_s) and the fastest (T_r) timescales, whereas the $\Delta\beta$ (yawing moment) equation operates at the intermediate T_f timescale. This complicates matters significantly. In terms of the dynamic modes, the $\Delta\beta$ (yawing moment) equation represents a second-order dynamics that is called the Dutch roll mode. The $\Delta\mu$ (rolling moment) equation must split into two first-order dynamics—a fast mode at the timescale T_r called the roll (rate) mode, and a slow mode at the timescale T_s called the spiral mode.

Next, we shall derive first-cut approximations to these three modes.

6.10 LATERAL-DIRECTIONAL DYNAMIC MODES

Since the lateral-directional equations of Table 6.1 involve three timescales, extracting the individual modes from these equations can be a little complicated. We shall follow a simplified, no-frills approach and cut a few corners, but still emerge with the correct solution. For those interested, a more detailed solution is provided in Ref. [1].

6.10.1 Roll (Rate) Mode

First, we extract the roll mode at the fastest timescale, T_r, from the $\Delta\mu$ equation in Table 6.1. The variable of interest is $\Delta\dot{\mu}$.

$\Delta\dot{\mu}$ is split into two components—one that responds at the fastest timescale T_r and another that changes as per the intermediate timescale T_f. With this split, we re-arrange the $\Delta\mu$ equation as follows:

$$\underbrace{\left\{\Delta\ddot{\mu}_r = \mathcal{L}_{p2}\Delta\dot{\mu}_r\right\}}_{\textit{Roll mode}}$$
$$+\underbrace{\left\{-\mathcal{L}_{r1}\Delta\dot{\beta}+\left[\mathcal{L}_{\beta}+\left(\frac{g}{V^*}\right)Y_{\beta}\mathcal{L}_{r2}\right]\Delta\beta+\mathcal{L}_{p2}\Delta\dot{\mu}_f+\left(\frac{g}{V^*}\right)\mathcal{L}_{r2}\Delta\mu\right\}}_{\textit{Equals zero}} \tag{6.39}$$

where, $\Delta\dot{\mu}_r$ is the component at the faster timescale T_r and $\Delta\dot{\mu}_f$ is the component at the timescale T_f.

The first set of braces contains the fastest dynamics at the timescale T_r, whereas at this timescale the terms in the second set of braces are collectively equal to zero. Thus, the roll (rate) mode dynamics is given by the first-order equation (where the subscript 'r' has been dropped):

$$\Delta\ddot{\mu} = \mathcal{L}_{p2}\Delta\dot{\mu} \tag{6.40}$$

which we can analyse following the methods of Chapter 2. In particular, the stability of the roll mode is given by the sign of the derivative $\mathcal{L}_{p2}$, which corresponds to the factor 'a' in Equation 2.1. The factor 'a' here is called the roll eigenvalue, λ_r. Hence, the requirement for the stability of the roll mode is:

$$\lambda_r = \mathcal{L}_{p2} < 0 \tag{6.41}$$

We shall discuss the physics behind the derivative $\mathcal{L}_{p2}$ in Chapter 7. Usually, $\mathcal{L}_{p2}$ is almost always negative, so the roll mode is mostly guaranteed to be stable.

Example 6.3

The roll rate response for an airplane is given in Figure 6.10. From the response, determine the roll eigenvalue, λ_r.

From the time history in Figure 6.10, the time-to-half amplitude can be found to be $t_{1/2} \sim 1.7$ seconds. Thus, $\lambda_r = \mathcal{L}_{p2} = -\dfrac{0.693}{1.7} = -0.407/s$.

The second set of braces in Equation 6.39 can be solved to give:

$$\Delta\dot{\mu}_f = \left(\frac{1}{\mathcal{L}_{p2}}\right)\left\{\mathcal{L}_{r1}\Delta\dot{\beta}-\left[\mathcal{L}_{\beta}+\left(g/V^*\right)Y_{\beta}\mathcal{L}_{r2}\right]\Delta\beta-\left(g/V^*\right)\mathcal{L}_{r2}\Delta\mu\right\} \tag{6.42}$$

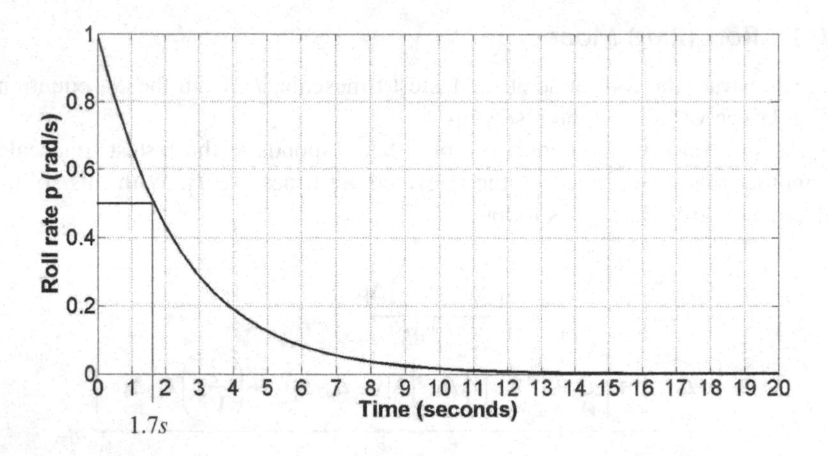

FIGURE 6.10 Roll rate response for Example 6.3.

Technically, this is called the 'residual'—the component of $\Delta\dot{\mu}$ that remains after the faster dynamics at the timescale T_r is complete and which now varies as per the next, intermediate timescale T_f.

6.10.2 Dutch Roll Mode

We move on to the next fastest mode at the timescale T_f, called the Dutch roll. The dominant variable for this mode is $\Delta\beta$ and we examine the yaw equation in Table 6.1 for $\Delta\beta$.

$$\Delta\ddot{\beta} + \left[-N_{r1} - \left(\frac{g}{V^*}\right)Y_\beta\right]\Delta\dot{\beta} + \left[N_\beta + \left(\frac{g}{V^*}\right)Y_\beta N_{r2}\right]\Delta\beta + \left[-\left(\frac{g}{V^*}\right)\right]\Delta\dot{\mu}_f + \left(\frac{g}{V^*}\right)N_{r2}\Delta\mu = 0 \tag{6.43}$$

where, as indicated, the $\Delta\dot{\mu}$ in Equation 6.43 is the residual $\Delta\dot{\mu}_f$ from the roll dynamics. So, we insert the expression for $\Delta\dot{\mu}_f$ from Equation 6.42 in the yaw Equation 6.43 to obtain:

$$\begin{aligned}\Delta\ddot{\beta} &+ \left[-N_{r1} - \left(\frac{g}{V^*}\right)Y_\beta\right]\Delta\dot{\beta} + \left[N_\beta + \left(\frac{g}{V^*}\right)Y_\beta N_{r2}\right]\Delta\beta \\ &+ \left[-\left(\frac{g}{V^*}\right)\right]\left(\frac{1}{\mathcal{L}_{p2}}\right)\left\{\mathcal{L}_{r1}\Delta\dot{\beta} - \left[\mathcal{L}_\beta + \left(\frac{g}{V^*}\right)Y_\beta\mathcal{L}_{r2}\right]\Delta\beta - \left(\frac{g}{V^*}\right)\mathcal{L}_{r2}\Delta\mu\right\} \\ &+ \left(\frac{g}{V^*}\right)N_{r2}\Delta\mu = 0\end{aligned} \tag{6.44}$$

Rearranging terms and dropping terms of higher order in (g/V^*), we have:

$$\Delta\ddot{\beta} + \left[-N_{r1} - \left(\frac{g}{V^*}\right)Y_\beta - \left(\frac{g}{V^*}\right)\left(\frac{\mathcal{L}_{r1}}{\mathcal{L}_{p2}}\right)\right]\Delta\dot{\beta} + \left[N_\beta + \left(\frac{g}{V^*}\right)Y_\beta N_{r2} + \left(\frac{g}{V^*}\right)\left(\frac{\mathcal{L}_\beta}{\mathcal{L}_{p2}}\right)\right]\Delta\beta + \left(\frac{g}{V^*}\right)N_{r2}\Delta\mu = 0 \tag{6.45}$$

Now, we split $\Delta\beta$ into two components—a faster one $\Delta\beta_f$ at the intermediate timescale T_f and a slower one $\Delta\beta_s$ at the slower timescale T_s. Then, rearranging the terms in Equation 6.45, we can write:

$$\underbrace{\Delta\ddot{\beta}_f + \left[-N_{r1} - \left(\frac{g}{V^*}\right)Y_\beta - \left(\frac{g}{V^*}\right)\left(\frac{\mathcal{L}_{r1}}{\mathcal{L}_{p2}}\right)\right]\Delta\dot{\beta}_f + \left[N_\beta + \left(\frac{g}{V^*}\right)Y_\beta N_{r2} + \left(\frac{g}{V^*}\right)\left(\frac{\mathcal{L}_\beta}{\mathcal{L}_{p2}}\right)\right]\Delta\beta_f}_{\textit{Dutch roll mode}}$$
$$+ \underbrace{\left\{\left[N_\beta + \left(\frac{g}{V^*}\right)Y_\beta \mathcal{L}_{r2} + \left(\frac{g}{V^*}\right)\left(\frac{\mathcal{L}_\beta}{\mathcal{L}_{p2}}\right)\right]\Delta\beta_s + \left(\frac{g}{V^*}\right)N_{r2}\Delta\mu\right\}}_{\textit{Equals zero}} = 0 \tag{6.46}$$

The first set of braces contains the second-order dynamics at timescale T_f in the variable $\Delta\beta$ of the Dutch roll mode:

$$\Delta\ddot{\beta} + \left[-N_{r1} - \left(\frac{g}{V^*}\right)Y_\beta - \left(\frac{g}{V^*}\right)\left(\frac{\mathcal{L}_{r1}}{\mathcal{L}_{p2}}\right)\right]\Delta\dot{\beta} + \left[N_\beta + \left(\frac{g}{V^*}\right)Y_\beta N_{r2} + \left(\frac{g}{V^*}\right)\left(\frac{\mathcal{L}_\beta}{\mathcal{L}_{p2}}\right)\right]\Delta\beta = 0 \tag{6.47}$$

From our discussion in Chapter 2, we can write the damping and frequency of the Dutch roll mode by inspection of Equation 6.47 as follows:

$$\omega_{nDR}^2 = N_\beta + \left(\frac{g}{V^*}\right)\left[Y_\beta N_{r2} + \left(\frac{\mathcal{L}_\beta}{\mathcal{L}_{p2}}\right)\right] \tag{6.48}$$

$$2\zeta_{DR}\omega_{nDR} = -N_{r1} - \left(\frac{g}{V^*}\right)\left[Y_\beta + \left(\frac{\mathcal{L}_{r1}}{\mathcal{L}_{p2}}\right)\right] \tag{6.49}$$

which is valid provided the right-hand side of Equation 6.48 is positive. However, note that the Dutch roll stiffness at high angles of attack and high Mach numbers may be an issue. This will be discussed in Chapter 7.

From the second set of braces in Equation 6.46, we have:

$$\Delta\beta_s = -\frac{\left(\frac{g}{V^*}\right)N_{r2}}{\left[N_\beta + \left(\frac{g}{V^*}\right)Y_\beta N_{r2} + \left(\frac{g}{V^*}\right)\left(\frac{\mathcal{L}_\beta}{\mathcal{L}_{p2}}\right)\right]}\Delta\mu = -\frac{\left(\frac{g}{V^*}\right)N_{r2}}{\omega_{nDR}^2}\Delta\mu \quad (6.50)$$

This is the residual after the Dutch roll dynamics has subsided. This will vary as per the slowest timescale T_s.

Example 6.4

A business-jet airplane has the following geometric and inertia characteristics:

$$W = 169921.24 \text{ N}, \; I_{xx} = 161032.43 \text{ kg} - \text{m}^2, \; I_{zz} = 330142.72 \text{ kg} - \text{m}^2,$$
$$S = 50.4 \text{ m}^2, \; b = 16.38 \text{ m}$$

The airplane is flying in cruise condition at sea level with a speed $V^* = 260$ *m/s* at an angle of attack $\alpha = 5°$. The various derivatives are given as:

$$C_{nr1} = -0.081/\text{rad}; \; C_{nr2} = -0.009/\text{rad}$$
$$C_{lp2} = -0.430/\text{rad};$$
$$C_{lr1} = 0.175/\text{rad}; \; C_{l\beta} = -0.03/\text{rad}$$

The Dutch roll mode characteristics are determined from a simulation of Equation 6.47 to be $\omega_{nDR} = 2.347$ *rad/s* and $\zeta_{DR} = 0.204$. Determine the derivatives $C_{Y\beta}$ and $C_{n\beta}$.

$$\bar{q}^* = \frac{1}{2}\rho V^{*2} = \frac{1}{2} \times 1.225 \times 260^2 = 41405.0 \text{ N/m}^2$$

$$\left(\frac{\bar{q}^* Sb}{I_{xx}}\right) = \frac{41405.0 \times 50.4 \times 16.38}{161032.43} = 212.27/s^2$$

$$\left(\frac{\bar{q}^* Sb}{I_{zz}}\right) = \frac{41405.0 \times 50.4 \times 16.38}{330142.72} = 103.54/s^2$$

$$\left(\frac{\mathcal{L}_{r1}}{\mathcal{L}_{p2}}\right) = \left(\frac{C_{lr1}}{C_{lp2}}\right) = -\frac{0.175}{0.43} = -0.407$$

$$N_{r1} = \left(\frac{\bar{q}^* Sb}{I_{zz}}\right) C_{nr1}\left(\frac{b}{2V^*}\right) = 103.54 \times (-0.081) \times \left(\frac{16.38}{2 \times 260}\right) = -0.264/s$$

$$N_{r2} = \left(\frac{\bar{q}^* Sb}{I_{zz}}\right) C_{nr2}\left(\frac{b}{2V^*}\right) = 103.54 \times (-0.009) \times \left(\frac{16.38}{2 \times 260}\right) = -0.03/s$$

$$\left(\frac{g}{V^*}\right) = \frac{9.81}{260} = 0.038/s$$

Using Equation 6.49:

$$2\zeta_{DR}\omega_{nDR} = -N_{r1} - \left(\frac{g}{V^*}\right)\left[Y_\beta + \left(\frac{\mathcal{L}_{r1}}{\mathcal{L}_{p2}}\right)\right]$$

$$2\times 0.204\times 2.347 = 0.264 - 0.038\times\left[Y_\beta - 0.407\right]$$

$$Y_\beta = \left(\frac{\bar{q}^* S}{W}\right)C_{Y\beta} = -17.845 \Rightarrow \left(\frac{41405.0\times 50.4}{169921.24}\right)C_{Y\beta} = -17.845$$

$$\Rightarrow C_{Y\beta} = -1.453/\text{rad}$$

Further, using Equation 6.48 and

$$\left(\frac{\mathcal{L}_\beta}{\mathcal{L}_{p2}}\right) = \left(\frac{C_{l\beta}}{C_{lp2}}\right)\cdot\left(\frac{2V^*}{b}\right)$$

$$= \left(\frac{-0.03}{-0.430}\right)\times\left(\frac{2\times 260}{16.38}\right) = 2.215/\text{s}$$

$$\omega_{nDR}^2 = N_\beta + \left(\frac{g}{V^*}\right)\left[Y_\beta N_{r2} + \left(\frac{\mathcal{L}_\beta}{\mathcal{L}_{p2}}\right)\right]$$

$$2.347^2 = N_\beta + 0.038\times\left[(-17.845\times -0.03) + 2.215\right]$$

$$N_\beta = \left(\frac{\bar{q}^* Sb}{I_{zz}}\right)C_{n\beta} = 5.404 \Rightarrow 103.54\ C_{n\beta} = 5.40 \Rightarrow C_{n\beta} = 0.052/\text{rad}$$

As a matter of curiosity, we can evaluate the residual sideslip angle:

$$\Delta\beta_s = -\frac{(g/V^*)N_{r2}}{\omega_{nDR}^2}\Delta\mu$$

$$= -\frac{0.038\times -0.03}{2.347^2}\Delta\mu = 2.069\times 10^{-4}\Delta\mu$$

6.10.3 Spiral Mode

To derive an approximation to the spiral mode, we go back to Equation 6.39 and examine the terms in the second set of braces:

$$\underbrace{\left\{-\mathcal{L}_{r1}\Delta\dot{\beta} + \left[\mathcal{L}_\beta + \left(\frac{g}{V^*}\right)Y_\beta\mathcal{L}_{r2}\right]\Delta\beta_s + \mathcal{L}_{p2}\Delta\dot{\mu} + \left(\frac{g}{V^*}\right)\mathcal{L}_{r2}\Delta\mu\right\}}_{\textit{Equals zero}} \quad (6.51)$$

which you will remember is what was left of the roll moment equation after the part representing the roll mode was detached. This may be written as below:

$$\Delta\dot{\mu}_s = \left(\frac{1}{\mathcal{L}_{p2}}\right)\left\{\mathcal{L}_{r1}\Delta\dot{\beta}_s - \left[\mathcal{L}_\beta + \left(\frac{g}{V^*}\right)Y_\beta\mathcal{L}_{r2}\right]\Delta\beta_s - \left(\frac{g}{V^*}\right)\mathcal{L}_{r2}\Delta\mu\right\} \tag{6.52}$$

where all the terms vary at the slowest timescale T_s. We ignore the $\Delta\dot{\beta}_s$ term in this equation, and use the expression for $\Delta\beta_s$ obtained from the residual in Equation 6.50 after the Dutch roll mode has been solved for. This gives (ignoring the subscript 's'):

$$\Delta\dot{\mu} = \left(\frac{1}{\mathcal{L}_{p2}}\right)\left\{\frac{\left[\mathcal{L}_\beta + (g/V^*)Y_\beta\mathcal{L}_{r2}\right]N_{r2}}{\left[N_\beta + (g/V^*)Y_\beta N_{r2} + (g/V^*)(\mathcal{L}_\beta/\mathcal{L}_{p2})\right]} - \mathcal{L}_{r2}\right\}\left(\frac{g}{V^*}\right)\Delta\mu$$

$$\Delta\dot{\mu} = \left(\frac{1}{\mathcal{L}_{p2}}\right)\left\{\frac{\left[\mathcal{L}_\beta + (g/V^*)Y_\beta\mathcal{L}_{r2}\right]N_{r2}}{\omega_{nDR}^2} - \mathcal{L}_{r2}\right\}\left(\frac{g}{V^*}\right)\Delta\mu \tag{6.53}$$

Rearranging terms and dropping terms in higher powers of (g/V^*) yields:

$$\Delta\dot{\mu} = \underbrace{\left(\frac{1}{\mathcal{L}_{p2}}\right)}_{1/\lambda_r}\left\{\frac{\left[\mathcal{L}_\beta N_{r2} - N_\beta\mathcal{L}_{r2}\right]}{\omega_{nDR}^2}\right\}\left(\frac{g}{V^*}\right)\Delta\mu \tag{6.54}$$

And since $\lambda_r = \mathcal{L}_{p2}$

$$\Delta\dot{\mu} = \left(\frac{g}{V^*}\right)\frac{\left[\mathcal{L}_\beta N_{r2} - N_\beta\mathcal{L}_{r2}\right]}{\lambda_r\omega_{nDR}^2}\Delta\mu \tag{6.55}$$

Following the condition on the factor 'a' in Equation 2.1, assuming $\lambda_r < 0$, for the spiral mode to be stable, we require:

$$\left[\mathcal{L}_\beta N_{r2} - N_\beta\mathcal{L}_{r2}\right] > 0 \tag{6.56}$$

which may or may not hold. As we shall soon see, the stability of the spiral mode is usually not a matter of much concern. We shall take up further study of these lateral-directional modes in Chapter 7.

Example 6.5

Determine the condition on the derivative $\mathcal{L}_{r2}$ so that the spiral mode may be stable for the airplane data given in Example 6.4.

Using the condition for stability of spiral mode given in Equation 6.56, we find that:

$$\mathcal{L}_{r2} < \left[\left(\frac{\mathcal{L}_\beta N_{r2}}{N_\beta} \right) = \left(\frac{C_{l\beta}/I_{xx}}{C_{n\beta}/I_{zz}} \cdot N_{r2} \right) \right]$$

Thus,

$$\mathcal{L}_{r2} < \left(\frac{-0.03 \times 330142.72 \times -0.03}{0.052 \times 161032.43} \right) \Rightarrow \mathcal{L}_{r2} < 0.0355$$

$$\left(\frac{\bar{q}^* S b}{I_{xx}} \right) C_{lr2} \left(\frac{b}{2V^*} \right) < 0.0355 \Rightarrow (212.27) C_{lr2} \left(\frac{16.38}{2 \times 260} \right) < 0.0355 \Rightarrow$$

$$C_{lr2} < 0.0053/\text{rad}$$

EXERCISE PROBLEMS

6.1 Construct some curved flight paths in the directional plane such that $\beta = 0$ is maintained during flight.

6.2 An airplane flying in cruise condition with velocity $V^* = 260\ m/s$ at an altitude of 11 km is supposed to follow a fixed straight line path that can be taken to be along the X^E axis of the Earth-fixed inertial coordinate system. The aircraft experiences a sidewind of velocity $v = 20\ m/s$ while heading along X^E. Determine the sideslip angle β and the wind orientation angle χ with respect to X^E axis.

6.3 We have seen four distinct timescales in this chapter including the fastest timescale $(b/2V^*)$. You may wonder whether the corresponding longitudinal timescale $(c/2V^*)$ has any significance. [*Hint:* consider the time taken by a signal to travel from the location of the bound vortices at wing quarter chord to the three-quarter chord point.]

6.4 For the airplane data given in Example 6.4, write a code to simulate Dutch roll and spiral motions. Verify the frequency and damping ratio for the Dutch roll motion given in Example 6.4 from your simulation.

6.5 It is desired to measure yaw stiffness and yaw damping derivatives for an airplane from wind-tunnel testing. The scaled-down model for the airplane is fixed in the tunnel so as to have only motion in yaw. Write down the appropriate governing equation of motion. The prototype executes oscillation in yaw under some perturbation from gust of wind released from the side-window while the airspeed in the wind-tunnel is $V^* = 15\ m/s$ and the airplane nose is directly pointing into the wind. Frequency and damping ratio of the motion is observed to be $\omega = 2.4\ rad/s$ and $\zeta = 0.2$. Other useful data given are: $S = 0.2m^2$, $b = 1m$, $I_{zz} = 0.11kg - m^2$. Determine the derivatives, $C_{n\beta}$ and C_{nr1}.

6.6 Determine different timescales of motion for the airplane model in Example 6.4 and the airplane model in Exercise 6.5 and study them comparatively. Is it possible to know, what would be an appropriate factor by which the airplane model in Example 6.4 ought to be scaled down so that the time response of the full scale and the scaled-down model are quantitatively same in yaw.

6.7 Equations 6.1 are true for any angle of attack in general as you will find in Chapter 8. Linearize the equations at $\alpha^* = 90°$.

REFERENCE

1. Raghavan, B. and Ananthkrishnan, N., Small-perturbation analysis of airplane dynamics with dynamic stability derivatives redefined, *Journal of Aerospace Sciences and Technologies*, 61(3), 2009, 365–380.

7 Lateral-Directional Dynamic Modes

In this chapter, we shall look closely at each of the three dynamic modes that we extracted in Chapter 6 from the small-perturbation lateral-directional dynamic equations.

7.1 ROLL (RATE) MODE

The fastest of the three modes is the roll mode, which was modelled by the first-order dynamical equation in $\Delta\dot{\mu}$:

$$\Delta\ddot{\mu} = \mathcal{L}_{p2}\Delta\dot{\mu} \tag{7.1}$$

As you can see, Equation 7.1 is actually written in terms of $\Delta\dot{\mu}$, the roll rate. Hence, to be correct, we should call it the *roll rate* mode, but it is often shortened to just the roll mode.

We have seen that the roll mode is stable, that is, perturbations in the roll rate die down, provided the coefficient $\mathcal{L}_{p2}$ is negative. An example of this was shown in Figure 6.10.

$\mathcal{L}_{p2}$ is the short form for the combination of terms as shown below:

$$\mathcal{L}_{p2} = \left(\frac{\overline{q}Sb}{I_{xx}}\right) C_{lp2} \left(\frac{b}{2V^*}\right)$$

which contains the fast timescale $(b/2V^*)$ and the aerodynamic derivative C_{lp2}. Thus, the roll mode is stable as long as:

$$C_{lp2} < 0$$

The derivative C_{lp2} is usually called the roll damping derivative.

It is important to note that although perturbations in roll rate $\Delta\dot{\mu}$ may die down, the roll mode does not deal with the roll angle $\Delta\mu$ itself, so the roll rate perturbation creates a roll angle that does not disappear immediately. Since the roll rate mode equation is of first order, a positive (right-wing down) roll rate leaves behind a positive roll angle perturbation, and a negative roll angle in the case of a left-wing down roll rate. We shall see later that the residual roll angle is then a perturbation for the slower Dutch roll and spiral modes.

7.2 ROLL DAMPING DERIVATIVE C_{lp2}

For most conventional airplanes, the main contributor to the roll damping derivative C_{lp2} is the wing. Let us evaluate the roll damping effect due to a wing. The following derivation is appropriate in the case of a large-aspect-ratio wing, as is seen on most civil airliners.

DOI: 10.1201/9781003096801-7

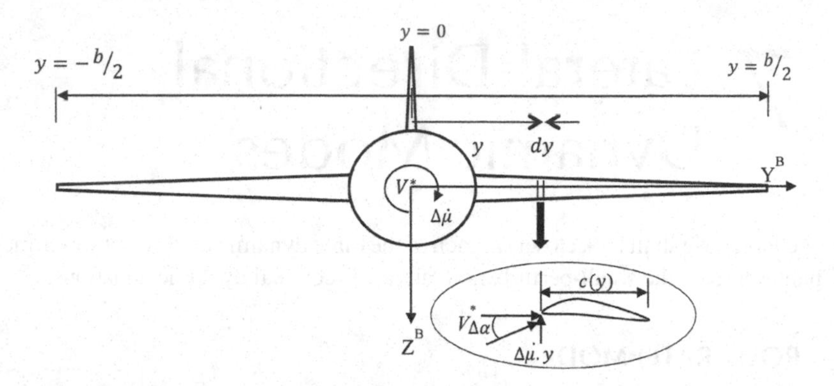

FIGURE 7.1 Rear view of an airplane and wing.

Consider the rear view of an airplane and wing shown in Figure 7.1 with the Y^B axis along the right wing and the Z^B axis in the downward direction. The fuselage centre line is at the origin of the axes ($y = 0$) and the right-wing tip is at $y = b/2$, where b is the wing span. The left-wing tip is at $y = -b/2$. The airplane velocity V^* is directed into the plane of the paper.

Imagine a small roll rate perturbation $\Delta\dot{\mu}$ in the positive sense as shown in Figure 7.1. As a result, every point on the wing (at station y) acquires a downward (right wing) or upward (left wing) velocity:

$$\Delta w(y) = \Delta\dot{\mu}.y$$

Imagine one such wing section as shown in the inset of Figure 7.1. The relative wind velocity due to the freestream, V^*, is the same at all sections. The downward/upward motion due to $\Delta\dot{\mu}$ creates a relative velocity $\Delta w(y) = \Delta\dot{\mu} \cdot y$ at that station (positive on the right wing and negative on the left wing). The result is a change in the angle of attack at that station, given by:

$$\Delta\alpha(y) = \Delta w(y)/V^* = \pm\Delta\dot{\mu} \cdot y/V^*$$

On the right down-going wing, this additional angle of attack is positive, so the wing sections add to the lift; whereas on the left up-going wing, the additional angle of attack, and hence the additional lift, is negative. The additional sectional lift coefficient is:

$$\Delta C_l(y) = C_{l\alpha} \cdot \Delta\alpha(y) = \pm C_{l\alpha} \cdot \Delta\dot{\mu} \cdot y/V^*$$

where, C_l is the sectional lift coefficient (not to be mistaken for the roll moment coefficient) and $C_{l\alpha}$ is the sectional (airfoil) lift curve slope, assumed to be the same for all sections.

Then, the additional lift at each section is:

$$l(y) = \bar{q} \cdot c(y) \cdot dy \cdot \Delta C_l(y) = \pm\bar{q} \cdot c(y) \cdot dy \cdot C_{l\alpha} \cdot \Delta\dot{\mu} \cdot y/V^*$$

This is positive (upwards) on the right wing and negative (downwards) on the left wing.

Considering a pair of stations at $\pm y$, the additional lift at these stations creates a net roll moment at the fuselage centre line ($y=0$), given by:

$$\Delta\mathcal{L}(\pm y) = -2l(y)\cdot y = -2\bar{q}\cdot c(y)\cdot dy\cdot C_{l\alpha}\cdot\Delta\dot{\mu}/V^*\cdot y^2$$

Note the negative sign because the rolling moment created is 'left-wing-down' and hence negative. The factor of two accounts for the equal and same-sense contribution from stations at $\pm y$. Since we have already considered a pair of span-wise stations at $\pm y$, the net rolling moment from the entire wing is obtained by integrating the incremental roll moments from $y=0$ to $y=b/2$:

$$\Delta\mathcal{L} = -2\bar{q}\cdot C_{l\alpha}\cdot\Delta\dot{\mu}/V^*\int_0^{b/2} c(y)\cdot y^2 dy$$

Subsequently, we can write the roll moment coefficient as:

$$\Delta C_l = \frac{\Delta\mathcal{L}}{\bar{q}Sb} = -2\left(\frac{C_{l\alpha}}{Sb}\right)\left(\frac{\Delta\dot{\mu}}{V^*}\right)\int_0^{b/2} c(y)\cdot y^2 dy$$

$$= -4\left(\frac{C_{l\alpha}}{Sb^2}\right)\left(\frac{\Delta\dot{\mu}b}{2V^*}\right)\int_0^{b/2} c(y)\cdot y^2 dy$$

Since the roll damping derivative is defined as (see Equation 6.20):

$$C_{lp2} = \left.\frac{\partial C_l}{\partial p_w(b/2V)}\right|_*$$

with $\Delta p_w = \Delta\dot{\mu}$, we can write:

$$C_{lp2} = \frac{\Delta C_l}{(\Delta\dot{\mu}b/2V^*)} = \left(\frac{-4C_{l\alpha}}{Sb^2}\right)\int_0^{b/2} c(y)\cdot y^2 dy \tag{7.2}$$

which is a generic expression for any high-aspect-ratio wing with a given span-wise distribution of chord length $c(y)$.

7.2.1 Special Case of Trapezoidal Wing

For the case of a trapezoidal wing as shown in Figure 7.2, which is common on many airplanes, we can explicitly evaluate the integral in Equation 7.2.

Given root chord c_r and tip chord c_t, the wing taper ratio λ is defined as:

$$\lambda = \frac{c_t}{c_r}$$

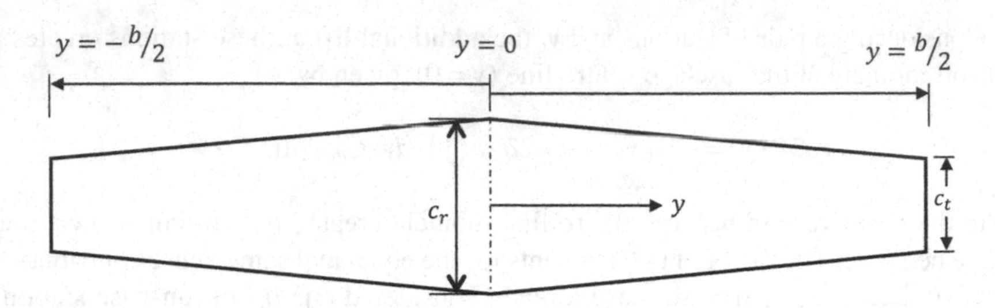

FIGURE 7.2 A trapezoidal wing.

The chord length $c(y)$ can be written as:

$$c(y) = c_r + \frac{2}{b}(c_t - c_r) \cdot y$$

And the wing area as:

$$S = \frac{b}{2} \cdot (c_t + c_r)$$

Hence, for a trapezoidal wing, we can evaluate Equation 7.2 as:

$$C_{lp2} = \left(\frac{-4C_{l\alpha}}{Sb^2}\right) \int_0^{b/2} \left\{c_r + \frac{2}{b} \cdot (c_t - c_r) \cdot y\right\} y^2 dy$$

$$= -\frac{8C_{l\alpha}}{(c_t + c_r)b^3} \int_0^{b/2} \left\{c_r y^2 dy + \frac{2}{b} \cdot (c_t - c_r) \cdot y^3 dy\right\}$$

$$= -\frac{8C_{l\alpha}}{(c_t + c_r)b^3} \left[c_r \frac{y^3}{3} + \frac{2}{b}(c_t - c_r)\frac{y^4}{4}\right]_0^{b/2}$$

$$= -\frac{C_{l\alpha}}{(c_t + c_r)(b^3/8)} \left[\frac{c_r}{3}\left(\frac{b}{2}\right)^3 + (c_t - c_r)\frac{1}{4} \cdot \left(\frac{b}{2}\right)^3\right]$$

$$= -\frac{C_{l\alpha}}{(c_t + c_r)} \left[\frac{c_r}{3} + (c_t - c_r)\frac{1}{4}\right]$$

$$= -\frac{C_{l\alpha}}{12} \cdot \frac{(c_r + 3c_t)}{(c_t + c_r)} = -\frac{C_{l\alpha}}{12} \cdot \frac{(1 + 3\lambda)}{(1 + \lambda)} \tag{7.3}$$

Homework Exercise: For many airplanes, the taper ratio λ is close to 0.4. Estimate C_{lp2} for $\lambda = 0.4$ and $C_{l\alpha} \approx 2\pi$ (comes out to ~ −0.82). If you do not have data for a conventional airplane, this is a good first guess for the value of C_{lp2}.

Special cases:

i. Rectangular wing, $\lambda = 1: C_{lp2} = -C_{l\alpha}/6$
ii. Delta (triangular) wing, $\lambda = 0: C_{lp2} = -C_{l\alpha}/12$

Clearly, this approach works if the wing has a large enough aspect ratio so that each span-wise section operates approximately like a two-dimensional airfoil section. On low-aspect-ratio wings, the flow is often too three-dimensional for this approach to be valid. In those cases, the broad principle of roll damping still holds, but this particular derivation of the roll damping derivative may no longer be sufficiently accurate.

7.2.2 Owing to Vertical Tail

A second source of roll damping on airplanes is the vertical tail, which can usually have a reasonably large span (height). The mechanism of creating roll damping due to the vertical tail is the same as that of the wing. See Figure 7.3 for a sketch.

Figure 7.3 shows the rear view of an airplane with a vertical tail. Consider a perturbation in roll rate, $\Delta\dot{\mu}$, in the positive sense (to the right) and a section of the vertical tail at height h from the fuselage centre line. As seen in the top view inset of Figure 7.3, there are two components of the relative velocity at this section: V^*, due to the freestream, and $\Delta\dot{\mu}h$, due to the roll rate. Together, they create an angle of attack, $\Delta\alpha$:

$$\Delta\alpha = \frac{\Delta\dot{\mu}h}{V^*}$$

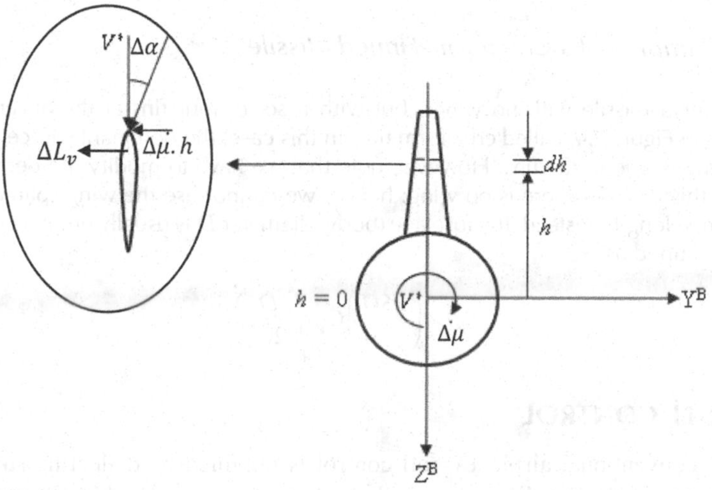

FIGURE 7.3 Rear view of an airplane with a vertical tail (top view inset).

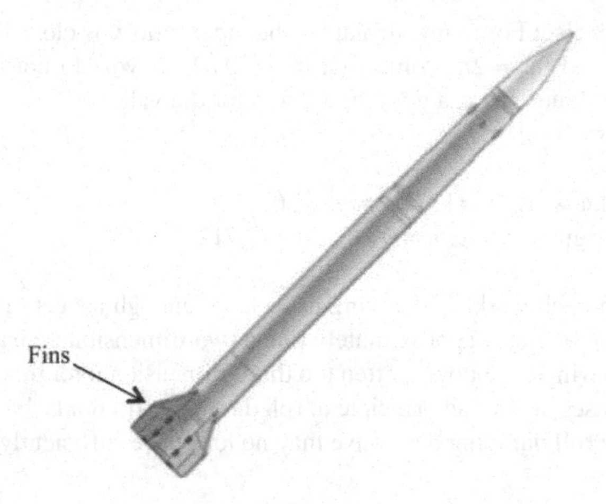

FIGURE 7.4 Missile with cruciform fins. (Reprinted with permission of the Aerospace Corporation, http://www.aerospaceweb.org.)

The resultant lift of the vertical tail section is actually a side force for the airplane that acts normal to the resultant relative wind, but for small angles $\Delta\alpha$ it can be taken to be approximately along the negative Y^B axis, as marked in Figure 7.3. Since this side force acts above the fuselage centre line (or the X^B axis), it creates a negative roll moment, that is, in the opposite sense to the perturbation in roll rate. Thus, the vertical tail also contributes to a negative C_{lp2} adding to the wing contribution.

Since the typical aspect ratio of a vertical tail (e.g., for civil airliners) is around 1.5, the calculations in Section 7.2 are perhaps not entirely appropriate to estimate the C_{lp2} effect due to the vertical tail. However, the principle is the same.

Example 7.1

Roll Damping of a Cruciform-Finned Missile

Consider a missile with no wings, but with a set of four fins at the aft end, as shown in Figure 7.4, called cruciform fins. In this case, the dominant source of roll damping is due to the fins. However, note that we have to modify the derivative $\mathcal{L}_{p2}$ in this case, as there is no wing; hence, we cannot use the wing span b as a reference length. Instead, the fuselage (body) diameter D is usually employed, and $\mathcal{L}_{p2}$ is defined as:

$$\mathcal{L}_{p2} \equiv \left(\frac{\bar{q}SD}{I_{xx}}\right) C_{lp2} \left(\frac{D}{2V^*}\right)$$

7.3 ROLL CONTROL

For most conventional airplanes, roll control is obtained by deflecting small flaps at the wing outboard trailing edges called ailerons, as described in Chapter 1. The ailerons are really a roll rate control device as we shall see below.

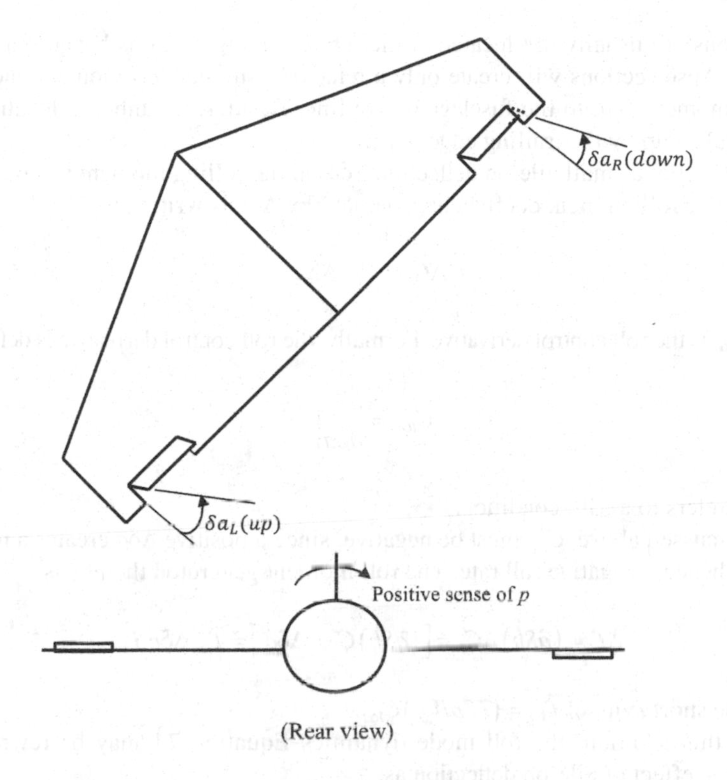

FIGURE 7.5 Sign convention for positive aileron deflection.

Ailerons are usually deflected in a pair. A positive aileron deflection, as indicated in Figure 7.5, is when the right aileron is deflected down and the left aileron deflected up. The sense of positive aileron deflection is the same as that of a positive roll rate marked by the arrow in Figure 7.5. When the right aileron is deflected up and the left aileron is deflected down, each is negative, and the net aileron deflection is negative as well.

The net aileron deflection is usually defined as:

$$\delta a = \frac{(\delta a_R + \delta a_L)}{2}$$

with the above sign convention.

Essentially, a downward deflected aileron adds to the camber of the airfoil sections over its span, which creates an additional lift. Likewise, an upward deflected aileron subtracts from the lift of the airfoil sections over its span. As shown in Figure 7.5, for a positive aileron deflection, the right aileron therefore adds to the lift of the right wing and the left aileron subtracts from the lift of the left wing. The difference creates a rolling moment to the left, which is in the negative sense. Thus, a positive aileron deflection causes a negative roll rate, and vice versa.

Ailerons are usually not located at the inboard wing sections because a change in lift at those sections will create only a relatively smaller roll moment due to the smaller moment arm to the fuselage centre line. Besides, the inboard locations are usually taken up by the trailing edge flaps.

The effect of a small aileron deflection $\Delta\delta a$ on the rolling moment is modelled as follows. The roll moment coefficient generated by $\Delta\delta a$ is written as:

$$\Delta C_l = C_{l\delta a}\Delta\delta a$$

where $C_{l\delta a}$ is the roll control derivative. Formally, the roll control derivative is defined as:

$$C_{l\delta a} = \left.\frac{\partial C_l}{\partial \delta a}\right|_*$$

where * refers to a trim condition.

As discussed above, $C_{l\delta a}$ must be negative, since a positive $\Delta\delta a$ creates a negative ΔC_l and hence a negative roll rate. The roll moment generated thereby is:

$$\Delta\mathcal{L} = (\bar{q}Sb)\Delta C_l = \left[(\bar{q}Sb)C_{l\delta a}.\Delta\delta a\right] = \mathcal{L}_{\delta a}\Delta\delta a.I_{xx}$$

where the short symbol $\mathcal{L}_{\delta a} = (\bar{q}Sb/I_{xx})C_{l\delta a}$.

With this addition, the roll mode dynamics Equation 7.1 may be rewritten to include the effect of aileron deflection as:

$$\Delta\ddot{\mu} = \mathcal{L}_{p2}\Delta\dot{\mu} + \mathcal{L}_{\delta a}\Delta\delta a \quad (7.4)$$

which is still a first-order dynamical system, but with a forcing term $\mathcal{L}_{\delta a}\Delta\delta a$.

Example 7.2

Roll Response to Step Input in Aileron

A step input as shown in Figure 7.6 means that the ailerons are deflected, either positive or negative, and held that way. This results in what appears from Figure 7.6 to be a steady roll rate. With a steady roll rate, the airplane will continue to roll, building up roll angle going on to a full 360° roll and further. That is why the aileron is called a roll rate control, not a roll angle control—a particular value of aileron gives a steady value of roll rate, not a fixed value of a roll angle.

However, in practice, once the airplane starts banking to one side due to the roll rate, the effect of gravity will induce a sideslip and set off the slower modes, Dutch roll and spiral. Thus, beyond a small value of the roll angle $\Delta\mu$, the solution of Equation 7.4 is not strictly meaningful. Nevertheless, Equation 7.4 provides some useful information, so it is still studied. But, since the effect of gravity is ignored in studying the solutions of Equation 7.4, the steady-state solutions of Equation 7.4 are called *pseudo*-steady states to distinguish them from real airplane steady states that may actually be observed in flight.

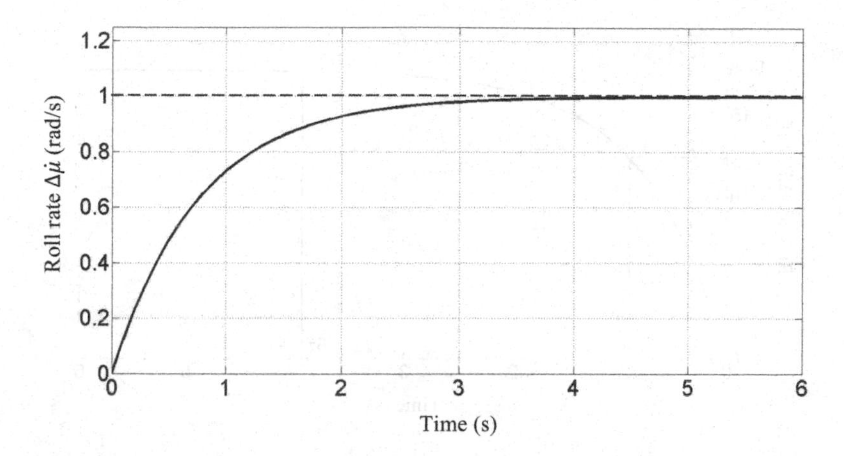

FIGURE 7.6 First-order roll rate response (solid line) to step demand in roll rate (dashed line).

The *pseudo*-equilibrium or *pseudo*-steady state of Equation 7.4 is, as discussed in Chapter 2, obtained by setting the time derivative terms to zero. Thus,

$$\mathcal{L}_{p2}\Delta\dot{\mu}^* + \mathcal{L}_{\delta a}\Delta\delta a = 0 \tag{7.5}$$

where * refers to the *pseudo*-steady state value. Writing out the long forms of the terms in $\mathcal{L}_{p2}$ and $\mathcal{L}_{\delta a}$, we obtain the non-dimensional pseudo-steady state roll rate for a step aileron deflection $\Delta\delta a$ as:

$$\Delta\dot{\mu}^*\left(b/2V^*\right) = -\left(C_{l\delta a}/C_{lp2}\right)\Delta\delta a \tag{7.6}$$

The factor $\Delta\dot{\mu}^*\left(b/2V^*\right)$ is a useful measure of roll performance. For instance, a requirement of $\Delta\dot{\mu}^*\left(b/2V^*\right)|_{max} > 0.09$ may be placed in the case of a fighter aircraft.

Example 7.3

Roll Performance of F-104A at Sea Level

Use the following data:

$$V^* = 87.0\text{ m/s};\quad S = 18.0\text{ m}^2;\quad b = 6.7\text{ m};\quad I_{xx} = 4676.0\text{ kg–m}^2;$$

$$C_{lp2} = -0.285/\text{rad};\quad C_{l\delta a} = -0.039/\text{rad}$$

Evaluate the response to a −5° step aileron input.

$$\bar{q} = 0.5\times 1.225\times 87.0^2 = 4636.01\text{ N/m}^2$$

$$\left(b/2V^*\right) = 6.7/\left(2\times 87.0\right) = 0.0385\text{ s}$$

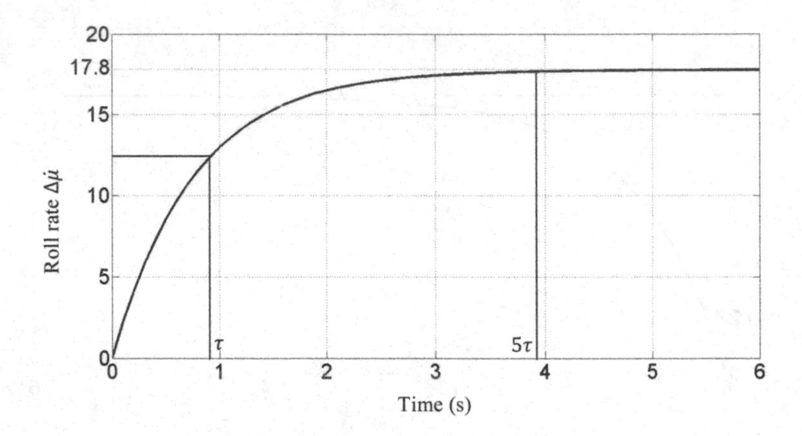

FIGURE 7.7 Roll response to –5° aileron step input in Example 7.3.

Non-dimensional pseudo-steady roll rate:

$$\Delta\dot{\mu}^*\left(b/2V^*\right) = -\left(C_{l\delta a}/C_{lp2}\right)\Delta\delta a = -(-0.039/-0.285)\times -5° = 0.684°$$

which gives the steady value of the roll-rate, $\Delta\dot{\mu}^* = 0.684°\times\dfrac{2\times 87.0}{6.7} = 17.76°/s$.

The time response appears as shown in Figure 7.7, steadily building up to the pseudo-steady state value. The time constant works out to:

$$1/\tau = -\left(\bar{q}Sb/I_{xx}\right)\left(b/2V^*\right)C_{lp2}$$
$$= -(4636.01\times 18.0\times 6.7/4676.0)\times 0.0385\times(-0.285) = 1.312$$

From which, $\tau = 0.762\ s$. This is the time taken for the response to build up to ~63% of the pseudo-steady state value. Approximately $5\tau = 3.81\ s$ is a good estimate of time to attain steady state as appears to be correct from the response in Figure 7.7.

7.4 AILERON CONTROL DERIVATIVE, $C_{l\delta a}$

We can work out a typical estimate for the aileron control derivative, $C_{l\delta a}$, for a conventional airplane wing.

Consider the airplane wing shown in Figure 7.8 with the aileron located between span-wise stations y_1 and y_2. The inset of Figure 7.8 shows a chord-wise cross section of the wing with the airfoil section and the trailing edge aileron flap.

The aileron effectiveness is measured by a parameter τ:

$$\tau = \frac{\Delta\alpha}{\Delta\delta a} = \frac{d\alpha}{d\delta a} \tag{7.7}$$

which indicates the effective change in the angle of attack $\Delta\alpha$ of the airfoil section for a trailing edge aileron flap deflection $\Delta\delta a$. τ, known as flap effectiveness parameter,

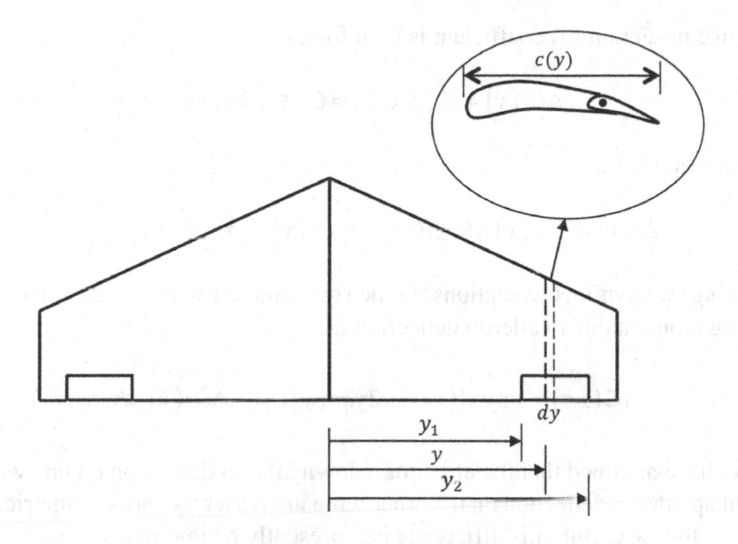

FIGURE 7.8 Span-wise location of aileron on wing.

is chiefly a function of the airfoil and aileron flap geometry, and is usually given as a function of the ratio of the aileron flap surface area to the wing surface (planform) area or the ratio of aileron flap chord to airfoil chord. In Figure 7.9, typical values of τ are shown as a function of the area ratio.

From Equation 7.7, the effect of an aileron deflection can be written in terms of the local angle of attack change at that section:

$$\Delta\alpha(y) = \tau \cdot \Delta\delta a(y)$$

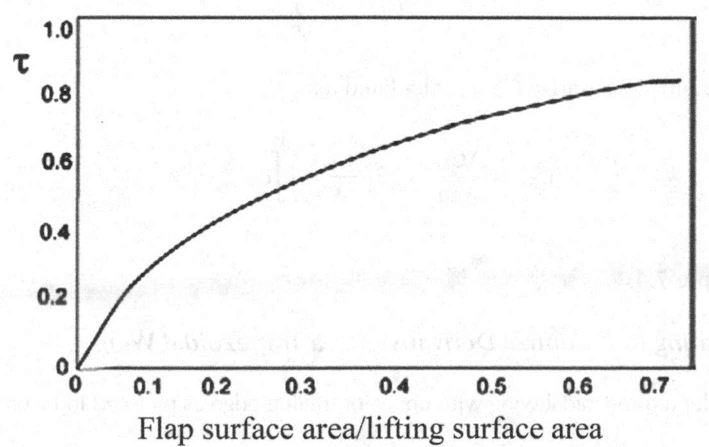

FIGURE 7.9 Flap effectiveness parameter. (Adapted from *Flight Stability and Automatic Control* by R.C. Nelson, Second Edition, McGraw Hill 2007, pp. 64.)

The change in section lift coefficient is then found as:

$$\Delta C_l(y) = C_{l\alpha}\Delta\alpha(y) = C_{l\alpha}\tau \cdot \Delta\delta a(y)$$

And the section lift as:

$$\Delta l(y) = \bar{q} \cdot c(y) \cdot dy \cdot \Delta C_l(y) = \bar{q}c(y)C_{l\alpha}\tau \cdot \Delta\delta a(y) \cdot dy$$

Considering two symmetric sections on the right and left wings at $\pm y$, we can write the rolling moment due to aileron deflection as:

$$\Delta\mathcal{L}(\pm y) = -2y\Delta l(y) = -2y\bar{q}c(y)C_{l\alpha}\tau \cdot \Delta\delta a(y) \cdot dy$$

where we have assumed that the effect of a down aileron deflection on one wing and that of an up aileron deflection on the other wing are perfectly anti-symmetric, which is not truly the case. But, this difference can presently be ignored.

The rolling moment coefficient due to these two symmetrically located sections is:

$$\begin{aligned}\Delta C_l(\pm y) &= \Delta\mathcal{L}(\pm y)/(\bar{q}Sb)\\ &= -2C_{l\alpha} \cdot (\tau/Sb) \cdot \Delta\delta a(y) \cdot c(y) \cdot y \cdot dy\\ &= -2C_{l\alpha} \cdot (\tau/Sb) \cdot \Delta\delta a \cdot c(y) \cdot y \cdot dy\end{aligned}$$

assuming $\Delta\delta a$ to be the same at all aileron sections (y), which is usually the case.

Over the entire span of the aileron, the rolling moment coefficient is then:

$$\Delta C_l = -2\frac{C_{l\alpha}\tau}{Sb} \cdot \Delta\delta a \cdot \int_{y_1}^{y_2} c(y)\, y\, dy$$

The roll control derivative $C_{l\delta a}$ is calculated as:

$$C_{l\delta a} = \frac{\Delta C_{l\alpha}}{\Delta\delta a} = -2\left(\frac{C_{l\alpha}\tau}{Sb}\right)\int_{y_1}^{y_2} c(y)\, y\, dy. \tag{7.8}$$

Example 7.4

Evaluating Roll Control Derivative for a Trapezoidal Wing

Consider a trapezoidal wing with unswept trailing edge as pictured in Figure 7.10.

The half-span $b/2 = 16.7$ m.
The tip and root chords $c_t = 3.9$ m and $c_r = 7.2$ m.
Hence, the taper ratio $\lambda = c_t/c_r = 0.54$.

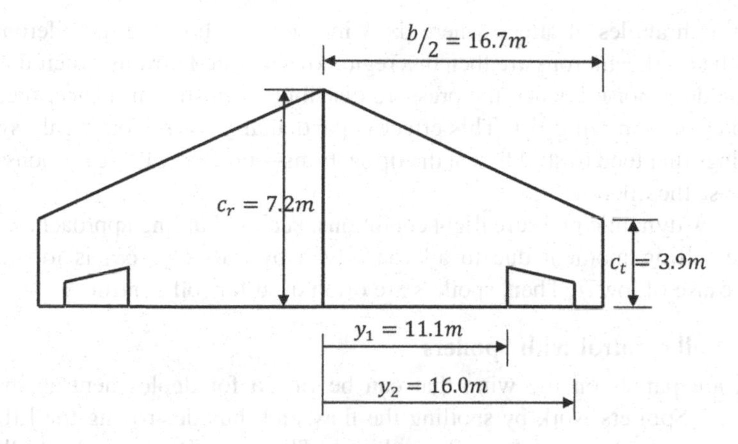

FIGURE 7.10 A trapezoidal wing with unswept trailing edge in Example 7.4.

The aileron spans between stations $y_1 = 11.1$ m and $y_2 = 16.0$ m, and the aileron chord is a fixed fraction of the wing chord at these sections, $c_a/c = 0.18$. For this chord fraction, the flap effectiveness parameter τ can be found from Figure 7.9 to be $\tau = 0.36$.

For the wing section, the lift curve slope $C_{l\alpha} = 4.44$/rad.

The wing area works out to be $S = 2 \times (1/2) \times 16.7 \times (7.2 + 3.9) = 185.37\ \text{m}^2$ (including both wing halves).

The wing chord can be written as the following function of y:

$$c(y) = c_r\left[1 + \frac{2y}{b}(\lambda - 1)\right]$$

Inserting all the numbers into Equation 7.8, we find:

$$C_{l\delta a} = -2\left(\frac{4.44 \times 0.36}{185.4 \times 16.7 \times 2}\right)\int_{11.1}^{16.0} 7.2\left[1 + 2 \times \frac{0.54 - 1}{16.7 \times 2}y\right]y\ dy = -0.155/\text{rad}$$

7.4.1 Other Roll Control Devices

There are conditions under which the ailerons are not very effective. Remember, the ailerons operate by changing the lift of the wing sections to whose trailing edge the ailerons are attached. Here are a few conditions under which aileron deflection does not significantly change the wing lift, and hence does not provide useful roll control.

a. At high (supersonic) Mach numbers—because of supersonic flow, trailing edge devices are not effective in changing the lift of the main lifting surface located upstream. The usual solution is to use asymmetric all-moving tail deflections for roll control.

b. At high angles of attack where the wing sections housing the aileron are stalled—the ailerons are then in a region of separated flow and their deflection does not influence the pressure distribution upstream; hence, there is no change in wing lift. This effect is particularly visible on highly swept wings that tend to stall first at the tip sections—precisely those sections that house the ailerons.

c. At low dynamic pressure flight conditions, such as landing approach, where the rolling moment due to aileron, given by $\overline{q}Sb \cdot C_{l\delta a} \cdot \delta a$, is too small because of low $\overline{q}$. Then, spoilers are often used for roll control.

7.4.1.1 Roll Control with Spoilers

Spoilers are panels on the wing that can be raised for deployment as shown in Figure 7.11. Spoilers work by spoiling the flow and thus destroying the lift on the wing sections over which they are deployed. They are therefore also called 'lift dumpers'. Usually, spoilers are deployed only on one wing when a roll is desired. Thus, for a roll to the left, the spoilers on the left wing are deployed. The left wing thus loses lift, while the lift on the right wing is maintained. This lift imbalance creates a roll to the left.

But, spoiler deployment also creates a huge drag. For example, when using spoilers on the left wing, the drag due to the left wing sharply increases, causing a drag differential between the left and right wings. This difference creates a yaw moment that yaws the airplane to the left. Thus, the use of spoilers creates a roll and yaw moment in the same sense.

FIGURE 7.11 Spoilers on the wing in action (https://upload.wikimedia.org/wikipedia/commons/1/11/Air_Jamaica_A321_landing_spoilers_opened.jpg).

Spoilers are effective irrespective of the flight dynamic pressure $\overline{q}$. But, the increased drag also slows down the airplane, which is usually undesirable. Due to this and structural reasons, spoilers are usually used only in low-speed flight, such as landing approach, where ailerons have poor effectiveness.

7.4.1.2 Roll Control by Differential Tail

Many military airplanes use differential horizontal tail deflections to provide roll control at supersonic speeds where the ailerons are not so effective. Deflecting the horizontal tails asymmetrically increases the lift on one tail and reduces the lift on the other. The differential lift creates a roll moment. However, because of the limited span of the horizontal tail, the moment arm is usually not so large, but it may be compensated by the relatively larger dynamic pressure $\overline{q}$ at higher speeds. When large tail deflections are used to produce a certain roll moment by creating a large lift differential, it creates a significant drag, especially at supersonic speeds. The increase in drag is approximately equal on both the horizontal tail halves, so the resultant yaw moment may not be very substantial.

7.4.1.3 Roll Control by Rudder

Just as the vertical tail contributes to roll damping, as discussed in Section 7.2.2, the rudder can also be used to provide roll control.

Firstly, let us state the sign convention for rudder deflection. As sketched in Figure 7.12, the top view shows the sign for a positive yaw rate by a clockwise arrow.

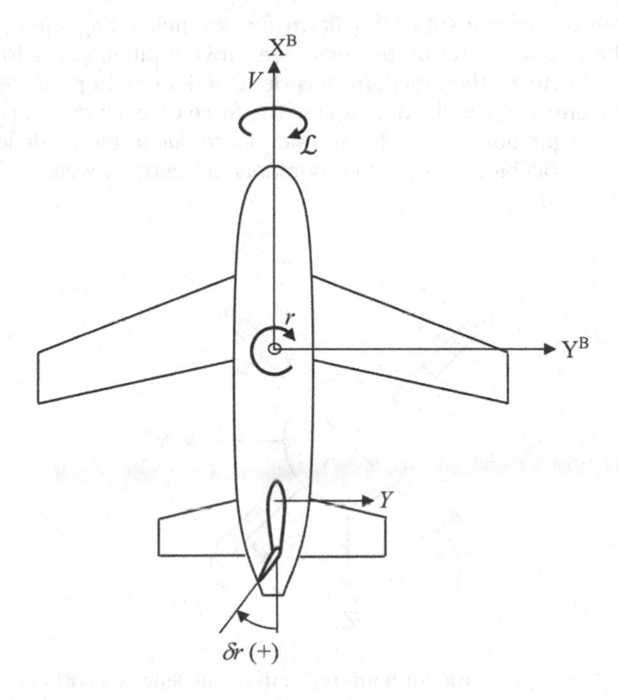

FIGURE 7.12 Sign convention for rudder deflection.

That is, a yaw rate to the right is positive. A rudder deflection in the same sense as the arrow is positive. Thus, a rudder deflection to the left is positive.

A positive rudder deflection creates a positive side force, marked by an arrow to the right in Figure 7.12. This force acting at a height above the fuselage centre line causes a positive roll moment, again to the right. We can define a rolling moment derivative due to rudder deflection as:

$$C_{l\delta r} = \left.\frac{\partial C_l}{\partial \delta r}\right|_*$$

And typically, $C_{l\delta r}$ will be positive, $C_{l\delta r} > 0$. The rudder as a roll control device is not hugely favoured for several reasons. The extent of the rolling moment created is limited by the vertical tail span (height). Rudder deflection adds to the drag. The rudder, which works by a similar principle of changing the lift of the vertical tail airfoil sections, may not be too effective at slow speeds and when the flow over the vertical tail is separated. This is often the case at high angles of attack when part or whole of the vertical tail may be blanketed in the fuselage wake.

Example 7.5

Roll Control by Cruciform Fins

Missiles often use only a set of cruciform fins as their lifting surfaces. In different combinations, the set of four fins then provide pitch, yaw and roll control. Figure 7.13 shows the cruciform fins deflected so as to produce a rolling moment. The arrows show the direction of the force on each fin. You can verify that in this configuration, the net lift and side force due to the fin deflection are zero, hence the pitching and yawing moments are zero as well, and the only result is a roll moment.

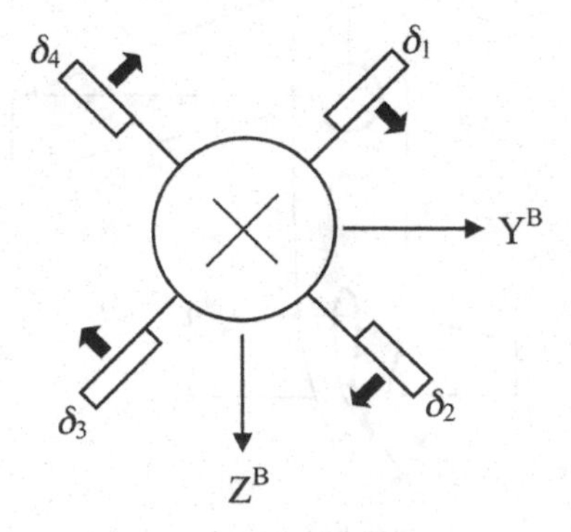

FIGURE 7.13 Rear view of cruciform fins on a missile in deflected position.

Example 7.6

Roll Control with Elevons

On tailless airplanes, the ailerons must act both as roll control as well as pitch control devices. Then, they are called *elevons*. For example, the elevons on the Concorde are shown in Figure 1.28. When symmetrically deflected, they act as elevators. In anti-symmetric deflection, they function as ailerons. Any combination of left and right elevon deflection can be split into a symmetric elevator component and an anti-symmetric aileron component.

Homework Exercise: Work out an expression for such a split.

7.5 YAW DUE TO ROLL CONTROL

The use of any roll control device to create a rolling moment also tends to create a yawing moment because of asymmetric drag increase between the left and right halves. We will look at the yaw due to roll control in the following cases.

7.5.1 Yaw due to Aileron

Consider a pair of ailerons as pictured in Figure 7.5 with the right aileron deflected down and the left aileron deflected up. This combination gives a roll to the left, as we have seen before. The deflected ailerons also add to the drag, which comes from two sources: one is the lift-induced drag that is roughly proportional to C_L^2, the other is called form drag. Since the down (right) aileron adds to C_L, whereas the up (left) aileron subtracts from C_L, the change in lift-induced drag is positive (increase) for the right wing and negative (decrease) for the left wing. The form drag increases on both wings, perhaps a little more in the case of the up-deflected aileron on the left wing. Adding the drag from these two sources together, usually the drag increase on the right wing (the one with the down-going aileron) and is often larger; hence, the airplane tends to yaw to the right due to this drag differential. Roll left, yaw right (or vice versa) is called *adverse yaw*. The other way round, roll and yaw in the same sense, is called *proverse yaw* or *pro yaw*.

From the piloting point of view, a small amount of adverse yaw is preferred because it requires the pilot to give a more natural combination of inputs. For example, to roll left, the stick is moved left. At the same time, to counter the adverse yaw, the left rudder pedal is pressed. This combination of left stick–left pedal (or right stick–right pedal for a right roll) is more natural. The left stick deflects the right aileron down and the left aileron up, giving a left roll. The left pedal moves the rudder to the left, giving a side force to the right, which creates a left yaw that counters the adverse (right) yaw due to a left roll.

Ailerons usually create too much adverse yaw, although the exact amount depends on the wing and aileron design and also on the flight condition. To overcome this and reduce the adverse yaw to a small quantity, the usual trick is to deflect the up-going (left aileron in our example above) proportionately more than the down-going aileron. This increases the drag on its wing half

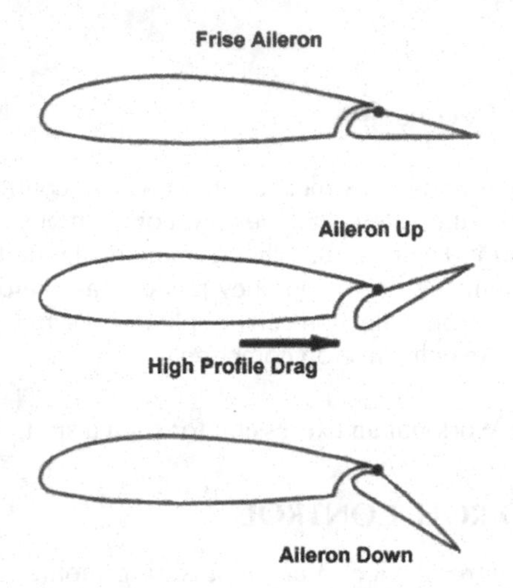

FIGURE 7.14 Frise aileron. (Reprinted with permission of the Aerospace Corporation, http://www.aerospaceweb.org/question/dynamics/yaw/frise.jpg)

and the net yawing moment is reduced. As a general rule of thumb, one may approximately take:

$$|\delta a_{\text{up}}| = 1.5|\delta a_{\text{down}}|$$

Another option is to use what are called *Frise ailerons*, which look like as shown in Figure 7.14.

The nose of the up-deflected Frise aileron protrudes into the wing undersurface flow thus adding to the drag on the down-going wing. The main purpose of the Frise aileron is to avoid excessive adverse yaw due to aileron deflection.

7.5.2 Yaw due to Spoilers

Consider the case in Figure 7.15 where the spoiler on the right wing has been deployed. This causes a loss of lift on the right wing causing it to drop thereby creating a roll to the right. In the process, there is a huge increase of form drag on

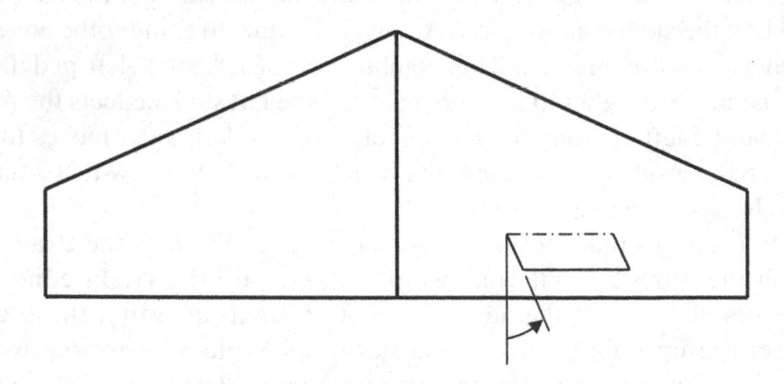

FIGURE 7.15 Spoiler in deflected position on the right wing.

the wing with the spoiler deployed; this causes the airplane to also yaw to the right. Thus, yaw due to spoilers is *pro yaw.* This yawing moment has to be countered by the use of rudder.

One option is to use a combination of ailerons and spoiler such that the net yawing moment is made quite small since ailerons usually give adverse yaw and spoilers always give pro yaw.

7.5.3 Yaw due to Differential Tail

Figure 7.16 shows a conventional airplane configuration with differential horizontal tail deflection—right at decreased angle of attack, hence lower lift; and left at an increased angle of attack, hence higher lift—giving a rolling moment to the right (positive). Notice that the vertical tail is located between and above the two horizontal tail halves. On the upper surface of the left tail, the higher lift means a lower pressure, that is, the suction mechanism that creates lift in subsonic flight. Likewise, the pressure on the upper surface of the right wing is higher. This pressure difference flanks the vertical tail on either side, as seen in Figure 7.16. Thus, the vertical tail sees a lower pressure to its left and a higher pressure to its right, and a pressure difference across a surface means lift. So, the vertical tail experiences a 'lift' force that acts to the left, a negative side force for the airplane. The moment due to this side force about the centre of gravity (CG) results in a positive (to the right) yawing moment. Thus, a roll to the right causes yaw to the right, which is *pro yaw.*

7.5.4 Yaw due to Rudder

The rudder is primarily meant to give a yawing moment, which we shall study later in this chapter. For now, imagine the rudder deflected to the left (positive) as shown in Figure 7.17.

This creates a side force to the right (positive) as marked in Figure 7.17. This side force creates a positive (to the right) rolling moment. It also results in a negative (to the left) yawing moment about the CG. Thus, a roll to the right comes with a yaw to the left in case of the rudder, that is, *adverse yaw.*

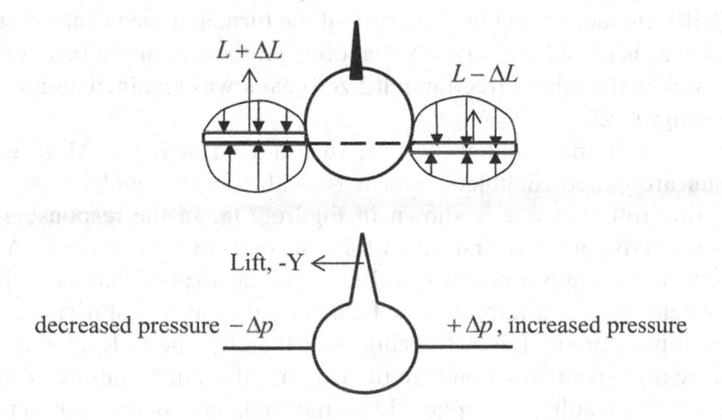

FIGURE 7.16 Rear view showing differential horizontal tail deflection.

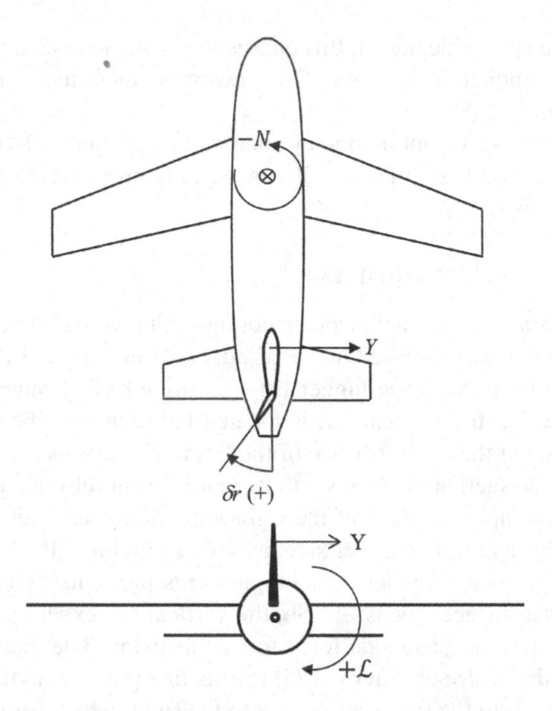

FIGURE 7.17 Yaw (top sketch) and roll (bottom sketch) due to positive rudder deflection.

7.6 AILERON INPUT FOR A BANK ANGLE

We have seen that a step aileron input gives a roll rate, as seen in the simulation of Equation 7.4 in Figure 7.6. However, a continuous roll, usually called a *barrel roll* or by some other similar name, is usually an aerobatic manoeuvre. Airplanes other than those in the aerobatic and combat categories would never need to perform such a manoeuvre. More normally, airplanes would use ailerons to bank to a certain angle to perform a turn manoeuvre, and again when the manoeuvre is done to return back to wings-level flight. For this purpose, one would first need to move the stick left/right (depending on the sense of the turn) and then centre it while the bank angle was held and the turn was ongoing. At the end of the turn, one would move the stick in the other direction until zero bank was regained and then centre it to hold wings level.

Let us run a sample simulation with such a kind of input. More generally, such inputs are called 'doublets' and a typical aileron doublet input and the corresponding roll response is shown in Figure 7.18. In the response shown in Figure 7.18, aileron pulse is first given for 2 seconds to initiate bank. A positive aileron deflection creates a negative roll rate and the airplane banks to the left as per the convention. The aileron is now held neutral so as to stabilize the airplane at a constant bank angle. Later, to bring back the airplane to level condition, an aileron pulse is given at 20 seconds in the negative direction, thus creating a positive roll rate. As a result, the airplane loses the bank and comes back to the wings level condition.

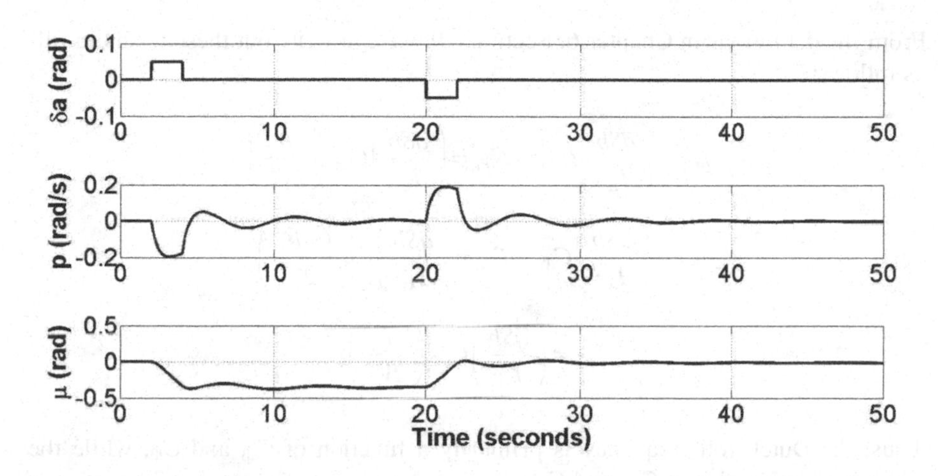

FIGURE 7.18 Roll response to aileron doublet input.

Homework Exercise: Set up a simulation model to study the effect of the aileron doublet input on roll response. Particularly, study the change in response with respect to the time interval between the two pulse inputs.

7.7 DUTCH ROLL MODE

The next mode, slower than the roll mode, is called the Dutch roll mode. We have seen in Chapter 6 that the Dutch roll dynamics is given by a second-order equation in the perturbed sideslip angle, $\Delta\beta$:

$$\begin{aligned}\Delta\ddot{\beta}+\left[-N_{r1}-\left(g/V^{*}\right)Y_{\beta}-\left(g/V^{*}\right)\mathcal{L}_{r1}/\mathcal{L}_{p2}\right]\Delta\dot{\beta}\\ +\left[N_{\beta}+\left(g/V^{*}\right)Y_{\beta}N_{r2}+\left(g/V^{*}\right)\mathcal{L}_{\beta}/\mathcal{L}_{p2}\right]\Delta\beta=0\end{aligned} \tag{7.9}$$

The Dutch roll dynamics is usually oscillatory with the damping and frequency usually given by the following expressions (reproduced from Chapter 6):

$$\omega_{nDR}^{2}=N_{\beta}+\left(\frac{g}{V^{*}}\right)\left[Y_{\beta}N_{r2}+\left(\frac{\mathcal{L}_{\beta}}{\mathcal{L}_{p2}}\right)\right]=\left\{N_{\beta}+\left(\frac{g}{V^{*}}\right)\left(\frac{\mathcal{L}_{\beta}}{\mathcal{L}_{p2}}\right)\right\}+\left(\frac{g}{V^{*}}\right)Y_{\beta}N_{r2} \tag{6.48}$$

$$2\zeta_{DR}\omega_{nDR}=-N_{r1}-\left(\frac{g}{V^{*}}\right)\left[Y_{\beta}+\left(\frac{\mathcal{L}_{r1}}{\mathcal{L}_{p2}}\right)\right]=\left\{-N_{r1}-\left(\frac{g}{V^{*}}\right)\left(\frac{\mathcal{L}_{r1}}{\mathcal{L}_{p2}}\right)\right\}-\left(\frac{g}{V^{*}}\right)Y_{\beta} \tag{6.49}$$

For simplicity, let us ignore the less significant Y_{β} term. Then, from the expressions in Equations 6.48 and 6.49, it appears that:

- The Dutch roll frequency is largely a function of N_{β} and $\mathcal{L}_{\beta}$.
- The Dutch roll damping is mainly dependent on N_{r1} and $\mathcal{L}_{r1}$.

From the definitions in Chapter 6, Section 6.9, we can write out these terms in full as follows:

$$N_\beta = \left(\frac{\bar{q}Sb}{I_{zz}}\right)C_{n\beta}; \quad N_{r1} = \left(\frac{\bar{q}Sb}{I_{zz}}\right)C_{nr1}\left(\frac{b}{2V^*}\right)$$

$$\mathcal{L}_\beta = \left(\frac{\bar{q}Sb}{I_{xx}}\right)C_{l\beta}; \quad \mathcal{L}_{r1} = \left(\frac{\bar{q}Sb}{I_{xx}}\right)C_{lr1}\left(\frac{b}{2V^*}\right)$$

$$\mathcal{L}_{p2} = \left(\frac{\bar{q}Sb}{I_{xx}}\right)C_{lp2}\left(\frac{b}{2V^*}\right)$$

Thus, the Dutch roll frequency is primarily a function of $C_{n\beta}$ and $C_{l\beta}$, while the damping is a function of C_{nr1} and C_{lr1}.

At first glance, one may be tempted to compare the second-order Equation 7.9 in $\Delta\beta$ for the Dutch roll mode with the second-order Equation 2.32 in $\Delta\alpha$ for the short-period mode and draw a parallel; for instance, between the derivatives $C_{n\beta}$ and C_{nr1} in the Dutch roll dynamics and the role played by the derivatives $C_{m\alpha}$ and C_{mq1}, respectively, for the short-period mode. There is some similarity—just as $C_{m\alpha}$ was responsible for short-period (pitch) stiffness, so $C_{n\beta}$ also contributes to Dutch roll (yaw) stiffness, and, the role of C_{nr1} in yaw damping is similar to that of C_{mq1} for pitch damping. However, note that in case of the Dutch roll mode, there are additional terms with the coefficient g/V^* that also contribute to yaw stiffness and damping. This is because of the coupling between the directional and lateral motions—sideslip creates a bank angle (through the $C_{l\beta}$ mechanism that we will study in detail later in this chapter) and a bank angle creates a sideslip by the effect of gravity (that we have already discussed in Chapter 6). Thus, to understand and analyse the Dutch roll dynamics, we need to consider both the directional and lateral motions, which makes it a little more involved than the pure α dynamics of the short period.

Applying the principles of stability for a second-order system, we know that the yaw stiffness (the ω_{nDR}^2 term) must be positive, which, ignoring the term in Y_β, requires:

$$N_\beta + \left(\frac{g}{V^*}\right)\left(\frac{\mathcal{L}_\beta}{\mathcal{L}_{p2}}\right) > 0$$

which, when written out in terms of the aerodynamic derivatives, appears as:

$$C_{n\beta} + \left\{\frac{\left(\frac{g}{V^*}\right)\left(\frac{2V^*}{b}\right)}{\left(\frac{\bar{q}Sb}{I_{zz}}\right)}\right\}\left(C_{l\beta}/C_{lp2}\right) > 0 \tag{7.10}$$

You can verify that the combination of terms in the braces $\{\cdot\}$ has no unit as both the numerator and the denominator have dimension of 1/time^2. In fact, this combination includes all the three timescales of interest to lateral-directional flight: $\frac{T_f^2}{T_s T_r}$. Since T_S ~ 10 seconds, T_f ~ 1 second, T_r ~ 0.1 second, the combination is of the order of 1. Let us use the symbol:

$$\varepsilon = \frac{T_f^2}{T_s T_r} \tag{7.11}$$

for convenience, where $\varepsilon \sim 1$. So, we can re-write Eq. 7.10 as:

$$C_{n\beta} + \varepsilon\left(C_{l\beta}/C_{lp2}\right) > 0 \tag{7.12}$$

as the approximate (because we ignored the Y_β term) condition for positive yaw stiffness. Note that $C_{n\beta} > 0$ contributes to yaw stiffness, but so does $C_{l\beta} < 0$ (since usually $C_{lp2} < 0$).

Along similar lines, we know that for the stability of the Dutch roll mode, yaw damping (the $2\zeta_{DR}\omega_{nDR}$ term) must also be positive, which requires:

$$N_{r1} + \left(\frac{g}{V^*}\right)\left(\frac{\mathcal{L}_{r1}}{\mathcal{L}_{p2}}\right) < 0$$

which, equivalently in terms of the aerodynamic derivatives, appears as:

$$C_{nr1} + \varepsilon\left(C_{lr1}/C_{lp2}\right) < 0 \tag{7.13}$$

Here, $C_{nr1} < 0$ would contribute to yaw damping, and so would $C_{lr1} > 0$ (because C_{lp2} is usually negative).

For a summary of the stability requirements in lateral-directional flight so far, see Table 7.1.

We shall examine these derivatives more closely subsequently in this chapter, but first let us look at the Dutch roll motion.

The Dutch roll motion can be set off by a small perturbation either in the roll angle or in the sideslip. Since a roll causes a sideslip and a sideslip induces a roll, it does not matter which of them is the starting cause, the Dutch roll motion soon shows a combination of both sideslip and roll. An airplane in Dutch roll motion sways

TABLE 7.1
Summary of Lateral-Directional Stability Requirements

Mode	Stability Requirement
Roll	$C_{lp2} < 0$
Dutch roll	$C_{n\beta} + \varepsilon\left(C_{l\beta}/C_{lp2}\right) > 0$ and $C_{nr1} + \varepsilon\left(C_{lr1}/C_{lp2}\right) < 0$

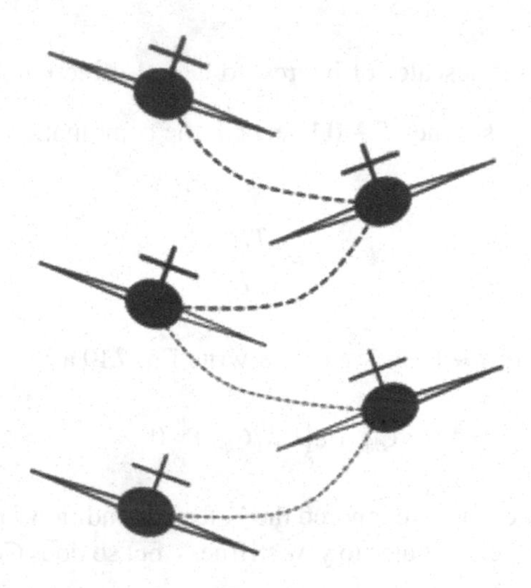

FIGURE 7.19 Dutch roll motion. (Adapted from: http://www.thephysicsofufos.com/kinema2a.jpg)

from side to side (even gracefully), somewhat like an ice skater. See the sequence in Figure 7.19. As the airplane sideslips to the right, it banks left (effect of $C_{l\beta}$) and the nose yaws right (effect of $C_{n\beta}$). Then, it recovers and starts a sideslip to the left—now the right wing falls (effect of $C_{l\beta}$) and the nose yaws left (effect of $C_{n\beta}$). As long as the Dutch roll mode is damped, the motion gradually reduces its amplitude with every cycle of oscillation. Typical requirements for lateral-directional mode parameters are discussed in Box 7.1.

BOX 7.1 LATERAL-DIRECTIONAL HANDLING OR FLYING QUALITIES[1]

The flying qualities of an aircraft are indicators of its performance in terms of stability and control characteristics. Pilots' rating of an airplane (1 for the best and 10 for the worst based on Cooper–Harper scale[1]) indicative of the flying qualities are based on dynamic characteristics of the aircraft and vary from Class or type (I–IV) of the aircraft to Category (A, B and C) of flight phases. Based on the dynamic characteristics of the aircraft in Dutch roll, roll and spiral modes, a pilot's opinion varies between Levels 1–3 as described below. The dynamic characteristics depend upon the frequency, damping and time-to-halve or time-to-double of the aircraft response in lateral-directional modes. For Class I (small light) and IV (high manoeuvrability, such as, fighter) airplanes, in Category A flight phase (non-terminal, rapid manoeuvring, demanding precision tracking and accurate flight-path control), lateral-directional flying qualities requirements are given in Table 7.2.

TABLE 7.2
Lateral-Directional Handling Qualities

	Spiral Mode	Roll Mode	Dutch Roll Mode		
	$T_{2,min}$ (Minimum Time-to-Double)	τ_{max} (Maximum Time Constant)	Min $\zeta\omega_n$ (rad/s)	Min ζ	Min ω_n (rad/s)
Level 1	12 s	1.0	0.35	0.19	1.0
Level 2	12 s	1.4	0.05	0.02	0.4
Level 3	4 s	10.0	—	0.02	0.4

Note: Level 1: Flying qualities adequate to accomplish mission flight phase.
Level 2: Flying qualities adequate to accomplish mission flight phase, but with increased pilot workload and degradation in mission effectiveness or both.
Level 3: Flying qualities such that the aircraft can be controlled safely, but with excessive pilot workload and mission effectiveness is inadequate.

Example 7.7

An unstable Dutch roll mode can lead to a limit cycle oscillatory motion called *wing rock*. An example of wing rock motion for an aircraft is shown via numerical simulation in Figure 7.20. In the plots shown in Figure 7.20, building up of transients in the bank angle ϕ and the sideslip angle β is seen around an equilibrium state representing a longitudinal trim condition, wings level and zero sideslip angle, $\phi^* = \beta^* = 0$. The growth of variables ϕ and β with time around the equilibrium state indicates that the equilibrium state is unstable. As a result, a stable oscillatory motion characterized by cyclic variations predominant in ϕ (roll) and β (sideslip) with constant amplitudes, ±15° and ±2.5°, respectively, and a constant time period of about 18 seconds eventually emerges. The motion appears as airplane rolling in a sequence from left wing down (right wing up) to left wing up (right wing down) and back to left wing down (right wing up), hence termed as *wing rock* motion.

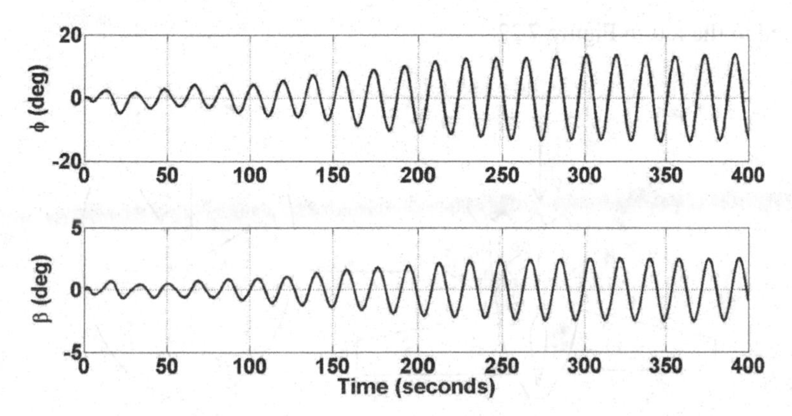

FIGURE 7.20 Wing rock motion.

7.8 DIRECTIONAL DERIVATIVES $C_{Y\beta}$ AND $C_{n\beta}$

In this section, we shall examine two of the three aerodynamic derivatives that influence the yaw stiffness—$C_{Y\beta}$ and $C_{n\beta}$. The third derivative, $C_{l\beta}$, will follow in the next section.

The main contributor to $C_{Y\beta}$ and $C_{n\beta}$ for conventional airplanes is the vertical tail. Consider the top view of an airplane with special reference to the vertical tail section as shown in Figure 7.21. The section of the vertical tail is a symmetric airfoil. When the airplane has a positive sideslip (to the right), there is a component of the relative airflow v coming from the right. When added vectorially to the freestream flow V^*, the resultant velocity vector approaches the vertical tail at an angle $\Delta\beta$, such that:

$$\Delta\beta \approx \upsilon/V^* \tag{7.14}$$

as shown in the inset of Figure 7.21. To the vertical tail airfoil section, the sideslip angle $\Delta\beta$ is effectively an angle of attack, so it responds with a lift, normal to the direction of the relative airflow, and of course a drag along the direction of the relative airflow. The lift and the drag on the vertical tail are also marked in Figure 7.21.

For small sideslip angles, we can assume that the lift and the drag on the vertical tail act in the direction of the body Y^B and X^B axes, respectively, as marked in Figure 7.22, instead of the directions normal and along the relative airflow. The error in making this assumption is usually small and it is quite justified within our study of airplane behavior under small perturbations (from an equilibrium state).

As seen in Figure 7.22, the 'lift' for the vertical tail is actually a side force for the airplane, whereas the drag on the vertical tail (also a drag for the airplane) is not relevant for the discussion in the section. A positive sideslip as marked in Figure 7.22 causes a negative (along the negative Y^B axis) side force. We can write this side force with the correct sign as follows.

Firstly, the 'lift' coefficient of the vertical tail, depending on the vertical tail lift-curve slope $C_{L\alpha_V}$ and the sideslip angle (its 'angle of attack'):

$$C_{LV} = C_{L\alpha_V}\Delta\beta$$

directed to the left in Figure 7.22.

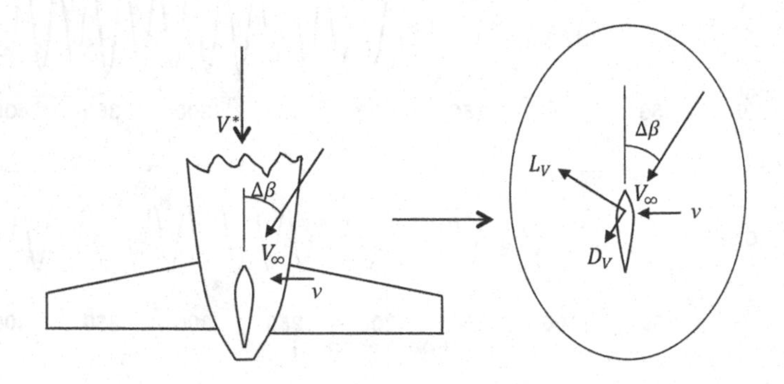

FIGURE 7.21 Lift and drag at the vertical tail due to sideslip.

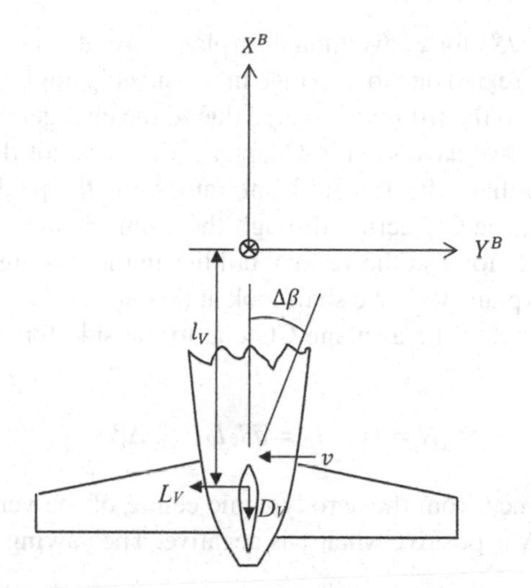

FIGURE 7.22 Lift and drag at the vertical tail due to sideslip approximated along airplane body axes.

So, the 'lift' on the vertical tail is:

$$L_V = \overline{q} S_V C_{LV} = \overline{q} S_V C_{L\alpha_V} \Delta\beta$$

where, S_V is the vertical tail planform area, and we use $\overline{q}$ for the dynamic pressure at the vertical tail as we did for the horizontal tail in Chapter 3.

We know this lift to be the (negative) side force for the airplane, so:

$$Y = -L_V = -\overline{q} S_V C_{L\alpha_V} \Delta\beta \tag{7.15}$$

The airplane side force coefficient due to the vertical tail is:

$$C_Y = Y / (\overline{q} S)$$

where, S now is the reference area we always use for the airplane, the wing planform area. Thus,

$$C_Y = -(S_V / S) C_{L\alpha_V} \Delta\beta$$

From our definition of $C_{Y\beta}$ in Chapter 6:

$$C_{Y\beta} = \left.\frac{\partial C_Y}{\partial \beta}\right|_*$$

It is easy to obtain that:

$$C_{Y\beta} = -(S_V / S) C_{L\alpha_V} \tag{7.16}$$

Typical values of (S_V/S) for conventional airplanes usually lie around 0.2. In that sense, the side force created due to a change in the sideslip angle at the vertical tail is similar in magnitude to the lift force created due to the change of the angle of attack at the horizontal tail. We have seen in Chapter 3 that it is not the lift force created by the horizontal tail that is itself significant, rather it is the pitching moment it can create about the airplane CG acting through the pitch moment arm. Likewise, the significance of the side force at the vertical tail lies in the yawing moment that it can produce about the airplane CG. We shall look at this next.

The yawing moment at the airplane CG due to the side force at the vertical tail can be found as:

$$N = L_V \times l_V = \bar{q} S_V l_V C_{L\alpha V} \Delta\beta$$

where, l_V is the distance from the aerodynamic centre of the vertical tail to the CG of the airplane, and N is positive when Y is negative. The yawing moment coefficient works out to:

$$C_n = N/(\bar{q}Sb) = (S_V l_V/Sb) C_{L\alpha V} \Delta\beta$$

The factor $(S_V l_V/Sb)$ is called the vertical tail volume ratio ($VTVR$) in analogy with the horizontal tail volume ratio ($HTVR$) that we defined for the horizontal tail. Then, the derivative $C_{n\beta}$ is defined as (from Chapter 6):

$$C_{n\beta} = \left.\frac{\partial C_n}{\partial \beta}\right|_*$$

and can be written as:

$$C_{n\beta} = (S_V l_V/Sb) C_{L\alpha V} = VTVR \times C_{L\alpha V} \tag{7.17}$$

Thus, under normal circumstance, for a vertical tail placed behind the airplane CG, we have:

$$C_{Y\beta} < 0, \quad \text{and} \quad C_{n\beta} > 0$$

Thus, a positive $C_{n\beta}$ will contribute to yaw stiffness (that is, keeping ω_{nDR}^2 positive).

Homework Exercise: Can you figure out why a vertical tail placed ahead of the airplane CG will not provide yaw stiffness, though in the case of the pitch dynamics (short-period mode), we could locate the horizontal tail ahead of airplane CG (called canard)?

7.8.1 Other Contributors to Yaw Stiffness

As you can guess, any part of the airplane that provides a vertical component of the surface to the cross-flow (v component of relative velocity) can potentially cause a side force and a consequent yaw moment. The fuselage (wing–body) with its attachments (engines, missiles, fuel tanks, etc.) are all sources of side force and yaw moment. Usually, the wing–body contribution to yaw moment due to sideslip, $C_{n\beta}$, is negative;

that is, it subtracts from the yaw stiffness provided by the vertical tail. This is similar to the contribution due to wing–body to the pitch stiffness that we saw in Chapter 3.

Example 7.8

Case of the Jaguar Airplane

Figure 7.23 shows the side view of the Jaguar. You can notice the large area presented to the cross-flow by the forward section of the fuselage. Another notable feature is the under-fuselage fuel tank. Most of it is ahead of the CG (approximate CG position marked on Figure 7.23). Hence, it subtracts from the yaw stiffness. If there are too many sources of negative $C_{n\beta}$, then the vertical tail needs to provide sufficient positive $C_{n\beta}$ to compensate for those sources. This means a larger *VTVR*—either more tail planform area S_V or a larger tail moment arm l_V. Since a larger l_V would mean a longer fuselage and that would upset too many calculations, it is easier to consider a larger S_V. The question is where to add the extra area to the vertical tail? One can either increase the chord or the span (height). In the case of increased chord, the additional area is closer to the airplane CG, so it is less effective than increasing the span (height). However, this is a common solution seen on many airplanes—it is called the dorsal fin (see Figure 7.24 for an example).

In the case of increased span, one must watch out for aeroelastic effects—the vertical tail may bend and/or twist in flight in such a way that the effective $C_{L\alpha V}$ may be reduced while you try to increase *VTVR*. Do take a look at the expression for $C_{n\beta}$ in Equation 7.17. An alternative could be to use two vertical tails.

Homework Exercise: Search the net or your library for airplanes with twin vertical tail and try to reason why this decision was taken.

Yet another way of providing vertical tail area is to add a vertical tail segment under the fuselage—this is called a ventral fin. There is a small one for the Jaguar in Figure 7.23. A larger ventral fin is found under the MiG-23/27 (see Figure 7.25). Of course, the trouble with the ventral fin is that it can scrape the ground during take-off and landing, so it needs to be folded away as long as the landing gears are down. Yet, there are several airplanes, mostly military, that sport a ventral fin, indicating that the yaw stiffness provided by the usual vertical tail is not adequate at least under certain flight conditions.

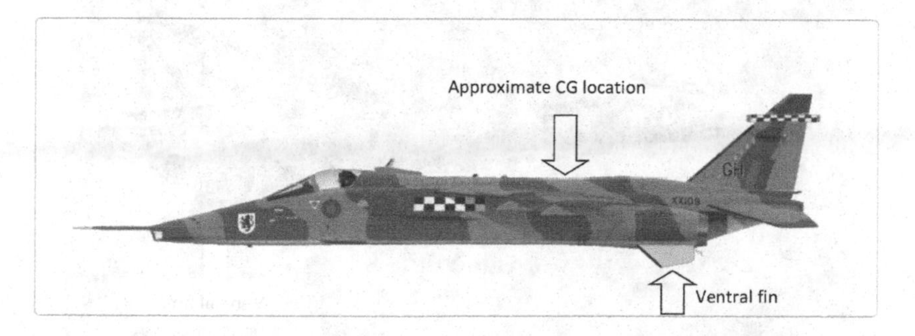

FIGURE 7.23 Ventral fin on Jaguar (https://upload.wikimedia.org/wikipedia/commons/6/67/SEPECAT_Jaguar.png).

FIGURE 7.24 Dorsal fin (https://www.aircraftrecognitionguide.com/pilatus-pc-12).

A look at Equation 7.12 shows that not all the contribution to yaw stiffness need come from $C_{n\beta}$; a negative $C_{l\beta}$ can provide the same effect as well. Earlier, two independent criteria were required to be satisfied for the lateral-directional dynamics—a positive $C_{n\beta}$ for yaw stiffness (called 'directional stability' earlier) and a negative $C_{l\beta}$ for what used to be called 'lateral stability'. Later, it was realized that yaw stiffness could be achieved with a combination of positive $C_{n\beta}$ and negative $C_{l\beta}$. An ad hoc combination, called $C_{n\beta,dyn}$, was proposed for this purpose (see Box 7.2). However, it is now recognized that the independent 'directional stability' and 'lateral stability' requirements are redundant, and that the $C_{n\beta,dyn}$ criterion is not appropriate for yaw stiffness. The correct statement of the yaw stiffness condition is as given by Equation 7.12.

FIGURE 7.25 Ventral fin on MIG-23/27 (https://military.wikia.org/wiki/Mikoyan_MiG-27).

BOX 7.2 $C_{n\beta,dyn}$ CRITERION

The role of $C_{l\beta}$ in providing yaw stiffness was understood some time ago and it was apparent that the condition for positive yaw stiffness should feature both $C_{n\beta}$ and $C_{l\beta}$, but the criterion in Equation 7.12 had not been derived. Instead, a different criterion called $C_{n\beta,dyn}$ (where 'dyn' stands for dynamic) containing both $C_{n\beta}$ and $C_{l\beta}$, as follows, emerged:

$$C_{n\beta,dyn} = C_{n\beta} \cos\alpha - (I_{zz}/I_{xx}) C_{l\beta} \sin\alpha \tag{7.18}$$

The $C_{n\beta,dyn}$ criterion had originally been proposed for a different purpose, but it began to be used as a condition for yaw stiffness. For example, $C_{n\beta,dyn}$ is also plotted in Figure 7.26 along with $C_{n\beta}$. Being a combination of $C_{n\beta}$ and negative $C_{l\beta}$, it behaves in a manner similar to the yaw stiffness condition of Equation 7.12. But, experience showed that the $C_{n\beta,dyn}$ criterion did not work consistently—for some airplanes, it gave a good approximation to the point of loss of yaw stiffness, but for others, it failed. We now know that the criterion in Equation 7.12 is the one that corresponds to the yaw stiffness. $C_{n\beta,dyn}$ still serves the purpose for which it was originally proposed—that of an indicator for wing rock onset. For more about $C_{n\beta,dyn}$ and the wing rock onset criterion, readers may refer to Ref. [2].

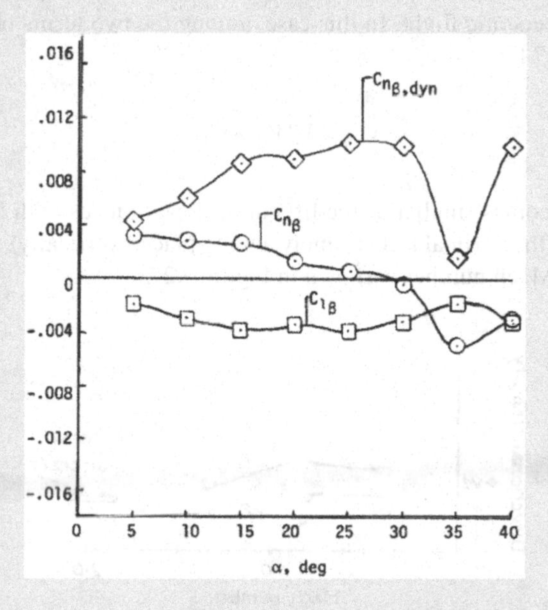

FIGURE 7.26 Variation of $C_{n\beta,dyn}$, $C_{n\beta}$ and $C_{l\beta}$ with angle of attack for an example airplane. (Nguyen, L.T. et al., Simulator Study of Stall/Post-Stall Characteristics of a Fighter Airplane with Relaxed Longitudinal Static Stability, NASA TP-1538, 1979.)

7.8.2 Loss of Vertical Tail Effectiveness

There are two main circumstances under which the vertical tail effectiveness is reduced or even totally lost. The first is in flight at high angles of attack when the fuselage wake may blanket parts (or even all) of the vertical tail. Those parts of the tail are then effectively lost and the effective tail area contributing to *VTVR* reduces; hence, there is a loss in $C_{n\beta}$, which by Equation 7.17 is $VTVR \times C_{L\alpha V}$. This is a common feature of most airplanes—the derivative $C_{n\beta}$ decreases with the angle of attack and eventually falls to zero around 25–35° angle of attack. The variation of $C_{n\beta}$ with the angle of attack for an example airplane is shown in Figure 7.26 in which $C_{n\beta}$ drops down to zero around 30° angle of attack.

In the early days, there was a fear that $C_{n\beta}$ becoming zero would mean loss of yaw stiffness (called 'directional static stability' then). But, we now know that the condition for positive yaw stiffness (even with Y_β neglected) has two terms:

$$C_{n\beta} + \varepsilon\left(C_{l\beta}/C_{lp2}\right) > 0 \tag{7.12}$$

So, even with $C_{n\beta} = 0$, the yaw stiffness condition in Equation 7.12 can be satisfied by a negative value of $C_{l\beta}$ (to be discussed in the next section).

For the example airplane data in Figure 7.26, $C_{l\beta}$ remains negative within the given range of the angle of attack. Thus, the criterion in Equation 7.12 is satisfied even at the point where $C_{n\beta} = 0$.

The other circumstance where the $C_{n\beta}$ contribution from the vertical tail is reduced is in supersonic flight. In this case, among the two terms on the right-hand side of Equation 7.17:

$$C_{n\beta} = VTVR \times C_{L\alpha V} \tag{7.17}$$

It is $C_{L\alpha V}$ that becomes smaller as the lift curve slope reduces with Mach number in supersonic flow (the vertical tail is simply a wing placed vertically). A typical variation of $C_{n\beta}$ with Mach number is shown in Figure 7.27.

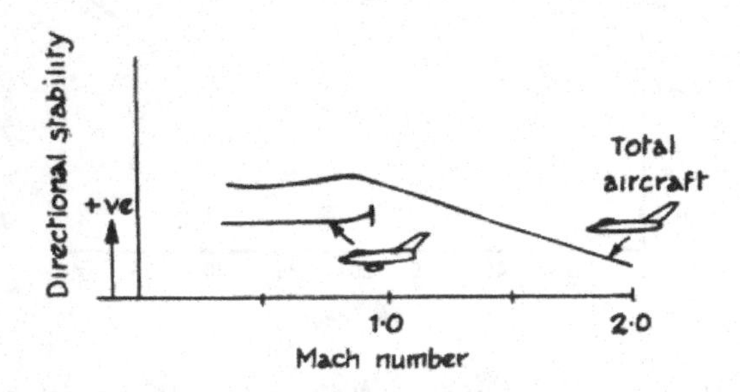

FIGURE 7.27 Variation of $C_{n\beta}$ with Mach number. (From Design for Air Combat by Ray Whitford, Jane's Information Group, 1987.)

7.9 LATERAL DERIVATIVE: $C_{l\beta}$

We have seen in Section 6.4 that an airplane that banks experiences a sideslip because of the effect of gravity that draws the airplane along the direction of the wing that has banked down. Likewise, an airplane that is side-slipping experiences a rolling motion. This is called the *dihedral effect* and is represented by the derivative $C_{l\beta}$, which is defined as:

$$C_{l\beta} = \left.\frac{\partial C_l}{\partial \beta}\right|_* \tag{7.19}$$

the rolling moment coefficient created per degree of the sideslip angle.

Example 7.9

An interesting incident occurred on a long-distance multi-engine airliner which lost power in one of its outer engines in flight. The resulting thrust differential created a yawing moment that induced a sideslip. The dihedral effect then caused a bank, which went unnoticed and continued to build up until the airplane literally fell out of the sky.

The main source of the *dihedral effect* is the wing dihedral, but there are other contributors too. We shall examine these next.

7.9.1 Wing Dihedral

Figure 7.28 shows a schematic of an airplane with: (a) wings inclined upwards, a positive dihedral configuration or simply dihedral, (b) wings with zero dihedral angle and (c) wings inclined downwards or an anhedral configuration. In the following, we develop expressions for roll moment caused due to wing dihedral and thereafter for the dihedral derivative, $C_{l\beta}$. We shall see how the wing dihedral angle contributes to the derivative $C_{l\beta}$.

Let us take the case of a wing configuration with dihedral angle Γ as shown (in the rear view of the airplane) in Figure 7.29. A positive sideslip implies that a relative wind from the right approaches the wing. Assuming the sideslip angle to be small, this side component of velocity is:

$$\upsilon = V \sin\beta \approx V\beta$$

Let us look at the resultant airflow over a section of each wing, as sketched below the right and left wings in Figure 7.29. The forward component of relative wind speed on each airfoil section is the same:

$$u = V \cos\beta \approx V$$

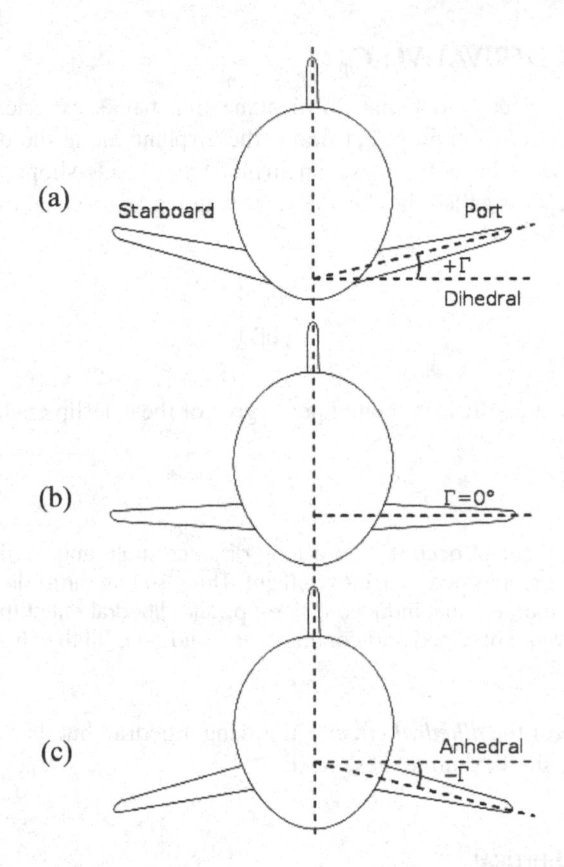

FIGURE 7.28 Wing dihedral (http://en.wikipedia.org/wiki/File:Dihedral_and_anhedral_angle_(aircraft_wing).svg).

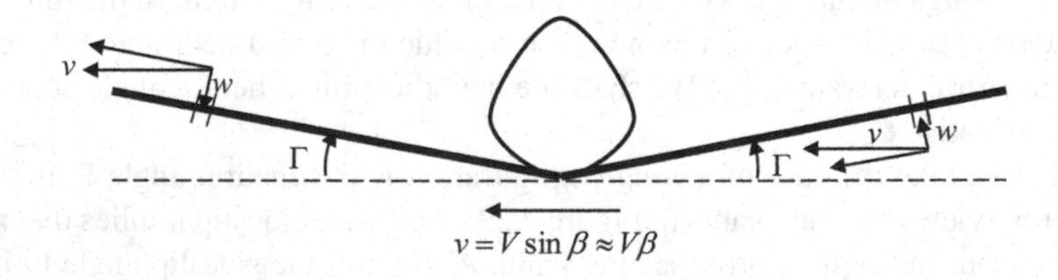

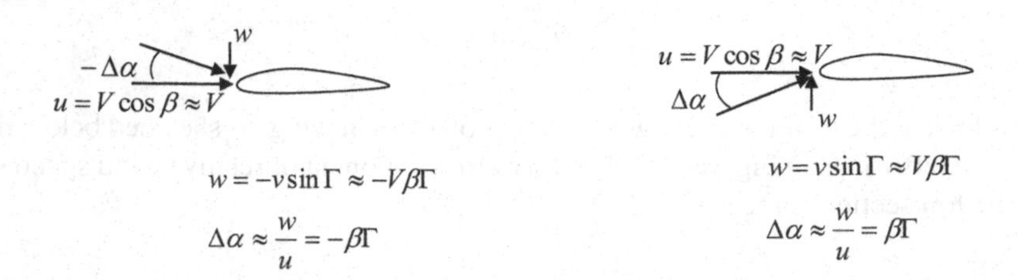

$$w = -v \sin \Gamma \approx -V\beta\Gamma$$

$$\Delta\alpha \approx \frac{w}{u} = -\beta\Gamma$$

$$w = v \sin \Gamma \approx V\beta\Gamma$$

$$\Delta\alpha \approx \frac{w}{u} = \beta\Gamma$$

FIGURE 7.29 Flow over an airplane with dihedral wing in a crosswind condition.

We can resolve the sideways component of the relative wind into two components—one along the plane of the wing and another normal to that plane. Owing to the dihedral angle Γ, the right wing (wind facing side or windward side) sees a normal component of wind given by:

$$w = \upsilon \sin\Gamma \approx \upsilon\Gamma = V\beta\Gamma$$

due to which the airfoil section sees an additional positive angle of attack given by:

$$\Delta\alpha = \frac{w}{u} = \beta\Gamma \tag{7.20}$$

For similar reasons, a wing section on the left (leeward) side of the wing sees a decrease in the angle of attack given by:

$$\Delta\alpha = \frac{w}{u} = -\beta\Gamma$$

This results in an increased lift over the right wing and a decreased lift over the left wing, causing a negative rolling moment given by:

$$\begin{aligned} \mathcal{L} &= -2\int_0^{b/2} \bar{q}\cdot\left(c(y)\cdot dy\right)\cdot C_l(y)\cdot y = -2\int_0^{b/2} q\cdot\left(c(y)\cdot dy\right)\cdot\left(C_{l\alpha}(y)\cdot\Delta\alpha\right)\cdot y \\ &= -2\int_0^{b/2} \bar{q}\cdot\left(c(y)\cdot dy\right)\cdot\left(C_{l\alpha}(y)\cdot\beta\Gamma\right)\cdot y \end{aligned} \tag{7.21}$$

For simplification, we assume that the wing is made up of the same airfoil section along its span, and the chord length is a constant, that is, $c(y) = c$. The airfoil section lift curve slope $C_{l\alpha}$ may also be taken to be a constant. Equation 7.21 then gets simplified to:

$$\bar{q}SbC_l = -2\bar{q}c\cdot C_{l\alpha}\cdot\beta\Gamma\int_0^{b/2} y\, dy = -2\bar{q}c\cdot C_{l\alpha}\cdot\beta\Gamma\cdot\frac{b^2}{8} \tag{7.22}$$

Solving Equation 7.22 for C_l results in (using $S = c.b$):

$$C_l = -C_{l\alpha}\cdot\beta\Gamma\cdot\frac{1}{4}$$

and the derivative $C_{l\beta}$ for the dihedral effect due to wing dihedral may be evaluated as:

$$C_{l\beta} = \frac{\partial C_l}{\partial\beta} = -\frac{C_{l\alpha}\Gamma}{4} \tag{7.23}$$

Homework Exercise: Carry out a similar derivation for $C_{l\beta}$ for a trapezoidal wing with dihedral angle Γ.

For an anhedral angle, we can use negative Γ, a negative dihedral. Equation 7.23 is still valid and in case of an anhedral, $C_{l\beta}$ turns out to be positive. Hence, an anhedral angle detracts from yaw stiffness in Equation 7.12.

Homework Exercise: Can we say anything about the extreme dihedral of configurations such as the one shown in Figure 7.30?

Homework Exercise: Does the wing dihedral contribute to the airplane $C_{n\beta}$ derivative?

7.9.2 Other Sources of $C_{l\beta}$

Other than the above direct effect due to wing dihedral angle, there are several other sources that contribute to the dihedral effect $C_{l\beta}$: significant among them are the wing sweep angle, the wing-fuselage interference and the vertical tail. We go through these contributions to $C_{l\beta}$ individually and develop approximate methods to evaluate them.

FIGURE 7.30 Flying wing configuration (http://www.nasa.gov/centers/dryden/news/ResearchUpdate/Helios/Previews/index_prt.htm).

7.9.2.1 Wing Sweep

Figure 7.31 considers a wing with sweep back having a positive sweep angle, Λ. There is no dihedral angle provided. Imagine this wing placed in a flow with sideslip β; that is, having relative flow coming from the right. Aerodynamic theory tells us that it is the component of relative velocity normal to the wing leading edge that matters as far as lift and drag are concerned. Owing to the combined effects of Λ and β, the right wing sections see a normal component of the relative wind:

$$u_1 = V\cos(\beta - \Lambda) = V\cos(\Lambda - \beta)$$

whereas the normal component of the relative wind on the left wing is given by:

$$u_2 = V\cos(\beta + \Lambda)$$

Typically, Λ is somewhere between 25° and 60° for most airplanes, and β should be less than 5° in most circumstances. As a result, u_1 is usually larger than u_2, so the right wing 'sees' a larger relative (normal) velocity, which means higher lift (also drag) than the left one. This results in a net negative roll moment due to the differential lift and a positive yaw moment due to the differential drag. The rolling moment caused by the change in lift on an airfoil section of chord length $c(y)$ at a distance y from the reference centre line (as marked in Figure 7.31) on the right wing due to sideslip is:

$$\Delta\mathcal{L}_R = -\Delta L_R \cdot y = -\left\{\left(\frac{1}{2}\rho u_1^2\right)\cdot\left(c(y)\cdot dy\right)\cdot c_l(y)\right\}\cdot y$$

$$= -\left(\frac{1}{2}\rho V^2\right)\cos^2(\beta-\Lambda)\cdot c(y)\cdot dy\cdot c_l(y)\cdot y = -\bar{q}\cos^2(\beta-\Lambda)\cdot c(y)\cdot dy\cdot c_l(y)\cdot y$$

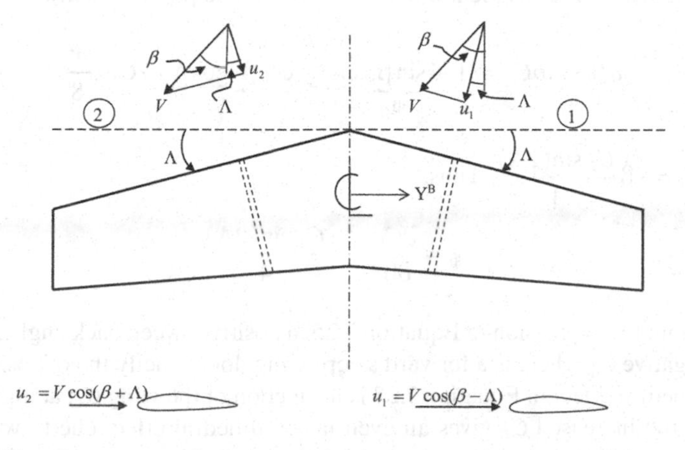

FIGURE 7.31 Components of relative wind velocity in a crosswind condition.

Integration over the whole right wing thus gives us the rolling moment caused by the lift over the right wing as:

$$\mathcal{L}_R = -\bar{q}\cos^2(\beta - \Lambda) \cdot \int_0^{b/2} c(y) \cdot dy \cdot c_l(y) \cdot y$$

Similarly, the rolling moment caused by the lift produced in the presence of sideslip over the left wing can be written as:

$$\mathcal{L}_L = \bar{q}\cos^2(\beta + \Lambda) \cdot \int_0^{b/2} c(y) \cdot dy \cdot c_l(y) \cdot y$$

Therefore, the net change in the rolling moment is given by:

$$\mathcal{L} = \mathcal{L}_L + \mathcal{L}_R = \bar{q}\left[\cos^2(\beta + \Lambda) - \cos^2(\beta - \Lambda)\right] \int_0^{b/2} \left(c(y) \cdot dy\right) \cdot c_l(y) \cdot y \quad (7.24)$$

Assuming identical airfoil sections with a constant lift curve slope over the whole wing span, Equation 7.24 can be rewritten as:

$$\bar{q}SbC_l = \bar{q}\left[\cos^2(\beta + \Lambda) - \cos^2(\beta - \Lambda)\right] c \cdot c_l \cdot \int_0^{b/2} y\, dy$$

$$= \bar{q}\left[\cos^2(\beta + \Lambda) - \cos^2(\beta - \Lambda)\right] c \cdot c_l \cdot \frac{b^2}{8}$$

which simplifies to (assuming the lift coefficient of airfoil sections $c_l = const$ to be same as the wing lift coefficient C_L, $S \approx b \cdot c$ and angle β to be small),

$$\bar{q}(c \cdot b)bC_l = \bar{q}[-4\underbrace{\sin\beta}_{\approx\beta}\ \sin\Lambda\ \underbrace{\cos\beta}_{\approx 1}\ \cos\Lambda]c \cdot C_L \cdot \frac{b^2}{8}$$

That is, $C_l = -\beta\dfrac{C_L \sin(2\Lambda)}{4}$. Thus,

$$C_{l\beta} = \frac{\partial C_l}{\partial \beta} = -\frac{C_L \sin(2\Lambda)}{4} \quad (7.25)$$

As seen from the expression in Equation 7.25, a positive sweep back angle Λ contributes to negative $C_{l\beta}$ whereas a forward swept wing does exactly the opposite.

The dihedral effect in Equation 7.25 is a function of the wing C_L; at higher angles of attack, the increased C_L gives an even larger dihedral effect due to wing sweep back, so much so that sometimes it can be too much. For compensation, the wing may have to be given an anhedral angle so as to keep the value of the $C_{l\beta}$ derivative in check. In contrast, if the increase in the angle of attack takes the wing past stall, then there may be a rapid loss of C_L and the negative $C_{l\beta}$ due to wing sweep may be drastically reduced.

7.9.2.2 Wing Vertical Position on Fuselage

Sketched in Figure 7.32 are three possible positions of the wing on the fuselage. Consider a flat wing (no curvature, no dihedral angle) and a circular or oval fuselage. Imagine the airplane side-slipping to the right (relative cross-flow coming from the right). In each case, the cross-flow has to pass over the right wing; one half each must turn up and down to pass over and under the fuselage, and then go over the left wing, as sketched.

The one in Figure 7.32a is called a mid-wing configuration. For the mid-wing configuration, the up and down components of the cross-flow with respect to each wing is symmetric, and hence there is no change in the forces created on either wing. There is no resultant change in the rolling moment either.

In the second case, a high-wing configuration (Figure 7.32b), the right wing encounters the up-going cross-flow near its root section, whereas around the left-wing root is a down-going cross-flow component. The up flow implies a slight increase in the local angle of attack at the right-wing root region, and likewise a slight decrease in the local angle of attack at the left-wing root due to the down flow. The resultant change in lift causes a negative rolling moment, thus a high-wing configuration contributes to negative $C_{l\beta}$ and supplements the dihedral effect due to wing dihedral. For similar reasons, a low-wing configuration (Figure 7.32c) gives positive $C_{l\beta}$ and detracts from the wing dihedral angle effect. However, note that the wing sections affected are close to the root, and hence the moment arm is not large. Also, the extent of the angle of attack change depends on the shape of the fuselage—vertical oval, circle or horizontal oval.

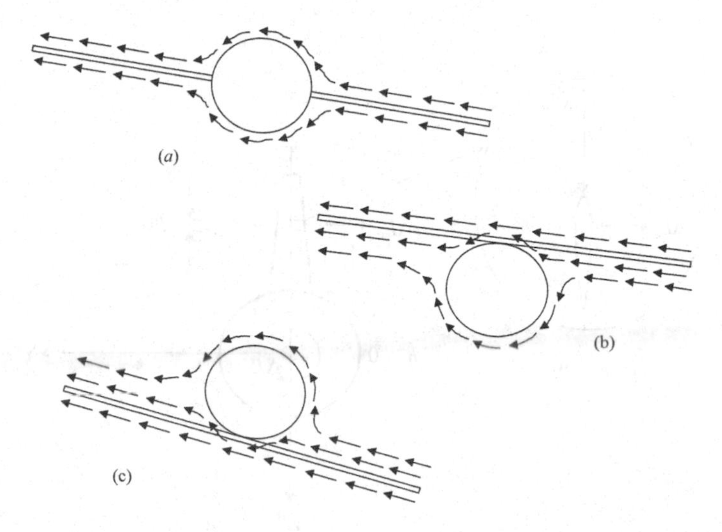

FIGURE 7.32 Rear view of fuselage with: (a) mid-wing, (b) high-wing and (c) low-wing configurations.

7.9.2.3 Vertical Tail

Refer Figure 7.33 for a sketch of the airplane and its vertical tail. We had found the side force created on the vertical tail due to a positive sideslip angle $\Delta\beta$ to be:

$$Y = -\bar{q}S_V C_{L\alpha V}\Delta\beta \tag{7.26}$$

If this side force acts at the vertical tail aerodynamic centre located at a height h_V above the fuselage centre line, it induces a negative (to the left) rolling moment given as:

$$\Delta\mathcal{L} = -\bar{q}S_V C_{L\alpha V}\cdot(\Delta\beta)\cdot h_V$$

The rolling moment coefficient can be obtained as:

$$\Delta C_l = \left(\frac{\Delta\mathcal{L}}{\bar{q}Sb}\right) = -\left(\frac{S_V h_V}{Sb}\right)C_{L\alpha V}\cdot(\Delta\beta) = -\left(\frac{S_V l_V}{Sb}\right)\cdot C_{L\alpha V}\cdot(\Delta\beta)\cdot(h_V/l_V)$$
$$= -VTVR\cdot C_{L\alpha V}\cdot(h_V/l_V)\cdot(\Delta\beta)$$

The derivative for the dihedral effect can then be found as:

$$C_{l\beta} = -VTVR\cdot C_{L\alpha V}\cdot(h_V/l_V) \tag{7.27}$$

It can be seen from Equation 7.27 that a vertical tail over the fuselage contributes a negative component to the dihedral derivative $C_{l\beta}$, while a vertical tail under the fuselage (ventral fin) has the opposite effect.

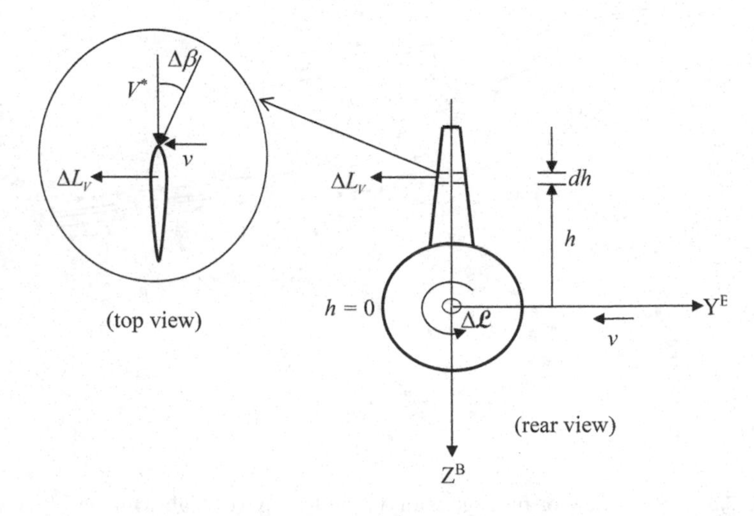

FIGURE 7.33 Rear view of an airplane with positive sideslip (velocity v).

Example 7.10

Why Do Some Airplanes Have Negative Dihedral (Anhedral) Angles?

Airplanes with a high wing (mounted high on the fuselage), with a large sweep back angle, and a tall vertical tail, may have too much (negative) $C_{l\beta}$ because all these contribute to the dihedral effect. Also note from the above discussion that the (negative) contribution of wing sweep to $C_{l\beta}$ increases with the angle of attack. All in all, too much negative $C_{l\beta}$ may give the airplane excessive yaw stiffness making it uncomfortable to fly (see Box 7.1 on handling qualities). There are flight conditions (e.g., landing in a crosswind) where we would like to hold wings level, but too much negative $C_{l\beta}$ can make this task difficult. Also, as we shall see later in this chapter, the same derivatives $C_{n\beta}$ and $C_{l\beta}$ also affect the third lateral-directional flight mode, called the *spiral* mode. So, some airplanes, usually military ones, have wings with a negative dihedral (also called *anhedral*) angle. The anhedral provides positive $C_{l\beta}$, reducing the excessively negative dihedral derivative. An example is the Harrier in Figure 7.34.

Homework Exercise: A parafoil such as the one shown in Figure 1.8 is an anhedral configuration, which should make it build up a bank angle in response to a sideslip. How does it maintain its stability in a crosswind condition? (Consider the effect on stability due to the underslung payload, which makes the CG fall well below the parafoil, in fact usually closer to the payload than the parafoil. This is sometimes called 'pendulum stability'.)

FIGURE 7.34 Wing anhedral on Harrier (http://4.bp.blogspot.com/–lkTiZpc400/T5TH4wEYTdI/AAAAAAAACJI/QQCzbEL7cG8/s1600/AV+8B+Harrier+II+Aircraft+1.jpg).

7.10 DAMPING DERIVATIVES: C_{nr1} AND C_{lr1}

Next, we study the derivatives C_{nr1} and C_{lr1} that together decide the yaw damping, according to Equation 7.13:

$$C_{nr1} + \varepsilon\left(C_{lr1}/C_{lp2}\right) < 0 \tag{7.28}$$

The derivatives C_{nr1} and C_{lr1} are defined (from Chapter 6) as follows:

$$C_{lr1} = \left.\frac{\partial C_l}{\partial\left(r_b - r_w\right)\left(b/2V^*\right)}\right|_* ; \quad C_{nr1} = \left.\frac{\partial C_n}{\partial\left(r_b - r_w\right)\left(b/2V^*\right)}\right|_* \tag{7.29}$$

They are the result of the difference between the body-axis yaw rate r_b and the wind-axis yaw rate r_w. For simplicity, let us consider the scenario in Figure 7.35 where the airplane continues to fly along a straight line path—along the direction marked by the velocity vector V^*, hence, $r_w = 0$. Let the airplane have a body-axis yaw rate r_b, positive as marked in Figure 7.35.

There are two main sources for these damping derivatives—the wing and the vertical tail.

7.10.1 WING CONTRIBUTION TO C_{nr1} AND C_{lr1}

Now consider a pair of stations on the wings at distance $+y$ (right wing) and $-y$ (left wing). The yaw rate r_b induces a forward velocity at the stations $\pm y$ as follows:

$$u = \mp r_b y \tag{7.30}$$

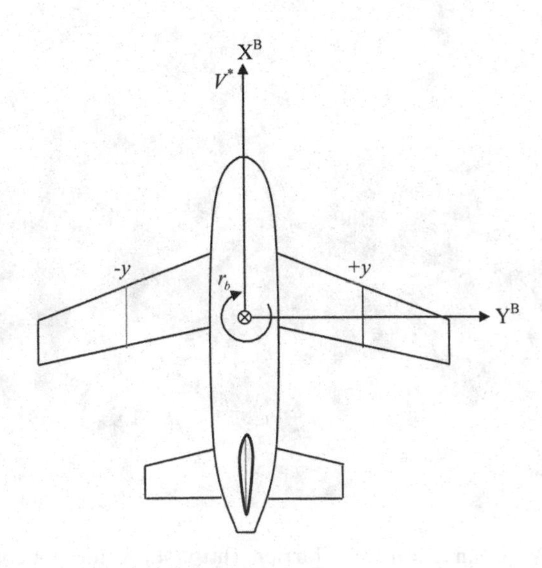

FIGURE 7.35 Airplane with a positive body-axis yaw rate r_b.

which is positive (along V^*) on the left wing and negative on the right wing. As a result, the effective relative wind velocity seen by these stations at $\pm y$ are as follows:

$$V = V^* + u = V^* \mp r_b y \tag{7.31}$$

Thus, wing sections on the left wing see a higher relative velocity and those on the right wing see a lower value of relative wind velocity. Since the aerodynamic forces on a wing section are a function of the relative wind velocity, the left wing sections have increased lift and drag, proportional to:

$$(V)^2 = \left(V^* + r_b y\right)^2 \tag{7.32}$$

whereas those on the right wing have a decreased lift and drag by:

$$(V)^2 = \left(V^* - r_b y\right)^2 \tag{7.33}$$

This difference in lift causes a rolling moment to the right (positive) and a similar difference in drag causes a yawing moment to the left (negative), in case of a positive (to right) yaw rate. Hence, the usual signs of the derivatives due to the wing are:

$$C_{nr1} < 0 \text{ and } C_{lr1} > 0 \tag{7.34}$$

Since C_{lp2} is usually negative, we can see from Equation 7.28 that both these derivatives with their usual signs contribute to yaw damping.

We can formally work out expressions for the wing contribution to C_{nr1} and C_{lr1} as follows.

A simple expression for the rolling moment due to the reduced lift on the right wing and increased lift on the left wing can be arrived at as:

$$\mathcal{L} = \frac{1}{2}\rho c_l c \int_0^{b/2} \left[\left(V^* + r_b y\right)^2 - \left(V^* - r_b y\right)^2\right] y \, dy$$

$$= \frac{1}{2}\rho c_l c \int_0^{b/2} \left[\left(V^{*2} + 2V^* r_b y + r_b^2 y^2\right) - \left(V^{*2} - 2V^* r_b y + r_b^2 y^2\right)\right] y \, dy$$

$$= \frac{1}{2}\rho c_l c\left(4V^* r_b\right) \int_0^{\frac{b}{2}} y^2 \, dy = \frac{1}{2}\rho c_l c\left(4V^* r_b\right)\left[\frac{y^3}{3}\right]_0^{\frac{b}{2}} = \frac{1}{2}\rho c_l c\left(4V^* r_b\right) \cdot \frac{b^3}{24}$$

assuming constant chord length c along the span of the wing and constant lift coefficient c_l of each section of the wing (a rectangular wing configuration).

Writing the roll moment in terms of the coefficient C_l, and assuming the wing lift coefficient C_L to be the same as the section lift coefficient c_l,

$$\frac{1}{2}\rho V^{*2} SbC_l = \frac{1}{2}\rho C_L \cdot c \cdot V^* \cdot r_b \cdot \frac{b^3}{6}$$

$$\frac{1}{2}\rho V^{*2}(c\cdot b)bC_l = \frac{1}{2}\rho C_L \cdot c \cdot V^{*2} \cdot \frac{b}{2V^*} \cdot r_b \cdot \frac{b^2}{3}$$

That is, $C_l = \frac{C_L}{3} r_b \frac{b}{2V^*}$, giving:

$$C_{lr1} = \frac{\partial C_l}{\partial\left(\frac{r_b b}{2V^*}\right)} = \frac{C_L}{3} \tag{7.35}$$

using Equation (7.29) with $r_w = 0$.

Homework Exercise: Similarly, derive an expression of C_{nr1} for a straight rectangular wing as $C_{nr1} = -(C_D/3)$. Show that these results hold for the general case of non-zero r_w.

7.10.2 Vertical Tail Contribution to C_{nr1} and C_{lr1}

A second source of the derivatives C_{nr1} and C_{lr1} is the vertical tail. Consider an identical scenario as in Figure 7.35, but focusing on the vertical tail as shown in Figure 7.36.

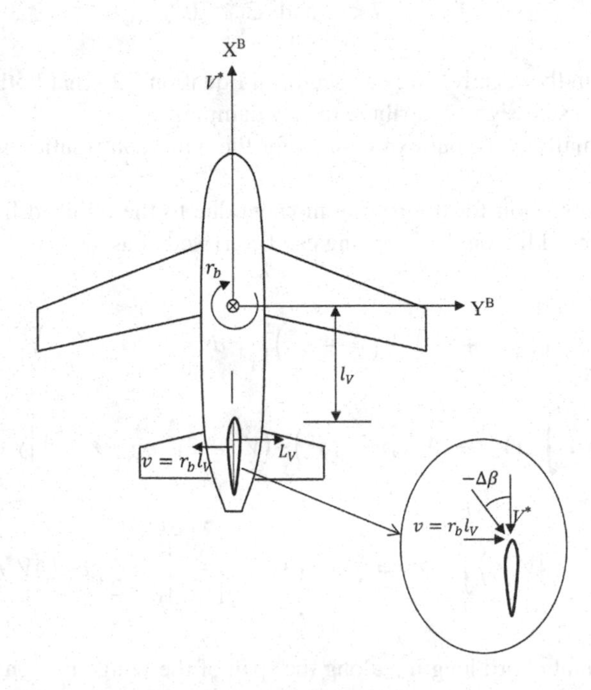

FIGURE 7.36 Sideways relative wind on vertical tail due to yaw rate r_b.

Here, the positive yaw rate r_b creates a sideways velocity for the vertical tail:

$$\upsilon = -r_b l_V \tag{7.36}$$

at its aerodynamic centre. This is equivalent to an induced wind velocity $\upsilon = r_b l_V$ at the vertical tail as shown in the inset to Figure 7.36. From the point of view of the vertical tail, the relative wind velocity is the vector sum of the freestream velocity V^* and this induced sideways velocity. The resultant of these two vectors, as marked in Figure 7.36, creates a sideslip angle given by:

$$\Delta\beta = -r_b l_V / V^* \tag{7.37}$$

which is negative when r_b is positive. This creates a positive side force (to the right) on the vertical tail as follows:

$$\begin{aligned} L_V &= \bar{q} S_V C_{L\alpha V} \Delta\alpha = \bar{q} S_V C_{L\alpha V} \Delta\beta = \bar{q} S_V C_{L\alpha V} \left(-r_b l_V / V^*\right) \\ Y &= \bar{q} S_V C_{L\alpha V} \left(r_b l_V / V^*\right) \end{aligned} \tag{7.38}$$

If the vertical tail aerodynamic centre is at a height h_V above the airplane X^B axis (approximately the centre line), then it will induce a positive rolling moment (to the right) as:

$$\mathcal{L} = \bar{q} S_V C_{L\alpha V} \left(r_b l_V / V^*\right) h_V \tag{7.39}$$

Thus, once again a positive yaw rate r_b induces a positive rolling moment, giving $C_{lr1} > 0$. That is, the vertical tail contribution is in the same sense as that of the wing and it also provides yaw damping.

To derive this effect formally, the rolling moment coefficient due to the vertical tail is:

$$C_l = (S_V/S) C_{L\alpha V} \left(r_b l_V / V^*\right)(h_V/b) = (S_V/S) C_{L\alpha V} \left(r_b b / 2V^*\right)(2 l_V / b)(h_V / b)$$

And hence using the formula for C_{lr1} in Equation 7.29, with $r_w = 0$:

$$\begin{aligned} C_{lr1} &= (S_V/S) C_{L\alpha V} (2 l_V / b)(h_V / b) = \left[(S_V/S)(l_V/b)\right] C_{L\alpha V} (2 h_V / b) \\ &= VTVR \cdot C_{L\alpha V} (2 h_V / b) \end{aligned} \tag{7.40}$$

which is obviously positive provided h_V is positive (i.e., vertical tail aerodynamic centre is above the fuselage centre line). Notice that a ventral fin would give C_{lr1} of the opposite sign, so even though it may help in yaw stiffness, it detracts from yaw damping as far as the C_{lr1} term in Equation 7.28 is concerned.

Going back to Equation 7.38, the side force also creates a yawing moment about the airplane CG and this can be found as:

$$N = -\bar{q} S_V C_{L\alpha V} \left(r_b l_V / V^*\right) l_V \tag{7.41}$$

A negative yawing moment for a positive yaw rate r_b implies that the derivative $C_{nr1} < 0$, which again is exactly in the same sense as the wing contribution. Thus, the vertical tail aids the wing in providing yaw damping.

The expression for C_{nr1} due to the vertical tail can be arrived at as follows:

$$C_n = -(S_V/S)C_{L\alpha V}\left(r_b l_V/V^*\right)(l_V/b) = -(S_V/S)C_{L\alpha V}\left(r_b b/2V^*\right)(2l_V/b)(l_V/b)$$

Applying the formula for C_{nr1} in Equation 7.29, with $r_w = 0$:

$$C_{nr1} = -\left(\frac{S_V}{S}\times\frac{l_V}{b}\right)C_{L\alpha V}\left(\frac{2l_V}{b}\right) = -VTVR\cdot C_{L\alpha V}\left(2l_V/b\right) \tag{7.42}$$

which clearly shows that $C_{nr1} < 0$ for a vertical tail located behind the airplane CG. Interestingly, a ventral fin also provides C_{nr1} in the same sense as a regular vertical tail (or a dorsal fin).

7.11 RUDDER CONTROL

The rudder, as we have seen in Section 7.4, is a small flap hinged to the rear of the vertical tail. A rudder deflected to the left as in Figure 7.17 is taken to be a positive deflection. A left rudder deflection creates a side force to the right (positive) as shown in Figure 7.17. The positive side force can then create a negative yawing moment about the airplane CG and (usually) a positive rolling moment about the X^B axis (fuselage centre line).

Thus, we can define three rudder derivatives, with the usual signs as follows:

$$C_{Y\delta r} = \left.\frac{\partial C_Y}{\partial \delta r}\right|_* > 0,\quad C_{l\delta r} = \left.\frac{\partial C_l}{\partial \delta r}\right|_* > 0,\quad C_{n\delta r} = \left.\frac{\partial C_n}{\partial \delta r}\right|_* < 0 \tag{7.43}$$

where * refers to the trim state at which these partial derivatives are evaluated.

To estimate these derivatives, we can model the effect of a deflected rudder in a manner similar to that adopted earlier. As in the case of the aileron, the flap effectiveness is measured by a parameter τ:

$$\tau = \frac{\Delta\beta}{\Delta\delta r} = \frac{d\beta}{d\delta r} \tag{7.44}$$

which indicates the effective change in sideslip $\Delta\beta$ of the vertical tail airfoil section for a trailing edge rudder flap deflection $\Delta\delta r$. Let us just assume that the rudder runs over the entire span (height) of the vertical tail.

Then, the 'lift' of the vertical tail due to a positive (to the left) rudder deflection $\Delta\delta r$ is:

$$\Delta C_{LV} = C_{L\alpha V}\Delta\beta = C_{L\alpha V}\,\tau\Delta\delta r$$

which is actually a (positive) side force for the airplane. The side force and resultant rolling and yawing moments come out to be:

$$\begin{aligned}\Delta Y &= \bar{q}S_V \Delta C_{LV} = \bar{q}S_V C_{L\alpha V} \tau \Delta\delta r \\ \Delta \mathcal{L} &= \bar{q}S_V C_{L\alpha V} \tau \Delta\delta r \cdot h_V \\ \Delta N &= -\bar{q}S_V C_{L\alpha V} \tau \Delta\delta r \cdot l_V\end{aligned} \tag{7.45}$$

where h_V is as marked in Figure 7.33 and l_V as in Figure 7.36.

The corresponding non-dimensional coefficients are:

$$\begin{aligned}C_Y &= (S_V/S) C_{L\alpha V} \tau \Delta\delta r \\ C_l &= \left(\frac{S_V}{S} \times \frac{l_V}{b}\right) . C_{L\alpha V} \tau \Delta\delta r \cdot (h_V/l_V) = VTVR \cdot C_{L\alpha V} \tau \Delta\delta r \cdot (h_V/l_V) \\ C_n &= -\left(\frac{S_V}{S} \times \frac{l_V}{b}\right) C_{L\alpha V} \tau \Delta\delta r = -VTVR \cdot C_{L\alpha V} \tau \Delta\delta r\end{aligned} \tag{7.46}$$

The derivatives in Equation 7.43 can then be evaluated to be:

$$\begin{aligned}C_{Y\delta r} &= (S_V/S) C_{L\alpha V} \tau > 0 \\ C_{l\delta r} &= VTVR \cdot C_{L\alpha V} \tau \left(\frac{h_V}{l_V}\right) > 0 \\ C_{n\delta r} &= -VTVR \cdot C_{L\alpha V} \tau < 0\end{aligned} \tag{7.47}$$

Of these, typically, only $C_{n\delta r}$ is really significant enough to be considered for further study in this text, though the others are not always completely negligible. By including the effect of $C_{n\delta r}$, we can rewrite the Dutch roll mode dynamics Equation 7.9 with the rudder forcing as:

$$\begin{aligned}&\Delta\ddot{\beta} + \left[-N_{r1} - (g/V^*) Y_\beta - (g/V^*) \mathcal{L}_{r1}/\mathcal{L}_{p2}\right] \Delta\dot{\beta} \\ &+ \left[N_\beta + (g/V^*) Y_\beta N_{r2} + (g/V^*) \mathcal{L}_\beta/\mathcal{L}_{p2}\right] \Delta\beta + N_{\delta r} \Delta\delta r = 0\end{aligned} \tag{7.48}$$

where the short symbol $N_{\delta r}$ stands for:

$$N_{\delta r} = (\bar{q}Sb/I_{zz}) C_{n\delta r} \tag{7.49}$$

In reality, we have seen that the yaw and the roll modes are coupled, and just as $\mathcal{L}_\beta$ and Y_β terms accompany N_β in the yaw stiffness term, with a proper derivation, one would expect $\mathcal{L}_{\delta r}$ and $Y_{\delta r}$ to appear with $N_{\delta r}$ beside the $\Delta\delta r$ term in Equation 7.48. We choose to ignore them on the understanding that the $\mathcal{L}_{\delta r}$ and $Y_{\delta r}$ effects are comparatively negligible, and this is usually justified for most conventional airplane configurations.

From Equation 7.48, we can see that a rudder input can excite the Dutch roll mode, and this is usually how it is excited during a flight test. But, unlike the longitudinal case, where an elevator input was used to initiate a pitch manoeuvre, there is almost no need for a yaw manoeuvre. The standard lateral-directional manoeuvre is a bank (and sometimes a continuous roll, for military airplanes), and usually the only concern as far as yaw is concerned is to hold a zero sideslip angle. So, the rudder is usually used in conjunction with the aileron to maintain zero sideslip in a lateral manoeuvre.

However, there are a few rare circumstances when the rudder is required to maintain a non-zero sideslip angle; these are called 'rudder trims'.

7.11.1 Crosswind Landing

Imagine an airplane coming in to land on a runway with the wind blowing from the right. As pictured in Figure 7.37, if the airplane's nose points along the runway, then there is a positive sideslip angle $\Delta\beta$ created due to the wind direction. The airplane's yaw stiffness will try to 'kill' this sideslip, that is, it will try to rotate the airplane nose into the wind so that $\Delta\beta = 0$. However, that will mean that the airplane is no longer aligned with the runway. To make a runway-aligned landing, the pilot must maintain the sideslip angle and resist the airplane's attempt to yaw into the wind by using the rudder.

Thus, the airplane must trim in a descending flight with wings level (zero bank angle) and a fixed sideslip angle depending on the extent of crosswind. It is easy to calculate the amount of rudder deflection needed to maintain this trim—the rudder moment needs to balance out the yaw stiffness moment that tries to realign the airplane's nose. From Equation 7.48, setting all time derivatives to zero at trim, we get:

$$\Delta\delta r = -\left[N_\beta + \left(g/V^*\right)Y_\beta N_{r2} + \left(g/V^*\right)\mathcal{L}_\beta/\mathcal{L}_{p2}\right]/N_{\delta r} * \Delta\beta \tag{7.50}$$

as the rudder deflection required to make a crosswind landing with sideslip.

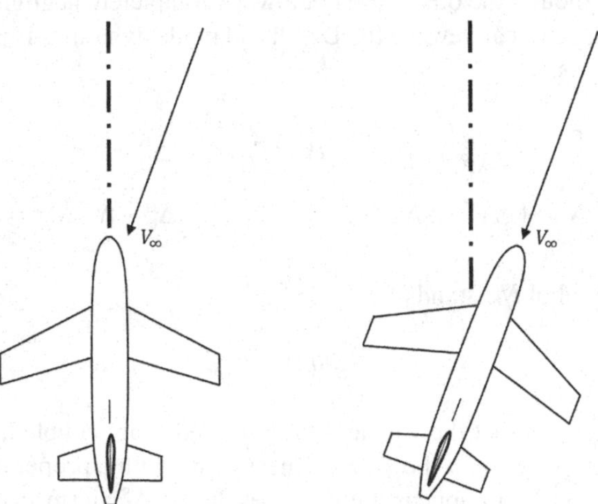

FIGURE 7.37 Landing in crosswind.

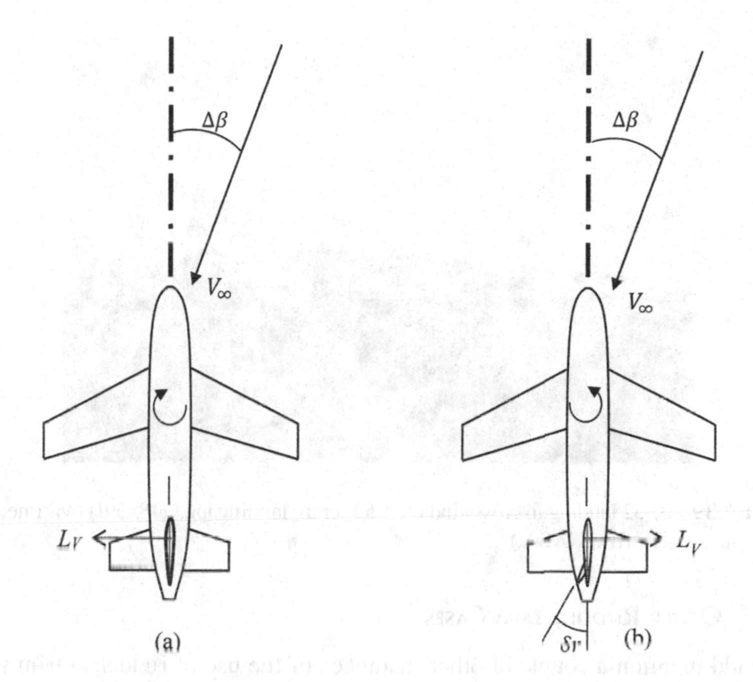

FIGURE 7.38 Vertical tail lift and yawing moment: (a) due to vertical tail alone and (b) due to left rudder deflection alone.

One should look at the forces acting on the vertical tail to understand this better (refer Figure 7.38). The vertical tail is 'lifting' due to two reasons—the constant positive sideslip causes a side force to the left acting at the tail aerodynamic centre (which is the one creating a positive yawing moment turning the airplane into the wind) and the left rudder produces a side force to the right, again acting at the tail aerodynamic centre (which must create an equal and opposite yawing moment). Equation 7.50 is a mathematical statement of these two yawing moments balancing each other, so that the airplane has no net yaw rate. At the vertical tail aerodynamic centre, the two side forces (due to sideslip and rudder, respectively) also cancel each other and do not create a net rolling moment either.

However, we know from Section 7.10 that the wing can create a rolling moment due to the $C_{l\beta}$ effect—a positive sideslip results in a roll to the left. Hence, when trimming at non-zero sideslip, the roll control (ailerons or spoilers) must also be used to null the net roll moment and maintain wings level. See Exercise Problem 7.11.

Example 7.11

The B-52 is shown in Figure 7.39 landing in a crosswind. While the aircraft wheels on ground are moving along the correct path on the runway, the aircraft body is oriented at an angle with respect to the runway. See Exercise Problem 7.1 for more.

FIGURE 7.39 B-52 landing in crosswind (XB-52_crab_landing.jpg (748×591) (wikimedia.org). (Photo courtesy: Loftin, NASA.)

7.11.2 Other Rudder Trim Cases

We should mention a couple of other instances of the use of rudder to trim the airplane. One of these is what is called a 'one engine inoperative' condition. This occurs when one of the engines in a multi-engine airplane either loses power or needs to be shut off due to a malfunction. With the other engine(s) still producing thrust, a net yawing moment is created about the airplane CG, which yaws the airplane and creates a sideslip. The rudder needs to create an opposing yawing moment to kill the yaw rate and maintain zero sideslip. Understandably, this requirement is more severe when one of the outer engines in a four-engine airplane fails and the airplane is in ascent after take-off when all the other engines are close to full power and the speed is still low so that the yawing moment produced by the rudder deflection is limited by the value of the dynamic pressure, $\overline{q}$.

Another example of the use of rudder to trim is to counteract the effect of propeller slipstream, especially on single-engine airplanes. The propeller slipstream passes over the fuselage and hits the vertical tail. The asymmetry in the flow field due to the slipstream induces a side force on the vertical tail. If left unchecked, it can cause an unrestricted yawing and rolling moment. Hence, the rudder has to be set such that the side force due to slipstream effects is cancelled and the airplane can fly in wings level, straight flight trim.

7.12 SPIRAL MODE

The last and the slowest of the lateral-directional modes is called the spiral mode and its dynamics is given by the following equation that we derived in Chapter 6:

$$\Delta\dot{\mu} = \left(\frac{g}{V^*}\right)\left\{\frac{\left[\mathcal{L}_\beta N_{r2} - N_\beta \mathcal{L}_{r2}\right]}{\lambda_r \omega_{nDR}^2}\right\}\Delta\mu \tag{6.55}$$

As you can see, it is a first-order dynamics in the bank angle $\Delta\mu$. When disturbed from a level flight trim state into a spiral mode, the airplane slowly banks to one side. One effect of this bank is to create a centripetal acceleration due to the $L \sin \Delta\mu$ component of the lift, as we pictured in Chapter 6 (Figure 6.3). This causes the airplane to also turn in the same sense, that is, it develops a $\Delta\chi$ at the same slow timescale T_s as the change in $\Delta\mu$. We have also seen in Chapter 6 (Section 6.4) that a bank angle causes the airplane to sideslip due to the effect of gravity, but that motion occurs at the faster Dutch roll timescale T_f, and assuming the Dutch roll dynamics to be stable, the sideslip will be killed at a much faster rate than the one at which $\Delta\mu$ and $\Delta\chi$ grow. Thus, the faster variables, roll rate (at timescale T_r) as well as the sideslip (at timescale T_f) are both quite small during the spiral motion and may be ignored.

A typical sketch of the airplane flight path in an unstable spiral motion is shown in Figure 7.40. A departure from the initial straight line path into a spiral downward trajectory can be observed from the plots. As the aircraft descends in a helical trajectory, the radius of turn decreases with both bank angle and yaw angle increasing monotonically.

Following the analysis of first-order dynamical systems in Chapter 2, we know that the condition for the spiral mode to be stable is:

$$\left(\frac{g}{V^*}\right)\left\{\frac{\left[\mathcal{L}_\beta N_{r2} - N_\beta \mathcal{L}_{r2}\right]}{\lambda_r \omega_{nDR}^2}\right\} < 0$$

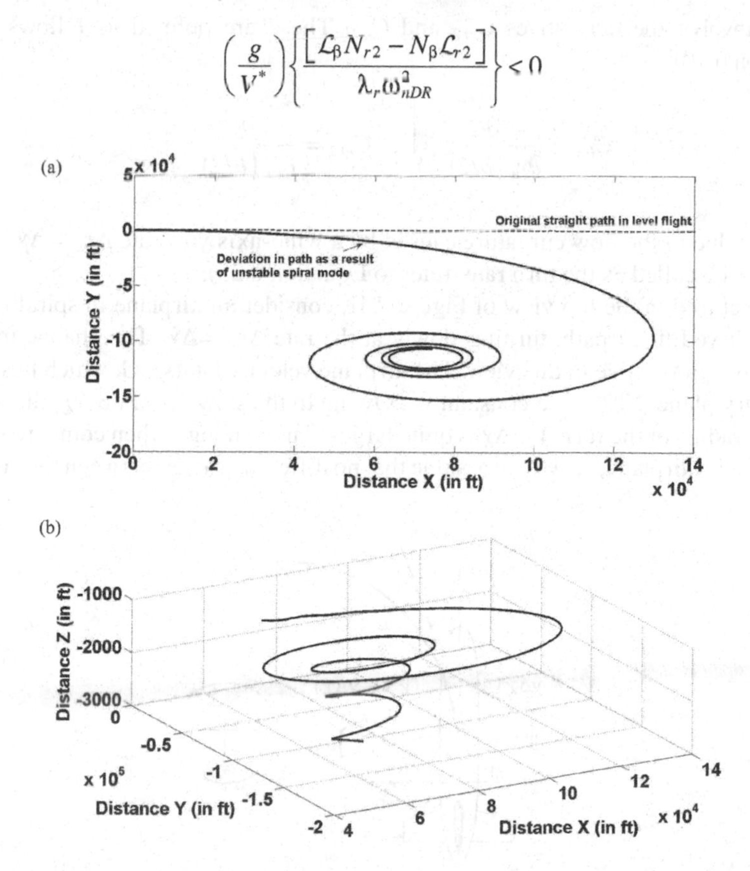

FIGURE 7.40 (a) Top view of a simulated spiral divergent path for F-18/HARV aircraft model, and (b) the three-dimensional view of the trajectory.

Since $\left(\frac{g}{V^*}\right) > 0$, and given $\lambda_r < 0$ and $\omega_{nDR}^2 > 0$, it reduces to:

$$\left[\mathcal{L}_\beta N_{r2} - N_\beta \mathcal{L}_{r2}\right] > 0 \tag{7.51}$$

To determine the sign of the expression in Equation 7.51, we need to figure out the signs of all the terms in it. Of these, we have already met $\mathcal{L}_\beta$ (usually negative) and N_β (usually positive). Now, we check out the other two terms that are derivatives with r_2.

7.12.1 C_{nr2} AND C_{lr2} DERIVATIVES

The N_{r2} and $\mathcal{L}_{r2}$ short symbols from Chapter 6 stand for the following:

$$N_{r2} = \left(\frac{\bar{q}Sb}{I_{zz}}\right) C_{nr2}\left(\frac{b}{2V^*}\right); \quad \mathcal{L}_{r2} = \left(\frac{\bar{q}Sb}{I_{xx}}\right) C_{lr2}\left(\frac{b}{2V^*}\right)$$

which involve the derivatives C_{nr2} and C_{lr2}. These are defined as follows (from Equation 6.20):

$$C_{lr2} = \left.\frac{\partial C_l}{\partial r_w\left(b/2V^*\right)}\right|_*; \quad C_{nr2} = \left.\frac{\partial C_n}{\partial r_w\left(b/2V^*\right)}\right|_*$$

Both are due to the flow curvature caused by a wind-axis yaw rate $\Delta r_w = \Delta\dot{\chi}$, which may also be called as the turn rate (refer to Equation 6.25).

As pictured in the top view of Figure 7.41, consider an airplane in spiral motion with a curved flight path, turning slowly at the rate $\Delta r_w = \Delta\dot{\chi}$. The change in bank angle $\Delta\mu$ is not visible in this view. The airplane velocity (at its CG, which lies in the symmetry plane, $X^B Z^B$) is a constant V^*. Owing to the slow turn rate $\Delta\dot{\chi}$, the instantaneous radius of the turn, $V^*/\Delta\dot{\chi}$ is quite large—large enough when compared to the length of the airplane, so we can assume that no difference is felt between the airplane

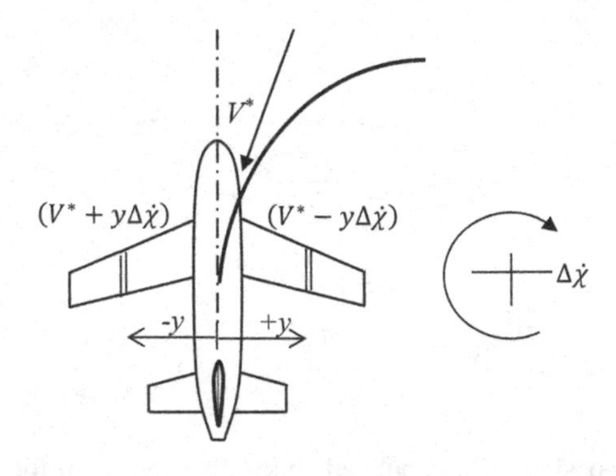

FIGURE 7.41 Relative wind on the two sides of the wing of an airplane in a curved flight path.

nose and its tail due to the flow curvature effect. However, for the turn rate to be the same across the span of the wing, the sections of the wing on the inside of the turn (right wing in Figure 7.41) must be travelling at a slower speed $(V^* - y\Delta\dot{\chi})$, where y is the coordinate along the Y^B axis as marked in Figure 7.41. Similarly, sections of the left wing, on the outside of the turn, must be travelling faster at $(V^* + y\Delta\dot{\chi})$. Since the aerodynamic forces are proportional to the square of the relative wind velocity, the left-wing sections experience a higher lift and drag than the right-wing sections whose lift and drag have reduced. In this particular case of a turn to the right, this causes a yaw to the left (due to the differential drag) and roll to the right (due to the differential lift). So, the derivatives have the following signs:

$$C_{nr2} < 0 \text{ and } C_{lr2} > 0$$

Effectively, the velocity difference generating the derivatives C_{nr2} and C_{lr2} above is very similar to that creating the C_{nr1} and C_{lr1} derivatives in Section 7.10.1 due to the wing. However, note that the mechanism is different in the two cases—the point about which the rotation or turn takes place is different. As a result, in the case of the derivatives C_{nr1} and C_{lr1}, there is another component that arises due to the action of the body-axis yaw rate Δr_b minus the wind-axis yaw rate Δr_w in creating a cross-flow at the vertical tail. In contrast, the wind-axis yaw rate $\Delta r_w = \Delta\dot{\chi}$ by itself does not create any effect at the vertical tail; so, there is no additional effect on the derivatives C_{nr2} and C_{lr2} due to the vertical tail. Hence, the derivatives C_{nr2} and C_{lr2} will be numerically different from C_{nr1} and C_{lr1} due to the absence of the vertical tail contribution.

Besides their numerical values (which are different, as we have just seen), the context of their usage and the factor they multiply with, in the equations, are also different. The derivatives C_{nr1} and C_{lr1} relate to the yaw damping of the Dutch roll motion where they enter Equation 7.9 multiplied by $(\Delta r_b - \Delta r_w) = \Delta\dot{\beta}$. In contrast, C_{nr2} and C_{lr2} relate to the spiral mode time constant and in deriving Equation 6.55 for the spiral mode they originally multiplied with $\Delta r_w = \Delta\dot{\chi}$ (see Chapter 6). Hence, the two sets of derivatives, C_{nr1} and C_{lr1}, and C_{nr2} and C_{lr2}, may not be used interchangeably.

Unfortunately in the past, this distinction was not recognized and a single set of derivatives (equivalent to) C_{nr1} and C_{lr1} was used both for the Dutch roll as well as the spiral mode. Consequently, the evaluation of the spiral mode time constant based on this usage was unreliable. Equally, the derivatives (equivalent to) C_{nr1} and C_{lr1} used to be multiplied in the Dutch roll analysis with Δr_b and not with the correct factor $(\Delta r_b - \Delta r_w) = \Delta\dot{\beta}$, leading to erroneous model of the Dutch roll dynamics.

7.12.2 Spiral Mode Stability

Rewriting Equation 7.51 in terms of the derivatives, we can state the spiral mode stability condition as:

$$[\underbrace{C_{l\beta}}_{-}\,\underbrace{C_{nr2}}_{-} - \underbrace{C_{n\beta}}_{+}\,\underbrace{C_{lr2}}_{+}] > 0 \tag{7.52}$$

with the nominal sign of each derivative shown below it. Each product is positive, and the expression in Equation 7.52 turns out to be the difference between two positive terms. In general, the difference can turn out to be either positive or negative—in either case, it is usually quite small and close to zero, so that the spiral motion either converges or diverges very slowly. Because of this slow response, spiral stability is usually not such a critical issue. The pilot (or the autopilot) can easily correct for the spiral motion provided the divergence is slower than a certain rate. For handling quality requirements on spiral mode divergence, refer to Box 7.1.

You may notice that the same derivatives $C_{l\beta}$ and $C_{n\beta}$ appear in the Dutch roll (yaw) stiffness criterion Equation 7.12 as well as spiral mode stability condition Equation 7.52. In the case of the Dutch roll, these two derivatives cooperate and their effects add up to contribute to the yaw stiffness. However, for the spiral mode, these derivatives are at loggerheads.

- $C_{n\beta}$ yaws the airplane nose into the turn and C_{lr2} banks the airplane further into the turn. Thus, this combination pushes the airplane further into the spiral motion.
- $C_{l\beta}$ lifts the banked wing up and C_{nr2} yaws the airplane nose out of the turn. Thus, this combination recovers the airplane from the spiral motion and tries to correct it.

Thus, too much yaw stiffness due to $C_{n\beta}$ can destabilize the spiral mode. This is a known fact from practice.

7.13 REAL-LIFE AIRPLANE DATA

As always, it is useful to be aware of how some of the aerodynamic derivatives vary for real airplanes, especially outside the 'linear regime' where the analytical approximations are not valid. Figure 7.42 shows plots of three key lateral-directional derivatives, $C_{l\beta}$, $C_{n\beta}$ and C_{lp2}, and $C_{n\beta} + \varepsilon\left(\frac{C_{l\beta}}{C_{lp2}}\right)$ for the F-18/HARV airplane as a function of the angle of attack α.

- The roll damping derivative C_{lp2} has a large negative value indicating stability of the roll (rate) mode. But there is a short stretch between $\alpha = 40–50°$ where it is close to zero.
- $C_{n\beta}$ is positive at small α, but it reduces to zero short of $\alpha = 30°$. Thereafter, it remains negative at higher α. This is possibly because the vertical tail remains submerged in the wing–body wake.
- $C_{l\beta}$ initially grows increasingly negative with α, no doubt the effect of wing sweep. It undergoes significant change in the $\alpha = 20–30°$ range, but remains negative all through.
- It is interesting to note from Figure 7.42b that in spite of $C_{n\beta}$ becoming negative beyond $\alpha = 30°$, Dutch-roll stiffness parameter $C_{n\beta} + \varepsilon\left(C_{l\beta}/C_{lp2}\right)$ is still positive.

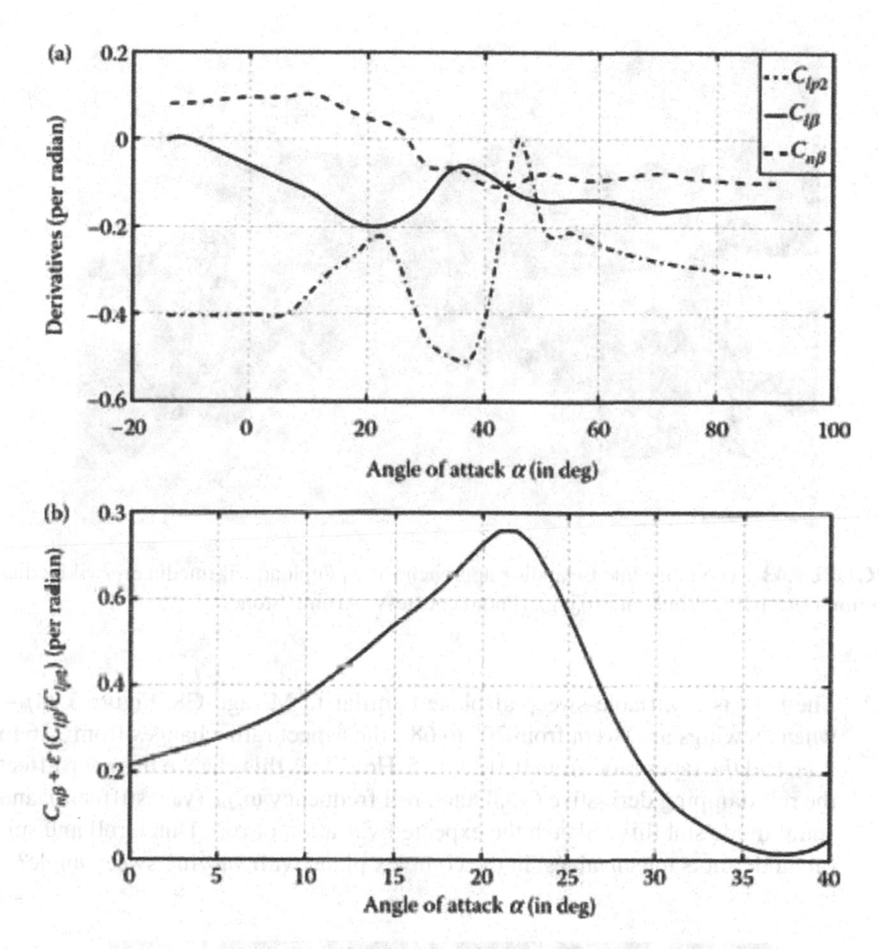

FIGURE 7.42 (a) Lateral-directional derivatives $C_{l\beta}$, $C_{n\beta}$ and C_{lp2}, and (b) plot of dutch-roll stiffness parameter $C_{n\beta} + \varepsilon\left(C_{l\beta}/C_{lp2}\right)$ for the F-18/HARV airplane (http://www.dfrc.nasa.gov/Research/HARV/Work/NASA2/nasa2.html).

EXERCISE PROBLEMS

7.1 The B-52 (see Figure 7.43) is a huge airplane with eight engines and manually operated controls, that is, no hydraulics. To keep hinge moments small, the rudder has a very small chord, only 10% of the vertical tail (fin) chord. Because of this, the B-52 has an unusual way of landing in crosswind, that is, wind coming not along the direction of the runway, but from either left or right. The landing gear could be rotated by a maximum of ±20° so that the wheels are pointed straight down the runway even when the airplane nose is pointed into the relative wind. Using a $C_n - \beta$ diagram with δr as parameter, explain: (i) How would the B-52 maintain zero-sideslip in cruising flight when one of the eight engines fails? (ii) Why did the landing gear rotation strategy for crosswind landing become necessary?

FIGURE 7.43 B-52 airplane in landing approach (https://upload.wikimedia.org/wikipedia/commons/c/cf/B52.climbout.arp.jpg). (Photo courtesy: Arping Stone.)

7.2 The F-14 is a variable-sweep airplane (similar to Mirage-G8, Figure 7.44)—when its wings are swept from 20° to 68°, the aspect ratio changes from 7.16 to 2.17, and the taper ratio from 0.31 to 0.25. How does this change in sweep affect the roll damping derivative C_{lp2}, Dutch roll frequency ω_{nDR} (yaw stiffness), and spiral mode stability? Sketch the expected variation of roll, Dutch roll and spiral mode roots (eigenvalues) in the complex plane with varying sweep angle?

FIGURE 7.44 Variable-sweep Mirage (https://commons.wikimedia.org/wiki/File:Dassault_Mirage_G8.jpg).

7.3 The variation of the derivative $C_{n\beta}$ (per degree) as a function of Mach number for the F-16 is as follows:

Ma	0.8	1.2	2.0
$C_{n\beta}$	0.004	0.006	0.002

Explain the possible reason for this variation using analytical expressions and graphs.

7.4 F-4 Phantom (see Figure 7.45)—a large anhedral angle (negative dihedral angle) of the horizontal tail in combination with a sharply defined positive dihedral in the outer wing panels are two conspicuous identifying features of the aircraft. Can you imagine why these were selected (Ref: http://www.hq.nasa.gov/office/pao/History/SP-468/ch11-5.htm)?

7.5 To roll the F-4 Phantom, the pilot had to use aileron below 12°, aileron and rudder combined between 12° and 16°, and rudder only above that point. Can you guess why?

7.6 The F-11 Tiger (shown in Figure 7.46) is noted for being the first jet aircraft to shoot itself down. How could this have happened?

7.7 For the Jaguar (shown in Figure 7.47), double slotted flaps are fitted along the whole wing trailing edge, leaving no space for ailerons. Then how is roll control achieved.

Answer: Through the use of spoilers operating together with the differential tail plane.

7.8 For a trapezoidal wing with span b and taper ratio $\lambda = c_t/c_r = 1/3$, estimate the roll damping derivative to be $C_{lp2} = -C_{L\alpha}/8$.

FIGURE 7.45 F-4 Phantom. (From www.airliners.net; photo courtesy: Geunwon.)

FIGURE 7.46 F-11 Tiger. (http://www.blueangels.org; Naval Fighters Number Forty Grumman F11F Tiger by Corwin "Corky" Meyer, Publication Date: December 1, 1997. With permission from William Swisher via Steve Ginter.)

7.9 For a trapezoidal wing with span b and taper ratio, $\lambda = c_t/c_r = 1/3$, derive the approximate formula for the rolling moment derivative due to yaw rate C_{lr1} as follows:

$$C_{lr1} = \left(\frac{1+3\lambda}{1+\lambda}\right)\cdot\frac{C_L}{6}$$

Hence, estimate C_{lr1} for such a wing with $b/c_r = 2.122$, $e = 0.8$, $C_{D0} = 0.02$, cruising at $(L/D)_{max}$.

7.10 In a similar manner as was done for C_{lp2}, for a wing with taper ratio, $\lambda = c_t/c_r$, show that:

$$C_{lr2} = -\left(\frac{1+3\lambda}{1+\lambda}\right)\cdot\frac{C_L}{12}$$

FIGURE 7.47 The Jaguar (https://www.airliners.net/photo/UK-Air-Force/Sepecat-Jaguar-GR3A/617357/L). (Photo courtesy: Kristof Jonckheere.)

7.11 An airplane is flying a straight-line, wings-level trim with $\beta \neq 0$ using a rudder deflection $\delta r = 10°$. Use the following data to estimate the aileron deflection δa required to maintain this straight and level trim flight:

$$C_{n\beta} = 0.195/\text{rad}, \quad C_{n\delta r} = -0.123/\text{rad}, \quad C_{n\delta a} = 0$$
$$C_{l\beta} = -0.28/\text{rad}, \quad C_{l\delta r} = 0.007/\text{rad}, \quad C_{l\delta a} = -0.053/\text{rad},$$

Assume reasonable values for missing data, if any.

7.12 Similar to downwash, wing tip vortices also create a sidewash effect over the vertical tail. The sidewash angle (σ) is a function of the sideslip angle and various other parameters of the aircraft. An empirical relation for this effect is given as:[3]

$$\left(1+\frac{d\sigma}{d\beta}\right)\eta_{VT} = 0.724 + 3.06\frac{S_V/S}{1+\cos\Lambda_{c/4}} + \frac{0.4z_w}{d_{f,\max}} + 0.009\,AR_w$$

where $\eta_V = (\bar{q}_V/\bar{q})$ is the ratio of dynamic pressures at the vertical tail and at the wing, which we can assume to be ~1, S_V/S is the ratio of the vertical tail area, including the part submerged in the fuselage, to the wing planform area, $\Lambda_{c/4}$ is the sweep of wing quarter chord line, z_w is the height of the wing root chord at quarter chord location from the fuselage reference line, $d_{f,max}$ is the fuselage maximum depth and AR_w is the aspect ratio of the wing. Calculate the contribution to the derivative $C_{n\beta}$ due to the vertical tail using the above formula for the following data: $AR_w = 6$, $S_V/S = 0.2$, $\Lambda_{c/4} = 20°$, $z_w = 0.0$, $l_V/b = 0.6$ [*Hint:* The angle of attack at the vertical tail, including the sidewash effect, is $\alpha_V = \beta + \sigma$].

BOX 7.3 WHY BIRDS AND AIRPLANES NEED NO VERTICAL TAIL

It has intrigued aeronautical scientists and engineers forever that the most adept flyers, the avians, have no need for a vertical tail. In contrast, almost every airplane designed over the last century of flight, but for a very small number of "tailless" ones,[5] sports a vertical tail; some have two, an odd one even has three. Ventral and dorsal fins to augment the vertical tail are commonly observed. The purpose of the vertical tail in its various forms is commonly explained as providing directional stability and damping in yaw. Additionally, the deflectable rudder surface at the trailing edge of the vertical tail is meant to provide directional control. This leads one to the obvious question: how do birds manage directional stability and control sans a vertical tail? Or is it that birds fly with some degree of lateral-directional instability, as some[6] have suggested?

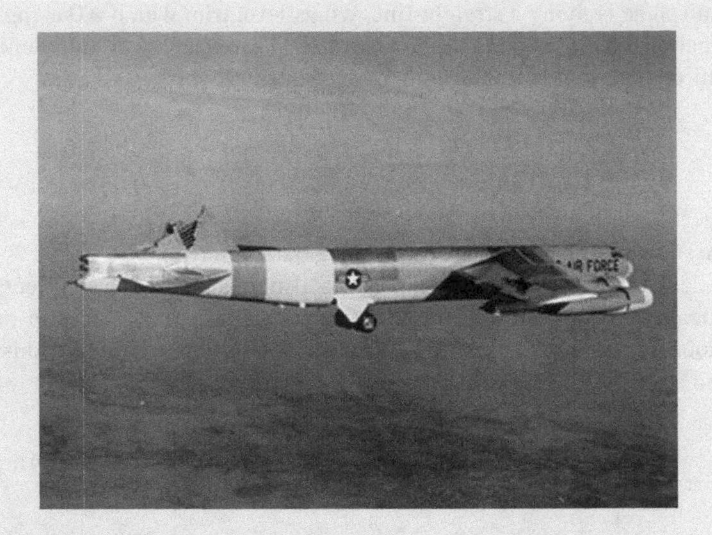

FIGURE 7.A1 A B-52 that managed to fly safely despite the almost total loss of its vertical tail (http://en.wikipedia.org/wiki/File:Boeing_B-52_with_no_vertical_stabilizer.jpg).

One point of view[7] is that it is a matter of scale; that the rationale for having a vertical tail disappears at the smaller scale corresponding to the flight of birds and 'mini'-scale airplanes. The implication being that 'full'-scale airplanes would generally continue to require a vertical tail to satisfy lateral-directional flight requirements. Yet, we have examples (see Figure 7.A1) of airplanes designed with a vertical tail that have managed to fly safely despite almost all of the vertical tail being accidentally lost. Hang gliders are another example of a system that does very well without the need for a vertical tail.[8]

The basis for most conclusions regarding the need or otherwise of a vertical tail for flight stems from an analysis of the literal approximation to the lateral-directional modes, particularly the Dutch roll and spiral modes. Based on the modal approximations derived herein, we can show that stability of the Dutch roll and spiral modes can be obtained in principle for any airplane without the need for a vertical tail.

Let us recall the three distinct timescales in lateral-directional flight which were identified as follows:

$$T_s = \left(V^*/g\right) \sim 10sec; \quad T_f = \sqrt{\frac{I_{zz}}{\bar{q}Sb}} \sim 1sec; \quad T_r = \left(b/2V^*\right) \sim 0.1sec \qquad (7.A1)$$

For a conventional airplane, the slow timescale T_s is of the order of 10 seconds and corresponds to the spiral mode. The intermediate timescale T_f is of the order of 1 second and is the scale at which the Dutch roll motion is observed.

The fast timescale T_r is of the order of 0.1 seconds corresponding to the roll (rate) mode. The clear distinction between these timescales is what allows their corresponding modes to be separated analytically from one another.

We also reproduce the approximations to the roll, Dutch roll and spiral mode parameters obtained previously:

Roll (rate) mode eigenvalue:

$$\lambda_r = \mathcal{L}_{p2} \tag{7.A2}$$

Dutch roll frequency and damping:

$$\omega_{nDR}^2 = N_\beta + \left(\frac{g}{V^*}\right)\left[Y_\beta N_{r2} + \left(\frac{\mathcal{L}_\beta}{\mathcal{L}_{p2}}\right)\right] \tag{7.A3}$$

$$2\zeta_{DR}\omega_{nDR} = -N_{r1} - \left(\frac{g}{V^*}\right)\left[Y_\beta + \left(\frac{\mathcal{L}_{r1}}{\mathcal{L}_{p2}}\right)\right] \tag{7.A4}$$

Spiral mode eigenvalue:

$$\lambda_s = \left(\frac{g}{V^*}\right)\frac{\left[\mathcal{L}_\beta N_{r2} - N_\beta \mathcal{L}_{r2}\right]}{\lambda_r \omega_{nDR}^2} \tag{7.A5}$$

Further, we have obtained stability criteria for the modes in terms of the aerodynamic derivatives (for simplicity, the term Y_β in Equations (7.A3) and (7.A4) has been neglected).

Roll (rate) mode:

$$C_{lp2} < 0 \tag{7.A6}$$

Dutch roll mode:

$$\text{Stiffness: } C_{n\beta} + \varepsilon\left(C_{l\beta}/C_{lp2}\right) > 0 \tag{7.A7}$$

$$\text{Damping: } C_{nr1} + \varepsilon\left(C_{lr1}/C_{lp2}\right) < 0 \tag{7.A8}$$

Spiral mode:

$$C_{l\beta}C_{nr2} - C_{n\beta}C_{lr2} > 0 \tag{7.A9}$$

Where ε is a dimensionless ratio of the various timescales as follows:

$$\varepsilon = T_f^2 / T_s T_r \tag{7.B1}$$

For a conventional airplane, since $T_f \sim 1$ *second*, $T_s \sim 10$ *seconds* and $T_r \sim 0.1$ *seconds*, as given in Equation 7.A1, $\varepsilon \sim 1$.

We can write down the typical signs of the various derivatives in Equations 7.A6 through 7.A9 and examine whether each stability criterion is usually met for a conventional airplane configuration. The usual sign of each derivative is marked under it in the following. Also, derivatives where the vertical tail has a predominant influence or a reasonable effect are marked with a bar over them.

Roll (rate) mode:

$$\underbrace{C_{lp2}}_{(-)} < 0 \tag{7.B2}$$

Dutch roll mode:

$$\text{Stiffness: } \underbrace{\overline{C_{n\beta}}}_{(+)} + \varepsilon\left(\underbrace{C_{l\beta}}_{(-)} / \underbrace{C_{lp2}}_{(-)}\right) > 0 \tag{7.B3}$$

$$\text{Damping: } \underbrace{\overline{C_{nr1}}}_{(-)} + \varepsilon\left(\underbrace{\overline{C_{lr1}}}_{(+)} / \underbrace{C_{lp2}}_{(-)}\right) < 0 \tag{7.B4}$$

Spiral mode:

$$\underbrace{C_{l\beta}}_{(-)}\underbrace{C_{nr2}}_{(-)} - \underbrace{\overline{C_{n\beta}}}_{(+)}\underbrace{C_{lr2}}_{(+)} > 0 \tag{7.B5}$$

Note that ε from Equation (7.B1) is of the order of 1.

- From Equation (7.B2), the roll mode stability is virtually guaranteed by the wing itself.
- Looking at Equation (7.B3), both terms contribute to the yaw stiffness—the first is mainly due to the vertical tail, the second is primarily from the wing. Even without a vertical tail (first term missing), with a sufficiently negative $C_{l\beta}$ (wing dihedral and sweep), yaw stiffness can be ensured.
- From Equation (7.B4), again each of the two terms is negative and ensures yaw damping. The vertical tail is a part contributor to the two derivatives with over-bars. Even without the vertical tail, the wing contribution to each of these two derivatives provides yaw damping.
- For the spiral mode eigenvalue in Equation (7.B5) alone, the two terms oppose each other. The net result is the difference between two positive terms and can go either way. Notably, the vertical tail contributes to only one of the four derivatives that go into Equation (7.B5). If the vertical tail is deleted, only the first term remains in Equation (7.B5) and it comfortably assures a stable spiral mode!

The conclusion is that the vertical tail is not necessary for the stability of any of the lateral-directional modes. As long as the wing can contribute sufficient $C_{l\beta}$, and a large enough C_{nr1}, C_{lr1}, yaw stiffness and damping can be managed. And without the vertical tail, spiral mode stability is no longer in question.

With no vertical tail, the rudder would also be absent, and the issue of alternative sources of directional control arises. Typically, the use of roll and pitch control devices does produce some yawing moment; hence they may be employed as surrogate yaw control devices. The use of asymmetric wing dihedral as a yaw control on airplanes without a vertical tail has been proposed.[9,10] The mechanism is apparently similar to that used by birds with their tail. Winglets, which are now ubiquitous on airplanes, could provide another source of yaw stability and control.[11]

For birds and small scale airplanes: First of all, we need to establish the scaling relation between the flight of 'full'-scale airplanes and that of birds and small-scale airplanes. The variety of flight over the range of Reynolds numbers is depicted in Figure 7.A2. When compared to 'full'-scale airplanes, we scale down the following variables by the respective order indicated to get to small scale or bird flight scale:

- Scale down by 1 order: flight velocity (V), wing chord (c), vehicle height (h).
- Scale down by 2 orders: wing span (b), vehicle length (l).

Thus, the Reynolds number which is defined as $Re = \rho Vl/\mu$, where ρ, μ are the air density and viscosity, respectively, scales down by 3 orders (1 from V, 2 from l). This agrees with Figure 7.A2 which shows general aviation and jet aircraft at $Re \sim 10^7 - 10^8$ and birds and model airplanes 3 orders lower at $Re \sim 10^4 - 10^5$. At this lower range of Reynolds number also, the aerodynamic model presented herein is applicable, hence the preceding lateral-directional dynamics equations may be used for at this scale as well.

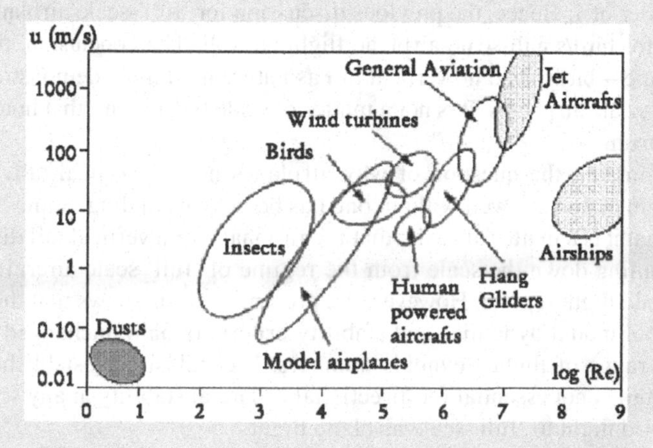

FIGURE 7.A2 Flight over the range of Reynolds numbers (from Ref. 12).

Next, examining the timescales in Equation (7.A1):

- The slow timescale T_s scales down by 1 order because of V.
- For the intermediate timescale T_f, we must consider the factors in the numerator and denominator. Each of their terms scales down by the order indicated in the parenthesis after them: I_{zz} (by 9 orders, since mass scales down as volume by 5 orders ($b.l.h$ ~ 2.2.1), and the two lengths perpendicular to the Z axis, wing span and vehicle length, scale down by 2 orders each). $\bar{q}$ (by 2 orders), b (by 2 orders) and S (by 3 orders, 2 for b and 1 for c). Thus, the numerator scales down by 9 orders and the denominator by 7, the difference is 2 orders, and after taking the square root, T_f scales down by 1 order.
- The fast timescale T_r scales down by 1 order (2 for b minus 1 for V).

Thus, the timescales at the bird flight and small airplane flight scale are arranged as:

$$T_s = \left(V^*/g\right) \sim 1sec; \quad T_f = \sqrt{\frac{I_{zz}}{\bar{q}Sb}} \sim 0.1sec; \quad T_r = \left(b/2V^*\right) \sim 0.01sec \qquad (7.B6)$$

Compared to the timescales for 'full'-scale airplanes in Equation 7.A1, each of the timescales in Equation 7.B6 is 1 order faster. However, the relative difference in timescales between the three of them is maintained. Hence the modal approximation based on the multiple timescale procedure holds, and the relations for the roll, dutch roll and spiral mode parameters in Equations 7.A2 through 7.A5 remain valid. Consequently, the stability criteria in Equations 7.A6 through 7.A9 also hold true for bird/small scale aircraft flight. As per the timescales in Equation 7.B6, the factor ε in Equation 7.B1 again works out to be of the order of 1. Hence, the previous discussion for 'full'-scale airplanes holds entirely for bird/small-scale airplane flight as well. The conclusion, therefore, is the same—birds need no vertical tail as nature has amply demonstrated, nor do small-scale airplanes. It is not a matter of scale but a verity that holds across the spectrum.

In summary, the question of why airplanes need a vertical tail, whereas birds seem to do very well without one has been longstanding. Some have suggested that it is a matter of scale; that the rationale for a vertical tail disappears when coming down the scale from the regime of 'full' scale aircraft to bird/small scale flight regime. However, the present analysis shows that the lateral-directional modal dynamics and stability criteria remain unchanged over the relevant range of flight Reynolds numbers. It conclusively establishes that a vertical tail is not essential for directional or lateral stability at any scale from that of bird flight to 'full'-scale airplane flight.

REFERENCES

1. MIL-F-8785C, Flying qualities of piloted airplanes, U.S. Department of Defense Military Specifications, November 5, 1980.
2. Ananthkrishnan, N., Shah, P. and Unnikrishnan, S., Approximate analytical criterion for aircraft wing rock onset, *Journal of Guidance, Control, and Dynamics*, 27(2), 2004, 304–307.
3. Nelson, R.C., *Flight stability and automatic control*. Second edition, McGraw Hill publication, Education (India) Private Limited, New Delhi, 2007.
4. Nguyen, L.T., Ogburn, M.E., Gilbert, W.P., Kibler, K.S., Brown, P.W. and Deal, P.L., Simulator Study of Stall/Post-Stall Characteristics of a Fighter Airplane with Relaxed Longitudinal Static Stability, NASA TP-1538, 1979.
5. Nickel, K. and Wohfahrt, M., *Tailless aircraft in theory and practice*, AIAA, Washington, 1994.
6. Paranjape, A.A., Dorothy, M.R., Chung, S.J. and Lee, K.D., A flight mechanics-centric review of bird-scale flapping flight, *International Journal of Aeronautical and Space Sciences*, 13(3), 2012, 267–281.
7. Sachs, G., Why birds and miniscale airplanes need no vertical tail, *Journal of Aircraft*, 44(4), 2007, 1159–1167.
8. Cook, M.V. and Spottiswoode, M., Modelling the flight dynamics of the hang glider, *The Aeronautical Journal*, 2006, Volume 109, Issue 1102, 1–20.
9. Paranjape, A.A., Chung, S.J., and Selig, M.S., Flight mechanics of a tailless articulated wing aircraft, *Bioinspiration & Biomimetics*, 6(2), 2011, 1–20.
10. Paranjape, A.A., Chung, S.J., Hilton, H.H. and Chakravarthy, A., Dynamics and performance of tailless micro aerial vehicle with flexible articulated wings, *AIAA Journal*, 50(5), 2012, 1177–1188.
11. Bourdin, P., Gatto, A. and Friswell, M.I., Aircraft control via variable cant-angle winglets, *Journal of Aircraft*, 45(2), 2008, 414–423.
12. Lissaman, P.B.S., Low-Reynolds-number airfoils, *Annual Review of Fluid Mechanics*, 15, 1983, 223–239.

8 Computational Flight Dynamics

8.1 AIRCRAFT EQUATIONS OF MOTION

Through Chapters 1–7, we tried to understand via simpler equations of motion (reduced-order models) how the stability and control characteristics of an aircraft are affected due to various factors related to aircraft sizing, configuration and control surface deflection. Dividing aircraft motions into two uncoupled sets, namely longitudinal and lateral-directional, and further based on the timescales, we examined what are known as aircraft dynamic modes through first- and second-order differential equations. We followed this approach for two reasons. Firstly, it is easy to work with decoupled first- or second-order equations (no numerical tool or computer coding required) and, secondly, from the simplified analyses, we could extract a lot of useful information that can be very helpful in the design and analysis of airplanes. And, of course, this is the best (perhaps the only) way to obtain a physical insight into the motion of airplanes.

In practice, as we have seen, for a comprehensive study of aircraft dynamics, an airplane must be modelled as a rigid body with six degrees of freedom (6DOF); certainly in many situations, all the modes of aircraft motion may become excited together at the same time. And disturbances (or manoeuvres) need not always be small in magnitude, so the decoupling of the various modes (which assumed small perturbations) may not always hold. To understand aircraft response to more general disturbances and also to design active flight control systems, it is important to work with the full-order 6DOF aircraft equations of motion. In the following, we first derive the 6DOF aircraft equations of motion using first principles, and then analyse some large-amplitude airplane manoeuvres.

8.2 DERIVATION OF AIRCRAFT EQUATIONS OF MOTION

As we discussed in Chapter 1, an aircraft, or for that matter any object in space, has 6DOF: three in translation and three in rotation. That is, assuming the aircraft to be a rigid body; ignoring the infinite degrees of freedom of small deformations that an aircraft body undergoes due to its elastic properties. Even though at times the structural deformations are large enough (for instance, wing tips of a Boeing 747 can have transverse deflections as large as 2 m) to cause notable changes in its translational and angular motion characteristics, in this chapter, we will consider the aircraft to be a rigid body. We will also ignore the effects of rotating parts, such as engine fan or propeller on aircraft motion.

With reference to Figure 8.1, let us assume an aircraft of mass m translating with velocity $\underline{V}_C$ and rotating with angular velocity $(\underline{\omega}_C)$ with respect to the inertial frame

DOI: 10.1201/9781003096801-8

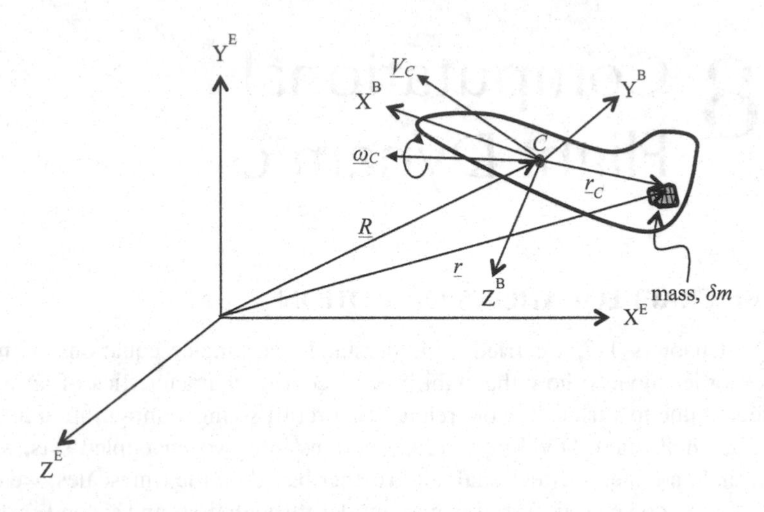

FIGURE 8.1 Aircraft as a point mass.

of reference $X^EY^EZ^E$, where C refers to the airplane centre of mass (CM), $\underline{R}$ is the position vector of the CM with respect to the origin of the $X^EY^EZ^E$ frame. The trajectory of the aircraft with respect to an inertial frame can be tracked by computing the vector $\underline{R}$. Let us make a few reasonable assumptions here before we proceed further, as follows:

- Earth is an inertial frame of reference.
- Earth is flat. Earth's curvature has no influence on the aircraft motion.
- Every part of the aircraft experiences the same acceleration due to gravity. This assumption also means that the centre of mass (CM) and centre of gravity (CG) of the aircraft coincide with each other.
- The aircraft is a rigid body. No relative motion between any two points on the airplane is considered, that is, $\dot{\underline{r}}_C = 0$. This allows us to arrive at a reasonable mathematical model.
- Shift in CG due to fuel depletion is considered to be insignificantly small.

Some more assumptions regarding the geometry of the aircraft will follow later. In Figure 8.1, the body-fixed axis system is shown to have its origin at the centre of mass C of the aircraft. The aircraft is symmetric about the X^BZ^B plane, which is also the longitudinal plane of the aircraft. The Y^B axis completes the right-handed orthogonal axis system.

8.2.1 Equations of the Translational Motion

Let us consider an element of mass δm at a distance $\underline{r}_C$ from the centre of mass C as shown in Figure 8.1. The velocity of this elemental mass as seen by an observer in the inertial frame of reference ($X^EY^EZ^E$) is:

$$\underline{V} = \underline{V}_C + \underline{\omega}_C \times \underline{r}_C \tag{8.1}$$

The linear momentum of the elemental mass δm is given by:

$$\delta m\underline{V} = \delta m\underline{V_C} + \delta m\left(\underline{\omega_C} \times \underline{r_C}\right) \tag{8.2}$$

Summing this over the whole body of the aircraft, the total linear momentum can be found as:

$$\begin{aligned} \sum \delta m\underline{V} &= \sum \delta m\underline{V_C} + \sum \delta m\left(\underline{\omega_C} \times \underline{r_C}\right) \\ &= \underline{V_C} \sum \delta m + \underline{\omega_C} \times \sum \delta m\underline{r_C} \end{aligned} \tag{8.3}$$

since the $\underline{V_C}$ and $\underline{\omega_C}$ experienced by all the elements of the airplane is the same. In the limit, when an elemental mass is assumed to be very small, that is, $\delta m \to 0$, the summation can be written as integral and therefore total linear momentum:

$$\underline{p} = \underline{V_C} \int \delta m + \underline{\omega_C} \times \int \underline{r_C}\, \delta m \tag{8.4}$$

Since all the distances are being measured on the aircraft with respect to its centre of mass (the origin of the body-fixed axis system), one can further simplify Equation 8.4 using $\int \underline{r_C} \delta m = 0$, which is the definition of the centre of mass, to:

$$\underline{p} = \underline{V_C} \int \delta m = m\underline{V_C} \tag{8.5}$$

The equations of translational motion of an aircraft can be written by differentiating Equation 8.5 with respect to time as:

$$\underline{F} = \frac{d\underline{p}}{dt} = m\frac{d\underline{V_C}}{dt} \tag{8.6}$$

where the rate of change d/dt is computed in the inertial frame. It is however convenient to calculate the rate of change with respect to the body frame.

For a rotating body, the rate of change of a vector $\underline{A}$ in body-axis frame as seen from an inertial frame of reference (subscript I) is given by:

$$\left.\frac{d\underline{A}}{dt}\right|_I = \left.\frac{d\underline{A}}{dt}\right|_B + \left(\underline{\omega_C} \times \underline{A}\right) \tag{8.7}$$

in terms of the rate of change in the body frame (subscript B), which has an angular velocity $\underline{\omega_C}$ with respect to the inertial frame. Thus, in the body frame, the force Equation 8.6 (describing the translational motion of the aircraft) appears as:

$$\underline{F} = m\left.\frac{d\underline{V_C}}{dt}\right|_I = m\left.\frac{d\underline{V_C}}{dt}\right|_B + m\left(\underline{\omega_C} \times \underline{V_C}\right) \tag{8.8}$$

Note that $\underline{V_C}$ and $\underline{\omega_C}$ are the translational and angular velocities of the body (equivalently, the body frame) with respect to the inertial (Earth-fixed) frame. In

Equation 8.8, only the derivative *d/dt* has been converted from the inertial to the body reference frame. We can now choose any coordinate system to express the vectors $\underline{F}$, $\underline{V}_C$ and $\underline{\omega}_C$ in terms of their components. This choice is independent of the frame of reference in which *d/dt* is expressed. The body-fixed axis is a standard choice to write the vector components so that $[u\ v\ w]'$ are components of the velocity vector $\underline{V}_C$ along the body-fixed axes and $[p\ q\ r]'$ are the components of angular velocity vector $\underline{\omega}_C$ about the body-fixed axes. The cross product of vector $\underline{\omega}_C$ can be expressed in matrix form as:

$$\underline{\omega}_C \times = \begin{bmatrix} 0 & -r & q \\ r & 0 & -p \\ -q & p & 0 \end{bmatrix} \tag{8.9}$$

Then, the force Equation 8.8 in the body-fixed axes system are:

$$\underline{F} = \begin{bmatrix} X \\ Y \\ Z \end{bmatrix} = m \begin{bmatrix} \dot{u} \\ \dot{v} \\ \dot{w} \end{bmatrix} + m \begin{bmatrix} 0 & -r & q \\ r & 0 & -p \\ -q & p & 0 \end{bmatrix} \begin{bmatrix} u \\ v \\ w \end{bmatrix} \tag{8.10}$$

or

$$\begin{aligned} \dot{u} &= rv - qw + \frac{X}{m} \\ \dot{v} &= pw - ru + \frac{Y}{m} \\ \dot{w} &= qu - pv + \frac{Z}{m} \end{aligned} \tag{8.11}$$

where X is the axial force (resultant force acting along the aircraft longitudinal axis X^B), Y is the side force (resultant force acting along the aircraft axis Y^B) and Z is the normal force (resultant force acting along the aircraft axis Z^B).

Figure 8.2 shows the various axes in use—Earth, Body, Wind and Stability (labelled E, B, W and S, respectively). More about the *Wind* and *Stability* axes later. Three different types of forces acting on the aircraft as shown in Figure 8.2 constitute the resultant force as follows:

$$\underline{F} = \underline{F}^{Gravitational} + \underline{F}^{Aerodynamic} + \underline{F}^{Propulsive} \tag{8.12}$$

The gravitational force or the weight of the airplane *mg* acts along the Z^E axis. Of the aerodynamic forces, *Lift 'L'* acts opposite to the Z^W axis, *Drag 'D'* is in the opposite direction to the X^S axis and *Side force 'Y'* is along the Y^B axis. Components of the aerodynamic force may be taken along X^B, Y^B, and Z^B axes. The propulsive force (thrust, *T*) may be assumed to act along X^B axis.

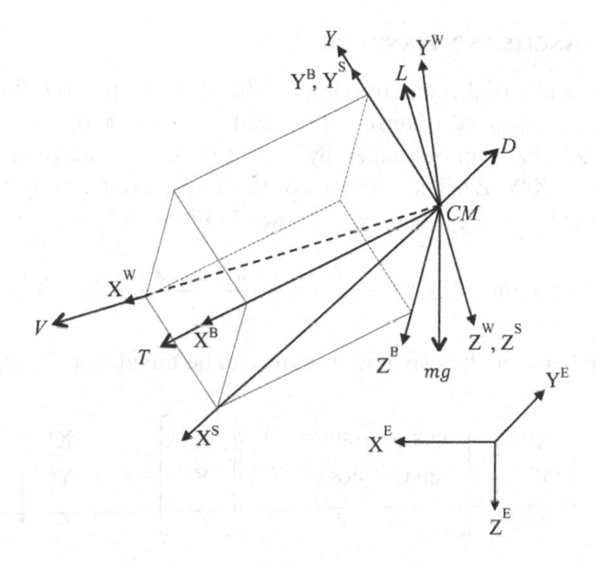

FIGURE 8.2 Forces acting on the aircraft (mg along Z^E, L along Z^W, D along X^S, Y along Y^B, T along X^B).

Thus, each of these forces may be conveniently expressed in a different axis system as:

$$\underline{F} = \begin{bmatrix} X \\ Y \\ Z \end{bmatrix} = \begin{bmatrix} 0 \\ 0 \\ mg \end{bmatrix}_E + \begin{bmatrix} X_A \\ Y_A \\ Z_A \end{bmatrix}_B + \begin{bmatrix} T \\ 0 \\ 0 \end{bmatrix}_B \tag{8.13}$$

where subscripts 'E' and 'B' stand for inertial (Earth-fixed) and body-fixed axis systems, respectively. X_A, Y_A and Z_A are the components of aerodynamic forces along the aircraft-fixed body axes.

We now need to transform the gravitational force to the body-fixed axis system. For that, we need to know the proper relations between different axis systems, defined by transformation matrices.

8.3 3–2–1 RULE

The 3–2–1 rule applies to the transformation of a vector from one orthogonal axis system to another. The rotation rule '3' refers to the (first) rotation of a frame of reference about its Z axis followed by '2', which refers to the rotation about the new Y axis (after the first rotation) and finally '1' refers to the third rotation of the axis system (after the second rotation) about the new X axis. Finally, the 3–2–1 rotations result in exact matching of the rotated frame of reference with the aircraft body-fixed axis system. In the following, we derive the relations between the vectors written in different axis systems through the rotation or transformation matrices.

8.3.1 Euler Angles and Transformation

With respect to an inertial frame of reference fixed at Earth, aircraft orientation (or attitude) can be determined in terms of the Euler angles, ϕ, θ, ψ. The Earth-fixed inertial frame $X^EY^EZ^E$ can be rotated by angles ϕ, θ, ψ to arrive at aircraft body-fixed axis system $X^BY^BZ^B$. The rotations of the Earth-fixed axis system follow the 3–2–1 rule in the following sequence as shown in Figure 8.3:

$$\text{First rotation: } X^E - Y^E - Z^E \xrightarrow{\text{rotation } \psi \text{ about } Z^E} X^1 - Y^1 - Z^1$$

The relation between the two axis systems can be found from Figure 8.3a as:

$$\begin{bmatrix} X^E \\ Y^E \\ Z^E \end{bmatrix} = \begin{bmatrix} \cos\psi & -\sin\psi & 0 \\ \sin\psi & \cos\psi & 0 \\ 0 & 0 & 1 \end{bmatrix} \begin{bmatrix} X^1 \\ Y^1 \\ Z^1 \end{bmatrix} = R_\psi \begin{bmatrix} X^1 \\ Y^1 \\ Z^1 \end{bmatrix} \tag{8.14}$$

$$\text{Second rotation: } X^1 - Y^1 - Z^1 \xrightarrow{\text{rotation } \theta \text{ about } Y^1} X^2 - Y^2 - Z^2$$

The second rotation results in the relation:

$$\begin{bmatrix} X^1 \\ Y^1 \\ Z^1 \end{bmatrix} = \begin{bmatrix} \cos\theta & 0 & \sin\theta \\ 0 & 1 & 0 \\ -\sin\theta & 0 & \cos\theta \end{bmatrix} \begin{bmatrix} X^2 \\ Y^2 \\ Z^2 \end{bmatrix} = R_\theta \begin{bmatrix} X^2 \\ Y^2 \\ Z^2 \end{bmatrix} \tag{8.15}$$

$$\text{Third rotation: } X^2 - Y^2 - Z^2 \xrightarrow{\text{rotation } \phi \text{ about } X^2} X^B - Y^B - Z^B$$

The third rotation follows the relation:

$$\begin{bmatrix} X^2 \\ Y^2 \\ Z^2 \end{bmatrix} = \begin{bmatrix} 1 & 0 & 0 \\ 0 & \cos\phi & -\sin\phi \\ 0 & \sin\phi & \cos\phi \end{bmatrix} \begin{bmatrix} X^B \\ Y^B \\ Z^B \end{bmatrix} = R_\phi \begin{bmatrix} X^B \\ Y^B \\ Z^B \end{bmatrix} \tag{8.16}$$

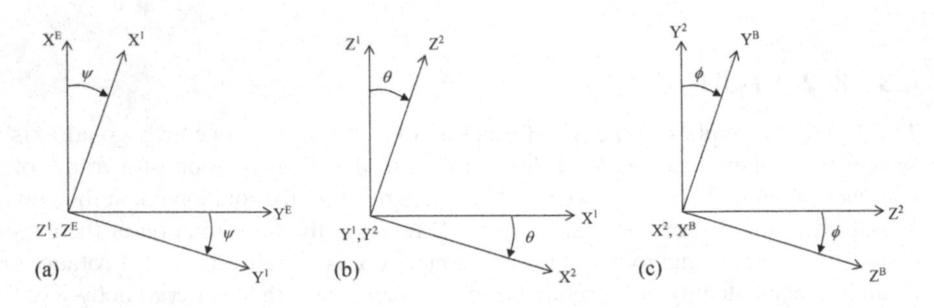

FIGURE 8.3 (a) Rotation '3' about the Z^E axis, (b) rotation '2' about Y^1 axis and (c) rotation '1' about X^2 axis.

In Equations 8.14–8.16, R_ψ, R_θ and R_ϕ are known as the rotation (or transformation) matrices which are orthogonal, that is $R_\psi^T R_\psi = R_\psi R_\psi^T = I$; $R_\theta^T R_\theta = R_\theta R_\theta^T = I$ and $R_\phi^T R_\phi = R_\phi R_\phi^T = I$, where I is the identity matrix. Thus, the relation between the axis systems $X^E Y^E Z^E$ and $X^B Y^B Z^B$ is given by the following consolidated transformation:

$$\begin{bmatrix} X^E \\ Y^E \\ Z^E \end{bmatrix} = R_\psi R_\theta R_\phi \begin{bmatrix} X^B \\ Y^B \\ Z^B \end{bmatrix} \tag{8.17}$$

Multiplying both sides of Equation 8.17 by $R_\phi^T R_\theta^T R_\psi^T$ and using the orthogonality condition conditions, we can further write Equation 8.17 as:

$$\begin{bmatrix} X^B \\ Y^B \\ Z^B \end{bmatrix} = R_\phi^T R_\theta^T R_\psi^T \begin{bmatrix} X^E \\ Y^E \\ Z^E \end{bmatrix} \tag{8.18}$$

Using the transformation in Equation 8.18, we are now in a position to write components of aircraft weight along the body-fixed axes as follows:

$$\begin{bmatrix} X^G \\ Y^G \\ Z^G \end{bmatrix} = R_\phi^T R_\theta^T R_\psi^T \begin{bmatrix} 0 \\ 0 \\ mg \end{bmatrix}$$

$$= \begin{bmatrix} 1 & 0 & 0 \\ 0 & \cos\phi & \sin\phi \\ 0 & -\sin\phi & \cos\phi \end{bmatrix} \begin{bmatrix} \cos\theta & 0 & -\sin\theta \\ 0 & 1 & 0 \\ \sin\theta & 0 & \cos\theta \end{bmatrix} \begin{bmatrix} \cos\psi & \sin\psi & 0 \\ -\sin\psi & \cos\psi & 0 \\ 0 & 0 & 1 \end{bmatrix} \begin{bmatrix} 0 \\ 0 \\ mg \end{bmatrix}$$

$$= \begin{bmatrix} -mg\sin\theta \\ mg\sin\phi\cos\theta \\ mg\cos\phi\cos\theta \end{bmatrix} \tag{8.19}$$

In a similar fashion, one can also describe the orientation of the relative wind (wind axis) with respect to the inertial frame using the angles μ, γ and χ in the sequence $\chi-\gamma-\mu$ (refer Figure 8.4). The relation turns out to be:

$$\begin{bmatrix} X^E \\ Y^E \\ Z^E \end{bmatrix} = R_\chi R_\gamma R_\mu \begin{bmatrix} X^W \\ Y^W \\ Z^W \end{bmatrix} \tag{8.20}$$

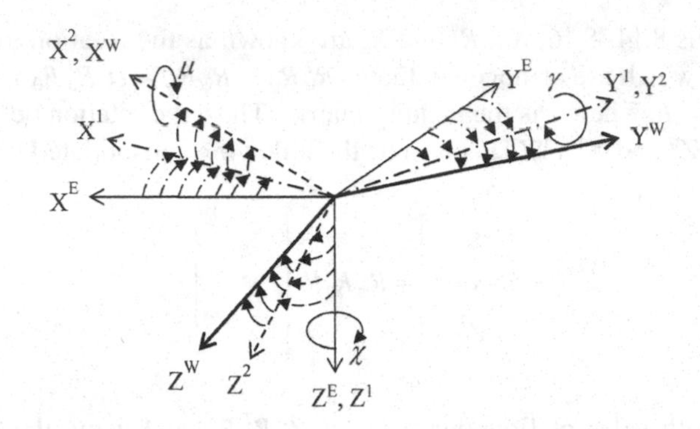

FIGURE 8.4 Rotations χ (about Z^E), γ (about Y^1) and μ (about X^2).

Where,

$$R_\chi = \begin{bmatrix} \cos\chi & -\sin\chi & 0 \\ \sin\chi & \cos\chi & 0 \\ 0 & 0 & 1 \end{bmatrix}; \quad R_\gamma = \begin{bmatrix} \cos\gamma & 0 & \sin\gamma \\ 0 & 1 & 0 \\ -\sin\gamma & 0 & \cos\gamma \end{bmatrix};$$

$$R_\mu = \begin{bmatrix} 1 & 0 & 0 \\ 0 & \cos\mu & -\sin\mu \\ 0 & \sin\mu & \cos\mu \end{bmatrix}$$

are rotation matrices that satisfy the orthogonality condition as well, that is,

$$R_\chi^T R_\chi = R_\chi R_\chi^T = I; \quad R_\gamma^T R_\gamma = R_\gamma R_\gamma^T = I; \quad R_\mu^T R_\mu = R_\mu R_\mu^T = I$$

Therefore,

$$\begin{bmatrix} X^W \\ Y^W \\ Z^W \end{bmatrix} = R_\mu^T R_\gamma^T R_\chi^T \begin{bmatrix} X^E \\ Y^E \\ Z^E \end{bmatrix} \tag{8.21}$$

A schematic summary of the rotations with 3–2–1 sequence for transformations between the Earth, aircraft body-fixed and wind-fixed coordinate systems used in flight dynamics is shown in Figure 8.5. The wind- to body-axis transformation will be discussed shortly.

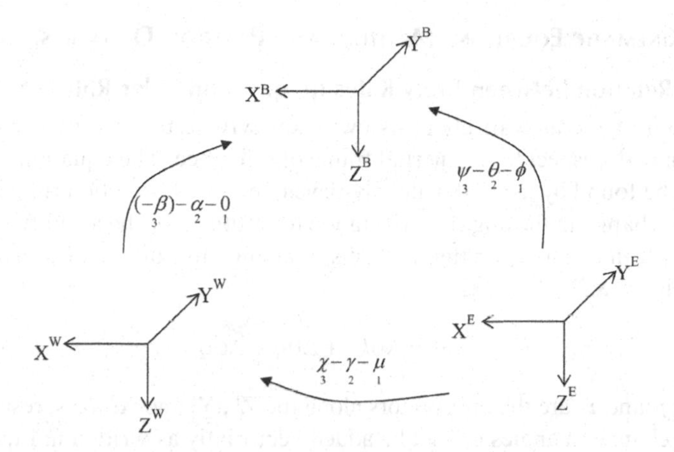

FIGURE 8.5 Schematic summary of rotations with sequence involved between the three coordinate systems used in flight dynamics.

Example 8.1

The components of the velocity vector in wind-fixed axes $\underline{V}_W = [V\ 0\ 0]'$ can now be found along the inertial axes as follows:

$$\begin{bmatrix} V_X \\ V_Y \\ V_Z \end{bmatrix} = \begin{bmatrix} \dot{x}_E \\ \dot{y}_E \\ \dot{z}_E \end{bmatrix} = R_\chi R_\gamma R_\mu \begin{bmatrix} V \\ 0 \\ 0 \end{bmatrix}$$

$$= \begin{bmatrix} \cos\chi & -\sin\chi & 0 \\ \sin\chi & \cos\chi & 0 \\ 0 & 0 & 1 \end{bmatrix} \begin{bmatrix} \cos\gamma & 0 & \sin\gamma \\ 0 & 1 & 0 \\ -\sin\gamma & 0 & \cos\gamma \end{bmatrix} \begin{bmatrix} 1 & 0 & 0 \\ 0 & \cos\mu & -\sin\mu \\ 0 & \sin\mu & \cos\mu \end{bmatrix} \begin{bmatrix} V \\ 0 \\ 0 \end{bmatrix}$$

$$= \begin{bmatrix} \cos\chi & -\sin\chi & 0 \\ \sin\chi & \cos\chi & 0 \\ 0 & 0 & 1 \end{bmatrix} \begin{bmatrix} \cos\gamma & 0 & \sin\gamma \\ 0 & 1 & 0 \\ -\sin\gamma & 0 & \cos\gamma \end{bmatrix} \begin{bmatrix} V \\ 0 \\ 0 \end{bmatrix}$$

$$= \begin{bmatrix} \cos\chi & -\sin\chi & 0 \\ \sin\chi & \cos\chi & 0 \\ 0 & 0 & 1 \end{bmatrix} \begin{bmatrix} V\cos\gamma \\ 0 \\ -V\sin\gamma \end{bmatrix}$$

$$= \begin{bmatrix} V\cos\gamma\cos\chi \\ V\cos\gamma\sin\chi \\ -V\sin\gamma \end{bmatrix} \tag{8.22}$$

Note that these relations were referred to in Chapter 6 (Equations 6.6 and 6.7) for calculating the cross range.

8.3.2 Kinematic Equations (Attitude and Position Dynamics)

8.3.2.1 Relation between Body Rates (p, q, r) and Euler Rates ($\dot{\phi}$, $\dot{\theta}$, $\dot{\psi}$)

The rotation of the aircraft about its own axes will result in time evolution of its orientation with respect to the inertial frame of reference. The equations of this evolution can be found by using the already-developed transformation rules as follows.

A small change in the angular orientation (or attitude) of the aircraft as seen from the inertial frame can be written in terms of change in Euler angles in vector form as (refer Figure 8.3):

$$\Delta\underline{\sigma} = \Delta\psi\hat{k}_E + \Delta\theta\hat{j}_1 + \Delta\phi\hat{i}_2 \tag{8.23}$$

where $\hat{k}_E, \hat{j}_1$ and $\hat{i}_2$ are the unit vectors along the Z^E, Y^1 and X^2 axes, respectively. In general, a change in angles cannot be added vectorially as written in Equation 8.23, but infinitesimal ones can be.

Taking the limit $\Delta t \rightarrow 0$ results in:

$$\lim_{\Delta t\rightarrow 0}\frac{\Delta\underline{\sigma}}{\Delta t} = \lim_{\Delta t\rightarrow 0}\left(\frac{\Delta\psi}{\Delta t}\hat{k}_E + \frac{\Delta\theta}{\Delta t}\hat{j}_1 + \frac{\Delta\phi}{\Delta t}\hat{i}_2\right)$$

or

$$\dot{\underline{\sigma}} = \dot{\psi}\hat{k}_E + \dot{\theta}\hat{j}_1 + \dot{\phi}\hat{i}_2 \tag{8.24}$$

The rate of change of angular orientation vector $\dot{\underline{\sigma}}$ is the angular velocity of the aircraft.

$$\underline{\omega}_C = \begin{bmatrix} p \\ q \\ r \end{bmatrix} = \begin{bmatrix} 0 \\ 0 \\ \dot{\psi} \end{bmatrix}_E + \begin{bmatrix} 0 \\ \dot{\theta} \\ 0 \end{bmatrix}_1 + \begin{bmatrix} \dot{\phi} \\ 0 \\ 0 \end{bmatrix}_2 \tag{8.25}$$

Making use of the transformation rules (Equations 8.14–8.16), we can write Equation 8.25 as:

$$\underline{\omega}_C = \begin{bmatrix} p \\ q \\ r \end{bmatrix} = R_\phi^T R_\theta^T R_\psi^T \begin{bmatrix} 0 \\ 0 \\ \dot{\psi} \end{bmatrix}_E + R_\phi^T R_\theta^T \begin{bmatrix} 0 \\ \dot{\theta} \\ 0 \end{bmatrix}_1 + R_\phi^T \begin{bmatrix} \dot{\phi} \\ 0 \\ 0 \end{bmatrix}_2$$

$$= \begin{bmatrix} 1 & 0 & 0 \\ 0 & \cos\phi & \sin\phi \\ 0 & -\sin\phi & \cos\phi \end{bmatrix} \begin{bmatrix} \cos\theta & 0 & -\sin\theta \\ 0 & 1 & 0 \\ \sin\theta & 0 & \cos\theta \end{bmatrix} \begin{bmatrix} \cos\psi & \sin\psi & 0 \\ -\sin\psi & \cos\psi & 0 \\ 0 & 0 & 1 \end{bmatrix} \begin{bmatrix} 0 \\ 0 \\ \dot{\psi} \end{bmatrix}$$

$$+ \begin{bmatrix} 1 & 0 & 0 \\ 0 & \cos\phi & \sin\phi \\ 0 & -\sin\phi & \cos\phi \end{bmatrix} \begin{bmatrix} \cos\theta & 0 & -\sin\theta \\ 0 & 1 & 0 \\ \sin\theta & 0 & \cos\theta \end{bmatrix} \begin{bmatrix} 0 \\ \dot{\theta} \\ 0 \end{bmatrix} + \begin{bmatrix} 1 & 0 & 0 \\ 0 & \cos\phi & \sin\phi \\ 0 & -\sin\phi & \cos\phi \end{bmatrix} \begin{bmatrix} \dot{\phi} \\ 0 \\ 0 \end{bmatrix} \tag{8.26a}$$

which is:

$$\underline{\omega}_C = \begin{bmatrix} p \\ q \\ r \end{bmatrix} = \begin{bmatrix} 1 & 0 & -\sin\theta \\ 0 & \cos\phi & \sin\phi\cos\theta \\ 0 & -\sin\phi & \cos\phi\cos\theta \end{bmatrix} \begin{bmatrix} \dot{\phi} \\ \dot{\theta} \\ \dot{\psi} \end{bmatrix} \tag{8.26}$$

Multiplying both sides of Equation 8.26a by $R_\theta R_\phi$ results in:

$$R_\theta R_\phi \begin{bmatrix} p \\ q \\ r \end{bmatrix} = R_\theta R_\phi R_\phi^T R_\theta^T R_\psi^T \begin{bmatrix} 0 \\ 0 \\ \dot{\psi} \end{bmatrix}_E + R_\theta R_\phi R_\phi^T R_\theta^T \begin{bmatrix} 0 \\ \dot{\theta} \\ 0 \end{bmatrix}_1 + R_\theta R_\phi R_\phi^T \begin{bmatrix} \dot{\phi} \\ 0 \\ 0 \end{bmatrix}_2$$

$$= R_\psi^T \begin{bmatrix} 0 \\ 0 \\ \dot{\psi} \end{bmatrix} + \begin{bmatrix} 0 \\ \dot{\theta} \\ 0 \end{bmatrix} + R_\theta \begin{bmatrix} \dot{\phi} \\ 0 \\ 0 \end{bmatrix}$$

$$= \begin{bmatrix} 0 \\ 0 \\ \dot{\psi} \end{bmatrix} + \begin{bmatrix} 0 \\ \dot{\theta} \\ 0 \end{bmatrix} + \begin{bmatrix} \cos\theta & 0 & \sin\theta \\ 0 & 1 & 0 \\ -\sin\theta & 0 & \cos\theta \end{bmatrix} \begin{bmatrix} \dot{\phi} \\ 0 \\ 0 \end{bmatrix} = \begin{bmatrix} \dot{\phi}\cos\theta \\ \dot{\theta} \\ \dot{\psi} - \dot{\phi}\sin\theta \end{bmatrix} \tag{8.27a}$$

$$\begin{bmatrix} \dot{\phi}\cos\theta \\ \dot{\theta} \\ \dot{\psi} - \dot{\phi}\sin\theta \end{bmatrix} = R_\theta R_\phi \begin{bmatrix} p \\ q \\ r \end{bmatrix}$$

$$= \begin{bmatrix} \cos\theta & 0 & \sin\theta \\ 0 & 1 & 0 \\ -\sin\theta & 0 & \cos\theta \end{bmatrix} \begin{bmatrix} 1 & 0 & 0 \\ 0 & \cos\phi & -\sin\phi \\ 0 & \sin\phi & \cos\phi \end{bmatrix} \begin{bmatrix} p \\ q \\ r \end{bmatrix}$$

$$= \begin{bmatrix} \cos\theta & \sin\theta\sin\phi & \sin\theta\cos\phi \\ 0 & \cos\phi & -\sin\phi \\ -\sin\theta & \cos\theta\sin\phi & \cos\theta\cos\phi \end{bmatrix} \begin{bmatrix} p \\ q \\ r \end{bmatrix} \tag{8.27b}$$

Thus,

$$\begin{bmatrix} \dot{\phi} \\ \dot{\theta} \\ \dot{\psi} \end{bmatrix} = \begin{bmatrix} 1 & \tan\theta\sin\phi & \tan\theta\cos\phi \\ 0 & \cos\phi & -\sin\phi \\ 0 & \sec\theta\sin\phi & \sec\theta\cos\phi \end{bmatrix} \begin{bmatrix} p \\ q \\ r \end{bmatrix} \tag{8.27}$$

Similarly, the wind-axis rates can be written as:

$$\underline{\omega}_W = \begin{bmatrix} p_w \\ q_w \\ r_w \end{bmatrix} = \begin{bmatrix} 0 \\ 0 \\ \dot{\chi} \end{bmatrix}_E + \begin{bmatrix} 0 \\ \dot{\gamma} \\ 0 \end{bmatrix}_1 + \begin{bmatrix} \dot{\mu} \\ 0 \\ 0 \end{bmatrix}_2 \tag{8.28}$$

where subscript '1' denotes an intermediate coordinate system after the first rotation and subscript '2' denotes the axis system after the second rotation. Then following similar lines as for the body-axis rates we obtain:

$$\begin{bmatrix} \dot{\mu} \\ \dot{\gamma} \\ \dot{\chi} \end{bmatrix} = \begin{bmatrix} 1 & \tan\gamma\sin\mu & \tan\gamma\cos\mu \\ 0 & \cos\mu & -\sin\mu \\ 0 & \sec\gamma\sin\mu & \sec\gamma\cos\mu \end{bmatrix} \begin{bmatrix} p_w \\ q_w \\ r_w \end{bmatrix} \tag{8.29}$$

$$\begin{bmatrix} p_w \\ q_w \\ r_w \end{bmatrix} = \begin{bmatrix} 1 & 0 & -\sin\gamma \\ 0 & \cos\mu & \sin\mu\cos\gamma \\ 0 & -\sin\mu & \cos\mu\cos\gamma \end{bmatrix} \begin{bmatrix} \dot{\mu} \\ \dot{\gamma} \\ \dot{\chi} \end{bmatrix} \tag{8.30}$$

8.3.2.2 Relation between Inertial Velocity and Body-Axis Velocity Components

The aircraft position at any instant can be determined with respect to the inertial frame of reference by using the following equations of kinematics:

$$\begin{bmatrix} \dot{x}_E \\ \dot{y}_E \\ \dot{z}_E \end{bmatrix} = \begin{bmatrix} \cos\psi & -\sin\psi & 0 \\ \sin\psi & \cos\psi & 0 \\ 0 & 0 & 1 \end{bmatrix} \begin{bmatrix} \cos\theta & 0 & \sin\theta \\ 0 & 1 & 0 \\ -\sin\theta & 0 & \cos\theta \end{bmatrix} \begin{bmatrix} 1 & 0 & 0 \\ 0 & \cos\phi & -\sin\phi \\ 0 & \sin\phi & \cos\phi \end{bmatrix} \begin{bmatrix} u \\ v \\ w \end{bmatrix}$$

$$\begin{bmatrix} \dot{x}_E \\ \dot{y}_E \\ \dot{z}_E \end{bmatrix} = \begin{bmatrix} \cos\psi\cos\theta & \cos\psi\sin\theta\sin\phi - \sin\psi\cos\phi & \cos\psi\sin\theta\cos\phi + \sin\psi\sin\phi \\ \sin\psi\cos\theta & \sin\psi\sin\theta\sin\phi + \cos\psi\cos\phi & \sin\psi\sin\theta\cos\phi - \cos\psi\sin\phi \\ -\sin\theta & \cos\theta\sin\phi & \cos\theta\sin\phi \end{bmatrix} \begin{bmatrix} u \\ v \\ w \end{bmatrix} \tag{8.31}$$

8.3.2.3 Relation between Body-Fixed and Wind-Fixed Coordinates

Another useful transformation is that between the wind-fixed axis system and the body-fixed axis system given by the following rotations:

$$\text{First rotation: } X^W - Y^W - Z^W \xrightarrow{\text{rotation } (-\beta) \text{ about } Z^W} X^S - Y^S - Z^S$$

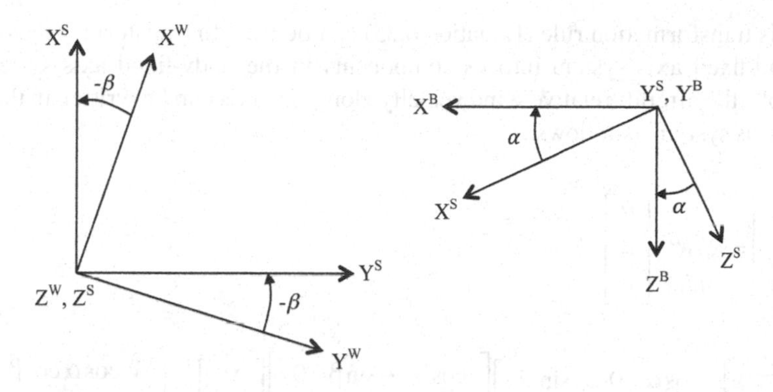

FIGURE 8.6 Rotations ($-\beta$) of wind-axis system and α of the stability-axis system.

The subscript 'S' above refers to what is called the *stability-axis* system. Following Figure 8.6, the relation between the two axis systems can be written as:

$$\begin{bmatrix} X^W \\ Y^W \\ Z^W \end{bmatrix} = \begin{bmatrix} \cos\beta & \sin\beta & 0 \\ -\sin\beta & \cos\beta & 0 \\ 0 & 0 & 1 \end{bmatrix} \begin{bmatrix} X^S \\ Y^S \\ Z^S \end{bmatrix} = R_{-\beta} \begin{bmatrix} X^S \\ Y^S \\ Z^S \end{bmatrix} \tag{8.32}$$

Second rotation: $X^S - Y^S - Z^S \xrightarrow{\text{rotation } \alpha \text{ about } Y^S} X^B - Y^B - Z^B$

$$\begin{bmatrix} X^S \\ Y^S \\ Z^S \end{bmatrix} = \begin{bmatrix} \cos\alpha & 0 & \sin\alpha \\ 0 & 1 & 0 \\ -\sin\alpha & 0 & \cos\alpha \end{bmatrix} \begin{bmatrix} X^B \\ Y^B \\ Z^B \end{bmatrix} = R_{\alpha} \begin{bmatrix} X^B \\ Y^B \\ Z^B \end{bmatrix} \tag{8.33}$$

The rotation matrices $R_{-\beta}$ and R_{α} satisfy the orthogonality conditions as well. Thus, the relation between the wind-fixed and body-fixed axes can be written using the following transformation rule:

$$\begin{bmatrix} X^W \\ Y^W \\ Z^W \end{bmatrix} = R_{-\beta} R_{\alpha} \begin{bmatrix} X^B \\ Y^B \\ Z^B \end{bmatrix} \tag{8.34}$$

or

$$\begin{bmatrix} X^B \\ Y^B \\ Z^B \end{bmatrix} = \begin{bmatrix} \cos\alpha & 0 & -\sin\alpha \\ 0 & 1 & 0 \\ \sin\alpha & 0 & \cos\alpha \end{bmatrix} \begin{bmatrix} \cos\beta & -\sin\beta & 0 \\ \sin\beta & \cos\beta & 0 \\ 0 & 0 & 1 \end{bmatrix} \begin{bmatrix} X^W \\ Y^W \\ Z^W \end{bmatrix} = R_{\alpha}^T R_{-\beta}^T \begin{bmatrix} X^W \\ Y^W \\ Z^W \end{bmatrix} \tag{8.35}$$

This transformation rule (Equation 8.35) can be used to transform a vector given in wind-fixed axis system into its components in the body-fixed axis system. For example, the aircraft relative wind velocity along X^W axis can be written in the body-fixed axis system as follows:

$$\begin{bmatrix} u \\ v \\ w \end{bmatrix} = \underbrace{R_\alpha^T R_{-\beta}^T}_{T_{BW}} \begin{bmatrix} V \\ 0 \\ 0 \end{bmatrix}$$

$$= \begin{bmatrix} \cos\alpha & 0 & -\sin\alpha \\ 0 & 1 & 0 \\ \sin\alpha & 0 & \cos\alpha \end{bmatrix} \begin{bmatrix} \cos\beta & -\sin\beta & 0 \\ \sin\beta & \cos\beta & 0 \\ 0 & 0 & 1 \end{bmatrix} \begin{bmatrix} V \\ 0 \\ 0 \end{bmatrix} = \begin{bmatrix} V\cos\alpha\cos\beta \\ V\sin\beta \\ V\sin\alpha\cos\beta \end{bmatrix}$$

where the transformation matrix:

$$T_{BW} = \begin{bmatrix} \cos\alpha & 0 & -\sin\alpha \\ 0 & 1 & 0 \\ \sin\alpha & 0 & \cos\alpha \end{bmatrix} \begin{bmatrix} \cos\beta & -\sin\beta & 0 \\ \sin\beta & \cos\beta & 0 \\ 0 & 0 & 1 \end{bmatrix}$$

$$= \begin{bmatrix} \cos\alpha\cos\beta & -\cos\alpha\sin\beta & -\sin\alpha \\ \sin\beta & \cos\beta & 0 \\ \sin\alpha\cos\beta & -\sin\alpha\sin\beta & \cos\alpha \end{bmatrix} \tag{8.36}$$

8.3.2.4 Relation between the Body-Axis and Wind-Axis Euler Angles

From Figure 8.5, we find that there are two equivalent ways of going from the Earth-fixed inertial axes to the airplane body-fixed axes. The first is a direct transformation through the set of three Euler angles ψ, θ, ϕ, as we have seen in Equation 8.17. Another way of reaching the same end state would be to first transform from the Earth-fixed inertial axes to the wind axes through the three Euler angles χ, γ, μ, as in Equation 8.20 and then from the wind axes to the body axes through the aerodynamic angles β, α as in Equation 8.34. That either route transforms the axes from the same initial state (Earth-fixed inertial) to the same end state (body-fixed) can be stated in matrix equality terms as given below:

$$R_\chi R_\gamma R_\mu R_{-\beta} R_\alpha = R_\psi R_\theta R_\phi$$

Or equivalently as given below:

$$R_\chi R_\gamma R_\mu = R_\psi R_\theta R_\phi R_\alpha^T R_{-\beta}^T \tag{8.37}$$

Equation 8.37, on expansion, results in nine relations between the body-axis Euler angles ψ, θ, ϕ, and the wind-axis Euler angles χ, γ, μ, in terms of the aerodynamic angles β,α:

$$\begin{bmatrix} c_\chi c_\gamma & c_\chi s_\gamma s_\mu - s_\chi c_\mu & c_\chi s_\gamma c_\mu + s_\chi s_\mu \\ s_\chi c_\gamma & s_\chi s_\gamma s_\mu + c_\chi c_\mu & s_\chi s_\gamma c_\mu - c_\chi s_\mu \\ -s_\gamma & c_\gamma s_\mu & c_\gamma c_\mu \end{bmatrix}$$

$$= \begin{bmatrix} c_\psi c_\theta & c_\psi s_\theta s_\phi - s_\psi c_\phi & c_\psi s_\theta c_\phi + s_\psi s_\phi \\ s_\psi c_\theta & s_\psi s_\theta s_\phi + c_\psi c_\phi & s_\psi s_\theta c_\phi - c_\psi s_\phi \\ -s_\theta & c_\theta s_\phi & c_\theta c_\phi \end{bmatrix} \begin{bmatrix} c_\alpha c_\beta & -c_\alpha s_\beta & -s_\alpha \\ s_\beta & c_\beta & 0 \\ s_\alpha c_\beta & -s_\alpha s_\beta & c_\alpha \end{bmatrix} \tag{8.38}$$

where, $c_{(\cdot)} = \cos(\cdot)$; $s_{(\cdot)} = \sin(\cdot)$. Equation 8.38 can be further expanded to:

$$\begin{bmatrix} c_\chi c_\gamma & c_\chi s_\gamma s_\mu - s_\chi c_\mu & c_\chi s_\gamma c_\mu + s_\chi s_\mu \\ s_\chi c_\gamma & s_\chi s_\gamma s_\mu + c_\chi c_\mu & s_\chi s_\gamma c_\mu - c_\chi s_\mu \\ -s_\gamma & c_\gamma s_\mu & c_\gamma c_\mu \end{bmatrix}$$

$$= \begin{bmatrix} \begin{array}{l} c_\psi c_\theta c_\alpha c_\beta \\ +s_\beta\left(c_\psi s_\theta s_\phi - s_\psi c_\phi\right) \\ +s_\alpha c_\beta\left(c_\psi s_\theta c_\phi + s_\psi c_\phi\right) \end{array} & \begin{array}{l} -c_\alpha s_\beta c_\psi c_\theta \\ +c_\beta\left(c_\psi s_\theta s_\phi - s_\psi c_\phi\right) \\ -s_\alpha s_\beta\left(c_\psi s_\theta c_\phi + s_\psi s_\phi\right) \end{array} & \begin{array}{l} -s_\alpha c_\psi c_\theta \\ +c_\alpha\left(c_\psi s_\theta c_\phi + s_\psi s_\phi\right) \end{array} \\ \begin{array}{l} s_\psi c_\theta c_\alpha c_\beta \\ +s_\beta\left(s_\psi s_\theta s_\phi + c_\psi c_\phi\right) \\ +s_\alpha c_\beta\left(s_\psi s_\theta s_\phi - c_\psi c_\phi\right) \end{array} & \begin{array}{l} -c_\alpha s_\beta s_\psi c_\theta \\ +c_\beta\left(s_\psi s_\theta s_\phi + c_\psi c_\phi\right) \\ -s_\alpha s_\beta\left(s_\psi s_\theta c_\phi - c_\psi s_\phi\right) \end{array} & \begin{array}{l} -s_\alpha s_\psi c_\theta \\ +c_\alpha\left(s_\psi s_\theta c_\phi - c_\psi s_\phi\right) \end{array} \\ \begin{array}{l} -s_\theta c_\alpha c_\beta \\ +c_\theta s_\phi s_\beta + c_\theta c_\phi s_\alpha c_\beta \end{array} & \begin{array}{l} s_\theta c_\alpha s_\beta \\ +c_\theta s_\phi c_\beta - c_\theta c_\phi s_\alpha s_\beta \end{array} & \begin{array}{l} s_\alpha s_\theta \\ +c_\theta c_\phi c_\alpha \end{array} \end{bmatrix} \tag{8.39}$$

Thus, for instance, by comparing the entries in the last row of the matrices in Equation 8.39 term wise, we obtain the following equalities:

$$\begin{aligned} \sin\gamma &= \cos\alpha\cos\beta\sin\theta - \sin\beta\sin\phi\cos\theta - \sin\alpha\cos\beta\cos\phi\cos\theta \\ \sin\mu\cos\gamma &= \sin\theta\cos\alpha\sin\beta + \sin\phi\cos\theta\cos\beta - \sin\alpha\sin\beta\cos\phi\cos\theta \\ \cos\mu\cos\gamma &= \sin\theta\sin\alpha + \cos\alpha\cos\phi\cos\theta \end{aligned} \tag{8.40}$$

There are nine such relations in all from Equation 8.39.

8.3.2.5 Relation between the Body-Axis and Wind-Axis Angular Rates

Equation 8.26 related the body-axis angular rates p, q, r to the rates of change of the body-axis Euler angles (defined with respect to the Earth-fixed inertial axis). Likewise, Equation 8.30 gives the wind-axis angular rates p_w, q_w, r_w in terms of the rates of change of the wind-axis Euler angles (again defined with respect to the Earth-fixed inertial axis).

In a similar fashion, we can write the difference between the body-axis and the wind-axis angular rates in terms of the relative rate of change of the aerodynamic angles, which define the transformation from the wind to the body axes. As we have seen from Equation 8.35, there are only two angles in question here: β and α. As seen in Figure 8.6, the first rotation by $-\beta$ takes place about the wind Z^W axis, which is also the stability Z^S axis (unit vector $\hat{k}_W$). The second rotation by α takes place about the stability Y^S axis, which also happens to be the body Y^B axis (unit vector $\hat{j}_B$).

Thus, the rate of rotation of the body axis with respect to the wind axis can be expressed in terms of the rate of change of angles as:

$$\begin{bmatrix} p_b \\ q_b \\ r_b \end{bmatrix}^B - \begin{bmatrix} p_w \\ q_w \\ r_w \end{bmatrix}^W = \left(-\dot{\beta}\right)\hat{k}_W + \dot{\alpha}\hat{j}_B$$

where the superscript 'B' indicates that the body-axis angular velocity is written in terms of its components along the body-fixed axes $X^BY^BZ^B$ and correspondingly the superscript 'W' indicates that the wind-axis angular velocity is in terms of the wind-axis components. Subscript 'b' is used to indicate body axis rates. To meaningfully subtract one from the other, we must ensure that the components of both vectors $\underline{\omega}_C$ and $\underline{\omega}_W$ are along the same axes, so we convert the wind-axis components of $\underline{\omega}_W$ to the body-axis by the transformation matrix T_{BW} of Equation 8.36. Likewise, the wind-axis unit vector $\hat{k}_W$ must also be transformed to the body axis.

$$\begin{bmatrix} p_b \\ q_b \\ r_b \end{bmatrix}^B - T_{BW}\begin{bmatrix} p_w \\ q_w \\ r_w \end{bmatrix}^W = T_{BW}\left(-\dot{\beta}\right)\hat{k}_W + \dot{\alpha}\hat{j}_B \tag{8.41}$$

Expanding Equation 8.41, where now all components are along the body axes, we get:

$$\begin{bmatrix} p_b - p_w \\ q_b - q_w \\ r_b - r_w \end{bmatrix}^B = T_{BW}\begin{bmatrix} 0 \\ 0 \\ -\dot{\beta} \end{bmatrix} + \begin{bmatrix} 0 \\ \dot{\alpha} \\ 0 \end{bmatrix}$$

$$= \begin{bmatrix} \cos\alpha\cos\beta & -\cos\alpha\sin\beta & -\sin\alpha \\ \sin\beta & \cos\beta & 0 \\ \sin\alpha\cos\beta & -\sin\alpha\sin\beta & \cos\alpha \end{bmatrix}\begin{bmatrix} 0 \\ 0 \\ -\dot{\beta} \end{bmatrix} + \begin{bmatrix} 0 \\ \dot{\alpha} \\ 0 \end{bmatrix} = \begin{bmatrix} \dot{\beta}\sin\alpha \\ \dot{\alpha} \\ -\dot{\beta}\cos\alpha \end{bmatrix} \tag{8.42}$$

The superscript 'B' refers to components along the body-fixed axes. This is the relation that was used in Chapter 6 (Equation 6.1).

8.3.3 Force Equations Summed Up

The aircraft equations of translational motion, Equation 8.11, and the forces in Equation 8.13 with components along the body-fixed axes can now be written component-wise as follows:

$$\begin{aligned}\dot{u} &= rv - qw + \frac{X^A}{m} + \frac{T}{m} - g\sin\theta \\ \dot{v} &= pw - ru + \frac{Y^A}{m} + g\sin\phi\cos\theta \\ \dot{w} &= qu - pv + \frac{Z^A}{m} + g\cos\phi\cos\theta\end{aligned} \tag{8.43}$$

In terms of the force coefficients:

$$\begin{aligned}\dot{u} &= rv - qw + \frac{\overline{q}SC_X}{m} + \frac{T}{m} - g\sin\theta \\ \dot{v} &= pw - ru + \frac{\overline{q}SC_Y}{m} + g\sin\phi\cos\theta \\ \dot{w} &= qu - pv + \frac{\overline{q}SC_Z}{m} + g\cos\phi\cos\theta\end{aligned} \tag{8.44}$$

In Equations 8.43 and 8.44, C_X, C_Y, C_Z represent the aerodynamic force coefficients.

Of course, we can take the components of the vectors $\underline{F}$, $\underline{V_C}$ and $\underline{\omega_C}$ along the wind axis as well, and often this may be more convenient.

8.3.3.1 Derivation of Force Equations in Wind-Fixed Axis System

Referring back to Equation 8.8, we can write the right-hand side in terms of wind-axis quantities thus:

$$\underline{F} = m\left.\frac{d\underline{V_C}}{dt}\right|_I = m\left.\frac{d\underline{V_C}}{dt}\right|_W + m\left(\underline{\omega_W} \times \underline{V_C}\right)$$

with $\underline{\omega_W}\times = \begin{bmatrix} 0 & -r_w & q_w \\ r_w & 0 & -p_w \\ -q_w & p_w & 0 \end{bmatrix}$

Expanding, we get:

$$m\left\{\begin{bmatrix}\dot{V}\\0\\0\end{bmatrix}+\begin{bmatrix}0\\r_wV\\-q_wV\end{bmatrix}\right\}=\begin{bmatrix}X_W\\Y_W\\Z_W\end{bmatrix}$$

That is,

$$\begin{aligned}\dot{V}&=\frac{X_W}{m}\\ r_wV&=\frac{Y_W}{m}\\ -q_wV&=\frac{Z_W}{m}\end{aligned}\tag{8.45}$$

The force components in wind axis appear as:

$$\begin{bmatrix}X_W\\Y_W\\Z_W\end{bmatrix}=T_{WE}\begin{bmatrix}0\\0\\mg\end{bmatrix}+\begin{bmatrix}X^A\\Y^A\\Z^A\end{bmatrix}+T_{WB}\begin{bmatrix}T\\0\\0\end{bmatrix}\tag{8.46}$$

Writing out the right-hand side terms one-by-one:

From Equation 8.21,

$$T_{WE}\begin{bmatrix}0\\0\\mg\end{bmatrix}=R_\mu^T R_\gamma^T R_\chi^T\begin{bmatrix}0\\0\\mg\end{bmatrix}=\begin{bmatrix}-mg\sin\gamma\\mg\cos\gamma\sin\mu\\mg\cos\gamma\cos\mu\end{bmatrix}\tag{8.46a}$$

$$\begin{bmatrix}X^A\\Y^A\\Z^A\end{bmatrix}=R_{-\beta}\bar{q}S\begin{bmatrix}-C_D\\C_Y\\-C_L\end{bmatrix}=\bar{q}S\begin{bmatrix}\cos\beta & \sin\beta & 0\\-\sin\beta & \cos\beta & 0\\0 & 0 & 1\end{bmatrix}\begin{bmatrix}-C_D\\C_Y\\-C_L\end{bmatrix}$$

$$=\bar{q}S\begin{bmatrix}-C_D\cos\beta+C_Y\sin\beta\\C_D\sin\beta+C_Y\cos\beta\\-C_L\end{bmatrix}\tag{8.46b}$$

where C_D, C_L are the drag and lift coefficients, and C_Y is the side force coefficient (see Figure 8.2 and the note on resolution of aerodynamic forces in Chapter 6). From Equation 8.36, with $T_{WB} = T_{BW}^T$:

$$T_{BW}^T \begin{bmatrix} T \\ 0 \\ 0 \end{bmatrix} = \begin{bmatrix} \cos\alpha\cos\beta & \sin\beta & \sin\alpha\cos\beta \\ -\cos\alpha\sin\beta & \cos\beta & -\sin\alpha\sin\beta \\ -\sin\alpha & 0 & \cos\alpha \end{bmatrix} \begin{bmatrix} T \\ 0 \\ 0 \end{bmatrix} = \begin{bmatrix} T\cos\alpha\cos\beta \\ -T\cos\alpha\sin\beta \\ -T\sin\alpha \end{bmatrix} \tag{8.46c}$$

So, collecting terms, we can write out Equation 8.46 in full as:

$$\begin{bmatrix} X_W \\ Y_W \\ Z_W \end{bmatrix} = \begin{bmatrix} -mg\sin\gamma \\ mg\cos\gamma\sin\mu \\ mg\cos\gamma\cos\mu \end{bmatrix} + \bar{q}S \begin{bmatrix} -C_D\cos\beta + C_Y\sin\beta \\ C_D\sin\beta + C_Y\cos\beta \\ -C_L \end{bmatrix} + \begin{bmatrix} T\cos\alpha\cos\beta \\ -T\cos\alpha\sin\beta \\ -T\sin\alpha \end{bmatrix}$$

And the wind-axis equations of translational motion can be written from Equation 8.45 as:

$$\dot{V} = -g\sin\gamma - \frac{\bar{q}S(C_D\cos\beta - C_Y\sin\beta)}{m} + \frac{T\cos\alpha\cos\beta}{m} \tag{8.47}$$

$$r_w V = g\cos\gamma\sin\mu + \frac{\bar{q}S(C_Y\cos\beta - C_D\sin\beta)}{m} + \frac{T\cos\alpha\sin\beta}{m} \tag{8.48}$$

$$-q_w V = g\cos\gamma\cos\mu - \frac{\bar{q}SC_L}{m} - \frac{T\sin\alpha}{m} \tag{8.49}$$

We have from Equation 8.30,

$$q_w = \dot{\gamma}\cos\mu + \dot{\chi}\sin\mu\cos\gamma$$
$$r_w = -\dot{\gamma}\sin\mu + \dot{\chi}\cos\mu\cos\gamma$$

Hence, the wind-axis equations of motion can be written as:

$$\begin{aligned} \dot{V} &= -g\sin\gamma - \frac{\bar{q}S(C_D\cos\beta - C_Y\sin\beta)}{m} + \frac{T\cos\alpha\cos\beta}{m} \\ V(-\dot{\gamma}\sin\mu + \dot{\chi}\cos\mu\cos\gamma) &= g\cos\gamma\sin\mu + \frac{\bar{q}S(C_Y\cos\beta - C_D\sin\beta)}{m} - \frac{T\cos\alpha\cos\beta}{m} \\ -V(\dot{\gamma}\sin\mu + \dot{\chi}\cos\mu\cos\gamma) &= g\cos\gamma\sin\mu - \frac{\bar{q}SC_L}{m} - \frac{T\sin\alpha}{m} \end{aligned} \tag{8.50}$$

When integrated, set of Equations (8.50) will give the velocity of the airplane and orientation angles γ, χ of the velocity vector with respect to the Earth. Thus, this constitutes the airplane's navigation equations.

An alternative way of dealing with Equations (8.48) and (8.49) is to convert the wind-axis angular rates p_w, q_w, r_w into the body-axis angular rates p, q, r by use of Equation (8.42) as below:

$$p_w^b = p_b^b - \dot{\beta}\sin\alpha$$
$$q_w^b = q_b^b - \dot{\alpha}$$
$$r_w^b = r_b^b + \dot{\beta}\cos\alpha$$

$$\begin{bmatrix} p_w^w \\ q_w^w \\ r_w^w \end{bmatrix} = T_{WB}\begin{bmatrix} p_w^b \\ q_w^b \\ r_w^b \end{bmatrix} = \begin{bmatrix} \cos\alpha\cos\beta & \sin\beta & \sin\alpha\cos\beta \\ -\cos\alpha\sin\beta & \cos\beta & -\sin\alpha\sin\beta \\ -\sin\alpha & 0 & \cos\alpha \end{bmatrix}\begin{bmatrix} p_b^b - \dot{\beta}\sin\alpha \\ q_b^b - \dot{\alpha} \\ r_b^b + \dot{\beta}\cos\alpha \end{bmatrix} \quad (8.51)$$

Thus,

$$q_w^w = -p_b^b\cos\alpha\sin\beta + \left(q_b^b - \dot{\alpha}\right)\cos\beta - r_b^b\sin\alpha\sin\beta$$
$$= -\left(p_b^b\cos\alpha + r_b^b\sin\alpha\right)\sin\beta + \left(q_b^b - \dot{\alpha}\right)\cos\beta$$

Therefore Equation (8.49) is,

$$\left(p_b^b\cos\alpha + r_b^b\sin\alpha\right)V\sin\beta - \left(q_b^b - \dot{\alpha}\right)V\cos\beta = g\cos\gamma\cos\mu - \frac{\bar{q}SC_L}{m} - \frac{T\sin\alpha}{m}$$

Rearranging terms yields:

$$-\left(q_b^b - \dot{\alpha}\right)V\cos\beta = g\cos\gamma\cos\mu - \frac{\bar{q}SC_L}{m} - \frac{T\sin\alpha}{m} - \left(p_b^b\cos\alpha + r_b^b\sin\alpha\right)V\sin\beta$$

Thus,

$$\dot{\alpha} = q_b - \frac{1}{\cos\beta}\left\{\left(p_b\cos\alpha + r_b\sin\alpha\right)\sin\beta - \frac{g}{V}\cos\gamma\cos\mu + \frac{\bar{q}SC_L}{mV} + \frac{T\sin\alpha}{mV}\right\} \quad (8.52)$$

where the superscripts on p, q, r are no longer needed.

Similarly, from Equation (8.51), we obtain:

$$r_w^w = -p_b^b\sin\alpha + \dot{\beta}\sin^2\alpha + r_b^b\cos\alpha + \dot{\beta}\sin^2\alpha = \left(-p_b^b\sin\alpha + r_b^b\cos\alpha\right) + \dot{\beta}$$

Then, Equation (8.48) appears as:

$$V\left[\left(-p_b^b \sin\alpha + r_b^b \cos\alpha\right) + \dot{\beta}\right] = g\cos\gamma\sin\mu + \frac{\bar{q}S\left(C_Y\cos\beta + C_D\sin\beta\right)}{m} - \frac{T\cos\alpha\sin\beta}{m}$$

Which yields,

$$\dot{\beta} = \left(p_b\sin\alpha - r_b\cos\alpha\right) + \frac{g}{V}\cos\gamma\sin\mu + \frac{\bar{q}S\left(C_Y\cos\beta + C_D\sin\beta\right)}{mV} - \frac{T\cos\alpha\sin\beta}{mV} \tag{8.53}$$

Equations (8.47), (8.52) and (8.53) is another way of writing the translational equations of motion in terms of the velocity V and the aerodynamic angles α, β – angles the wind-axis makes with the body-fixed axis.

Dropping the subscripts on p, q, r, which are body-fixed rates, and collecting the equations, one can arrive at the following set of force equations of aircraft motion in wind-fixed axes:

$$\begin{aligned}
\dot{V} &= -g\sin\gamma - \frac{\bar{q}S\left(C_D\cos\beta - C_Y\sin\beta\right)}{m} + \frac{T\cos\alpha\cos\beta}{m} \\
\dot{\alpha} &= q - \frac{1}{\cos\beta}\left\{\left(p\cos\alpha + r\sin\alpha\right)\sin\beta - \frac{g}{V}\cos\gamma\cos\mu + \frac{\bar{q}SC_L}{mV} + \frac{T\sin\alpha}{mV}\right\} \\
\dot{\beta} &= \left(p\sin\alpha - r\cos\alpha\right) + \frac{g}{V}\cos\gamma\sin\mu + \frac{\bar{q}S\left(C_Y\cos\beta + C_D\sin\beta\right)}{mV} - \frac{T\cos\alpha\sin\beta}{mV}
\end{aligned} \tag{8.54}$$

8.4 DERIVATION OF AIRCRAFT EQUATIONS OF MOTION (CONTD.)

8.4.1 Equations for the Rotational Motion

The equations for the rotational motion of an aircraft can be arrived at by taking the rate of change of angular momentum of the elemental mass δm shown in Figure 8.1. Angular momentum of the elemental mass δm about the center-of-mass C is given by:

$$\delta\underline{h} = \underline{r}_C \times \delta m\left(\underline{V}_C + \underline{\omega}_C \times \underline{r}_C\right)$$

Summing over the whole body of aircraft:

$$\begin{aligned}
\underline{h} &= \sum\delta\underline{h} = \sum\underline{r}_C \times \delta m\left(\underline{V}_C + \underline{\omega}_C \times \underline{r}_C\right) \\
&= \underbrace{\sum\underline{r}_C\delta m \times \underline{V}_C}_{1} + \underbrace{\sum\delta m\left\{\underline{r}_C \times \underline{\omega}_C \times \underline{r}_C\right\}}_{2}
\end{aligned} \tag{8.55}$$

The term '1' in Equation 8.55 is zero since we measure all distances from the centre of mass of the aircraft, so that $\sum \underline{r}_C \delta m = 0$, which is also $\int \underline{r}_C dm = 0$ in the limit $\delta m \to 0$, as before. Expanding the term '2' results in components of angular momentum vector about the aircraft-fixed axes as:

$$\begin{aligned}
\underline{h} &= \sum \delta m\left\{\left(x\hat{i} + y\hat{j} + z\hat{k}\right)\times\left[\left(p\hat{i} + q\hat{j} + r\hat{k}\right)\times\left(x\hat{i} + y\hat{j} + z\hat{k}\right)\right]\right. \\
&= \sum \delta m\left\{\left(x\hat{i} + y\hat{j} + z\hat{k}\right)\times\left[(qz - ry)\hat{i} + (rx - pz)\hat{j} + (py - qx)\hat{k}\right]\right. \\
&= \left[p\sum(y^2 + z^2)\delta m - q\sum xy\delta m - r\sum xz\delta m\right]\hat{i} \\
&\quad + \left[-p\sum yx\delta m + q\sum(x^2 + z^2)\delta m - r\sum yz\delta m\right]\hat{j} \\
&\quad + \left[-p\sum zx\delta m - q\sum zy\delta m + r\sum(x^2 + y^2)\delta m\right]\hat{k}
\end{aligned}$$

which in the limit $\delta m \to 0$ can be written as:

$$\begin{aligned}
\underline{h} &= \left[p\int(y^2 + z^2)\, dm - q\int xy\, dm - r\int xz\, dm\right]\hat{i} \\
&\quad + \left[-p\int yx\, dm + q\int(x^2 + z^2)dm - r\int yz\, dm\right]\hat{j} \\
&\quad + \left[-p\int zx\, dm - q\int zy\, dm + r\int(y^2 + x^2)dm\right]\hat{k} \\
&= \left[pI_{xx} - qI_{xy} - rI_{xz}\right]\hat{i} + \left[-pI_{yx} + qI_{yy} - rI_{yz}\right]\hat{j} + \left[-pI_{zx} - qI_{zy} + rI_{zz}\right]\hat{k}
\end{aligned} \tag{8.56}$$

Equation 8.56 can be rewritten in matrix form as:

$$\underline{h} = \begin{bmatrix} h_x \\ h_y \\ h_z \end{bmatrix} = \begin{bmatrix} I_{xx} & -I_{xy} & -I_{xz} \\ -I_{yx} & I_{yy} & -I_{yz} \\ -I_{zx} & -I_{zy} & I_{zz} \end{bmatrix}\begin{bmatrix} p \\ q \\ r \end{bmatrix} = \underline{\underline{I}} \cdot \underline{\omega}_C \tag{8.57}$$

where the moments of inertia terms are defined as:

$$I_{xx} = \int(y^2 + z^2)\, dm,\quad I_{yy} = \int(x^2 + z^2)\, dm,\quad I_{zz} = \int(x^2 + z^2)\, dm,$$

$$I_{xy} = I_{yx} = \int xy\, dm,\quad I_{xz} = I_{zx} = \int xz\, dm,\quad I_{yz} = I_{zy} = \int yz\, dm$$

Now, using the rule for converting the derivative *d/dt* from inertial (Earth) axis to body axis, as before:

$$\underline{M} = \left.\frac{d\underline{h}}{dt}\right|_I = \left.\frac{d\underline{h}}{dt}\right|_B + \underline{\omega}_C \times \underline{h} \tag{8.58}$$

Once again we can write the vectors $\underline{h}$, $\underline{\omega}_C$ in terms of their components in any coordinate system, but the body-axis coordinates is the obvious choice because that is the only one in which the moments of inertia remain constant (assuming of course no loss or redistribution of mass). Thus, the standard form of the equations for aircraft rotational motion is written in terms of angular momentum and angular velocity components along the body axes as below:

$$\underline{M} = \begin{bmatrix} M_x \\ M_y \\ M_z \end{bmatrix} = \begin{bmatrix} \dot{h}_x \\ \dot{h}_y \\ \dot{h}_z \end{bmatrix} + \begin{bmatrix} 0 & -r & q \\ r & 0 & -p \\ -q & p & 0 \end{bmatrix} \begin{bmatrix} h_x \\ h_y \\ h_z \end{bmatrix} \tag{8.59}$$

where $\underline{M}$ is the sum of external moments acting on the aircraft, and p, q, r are body-axis angular rates.

8.4.2 Symmetry of Aircraft

Most airplanes have a plane of symmetry—the longitudinal plane $X^B Z^B$, which means that the cross products of inertia, $I_{xy} = I_{yz} = 0$. Further, if we assume that the body-fixed axis system coincides with the principal axes of the aircraft (since we are free to choose the body-fixed axes as we please), then the cross moment of inertia terms would all be zero, that is, $I_{xz} = I_{xy} = I_{yz} = 0$. In that case, from Equation 8.57, $h_x = I_{xx} p$; $h_y = I_{yy} q$; $h_z = I_{zz} r$, and

$$\underline{M} = \begin{bmatrix} M_x \\ M_y \\ M_z \end{bmatrix} = \begin{bmatrix} I_{xx}\dot{p} \\ I_{yy}\dot{q} \\ I_{zz}\dot{r} \end{bmatrix} + \begin{bmatrix} 0 & -r & q \\ r & 0 & -p \\ -q & p & 0 \end{bmatrix} \begin{bmatrix} I_{xx} p \\ I_{yy} q \\ I_{zz} r \end{bmatrix} \tag{8.60}$$

where $\underline{M} = \underline{M}^{Aerodynamic} + \underline{M}^{Propulsive}$ is the sum of external moments acting on the aircraft.

$$\underline{M}^{Aerodynamic} = \begin{bmatrix} \mathcal{L} \\ M \\ N \end{bmatrix} \tag{8.61}$$

is the aerodynamic moment vector, where $\mathcal{L}$, M and N are rolling, pitching and yawing moments, respectively. $\underline{M}^{Propulsive}$ is the moment vector due to engine power. Note that the gravity terms will act at the centre of gravity (equivalent to the centre of mass here); hence, they do not contribute to the moment at the CG.

Assuming $\underline{M}^{Propulsive} = 0$, that is, propulsive forces pass through the centre of gravity of the aircraft, Equation 8.60 for the rotational motion of an aircraft appears as:

$$\dot{p} = \left(\frac{I_{yy} - I_{zz}}{I_{xx}}\right) qr + \frac{\mathcal{L}}{I_{xx}}$$
$$\dot{q} = \left(\frac{I_{zz} - I_{xx}}{I_{yy}}\right) pr + \frac{M}{I_{yy}} \tag{8.62}$$
$$\dot{r} = \left(\frac{I_{xx} - I_{yy}}{I_{zz}}\right) pq + \frac{N}{I_{zz}}$$

Equation 8.62 can be written in terms of moment coefficients as:

$$\dot{p} = \left(\frac{I_{yy} - I_{zz}}{I_{xx}}\right) qr + \frac{1}{2I_{xx}} \rho V^2 SbC_l$$
$$\dot{q} = \left(\frac{I_{zz} - I_{xx}}{I_{yy}}\right) pr + \frac{1}{2I_{yy}} \rho V^2 ScC_m \tag{8.63}$$
$$\dot{r} = \left(\frac{I_{xx} - I_{yy}}{I_{zz}}\right) pq + \frac{1}{2I_{zz}} \rho V^2 SbC_n$$

The various equations describing the aircraft dynamics and kinematics obtained thus far have been summarized in Tables 8.1 and 8.2.

In either case, there are 12 differential equations in all, each of first order in time. All 12 of them can be integrated simultaneously to determine aircraft velocity, trajectory, angular velocity and attitude at any instant of time. However, in studying the aircraft dynamics, the effect of the position variables x_E, y_E, z_E (ignoring density variations with altitude, which is not admissible in some cases) and the heading angle ψ may generally be ignored, which leaves behind eight first-order differential equations. These give rise to the various airplane dynamic modes that we have already described: short period (2), phugoid (2), roll rate (1), Dutch roll (2) and spiral (1)—the numbers in parentheses give the number of variables needed to describe each dynamic mode, clearly they sum up to 8.

Note: In the presence of wind, the dynamics equations need to be modified to account for change in relative wind speed of aircraft as $\underline{V} = \underline{V_E} - \underline{W}$, where $\underline{V_E}$ is the inertial velocity of aircraft. Appropriate components of wind $\underline{W}$ usually available in the Earth-fixed axis system can be determined using the rotation matrices presented earlier in this chapter. Effect of wind on the aircraft dynamics is via the aerodynamic forces and moments which are modified due to the dynamic pressure term via $V^2 = (u - W_{xB})^2 + (v - W_{yB})^2 + (w - W_{zB})^2$. W_{xB}, W_{yB} and W_{zB} are components of $\underline{W}$ along body fixed axes. Angle of attack and sideslip angle are now defined as $\alpha = tan^{-1}\left(\frac{w - W_{zB}}{u - W_{xB}}\right)$; $\beta = sin^{-1}\left(\frac{v - W_{yB}}{V}\right)$. Changes in both modal as well as overall

TABLE 8.1
Aircraft Equations of Motion with Components along Body Axes

Label	Equations
Translational dynamics	$\dot{u} = rv - qw + \frac{1}{2m}\rho V^2 SC_X + \frac{T}{m} - g\sin\theta$ $\dot{v} = pw - ru + \frac{1}{2m}\rho V^2 SC_Y + g\sin\phi\cos\theta$ $\dot{w} = qu - pv + \frac{1}{2m}\rho V^2 SC_Z + g\cos\phi\cos\theta$
Translational kinematics	$\begin{bmatrix} \dot{x}_E \\ \dot{y}_E \\ \dot{z}_E \end{bmatrix} = \begin{bmatrix} \cos\psi\cos\theta & \cos\psi\sin\theta\sin\phi - \sin\psi\cos\phi & \cos\psi\sin\theta\cos\phi + \sin\psi\sin\phi \\ \sin\psi\cos\theta & \sin\psi\sin\theta\sin\phi + \cos\psi\cos\phi & \sin\psi\sin\theta\cos\phi - \cos\psi\sin\phi \\ -\sin\theta & \cos\theta\sin\phi & \cos\theta\sin\phi \end{bmatrix} \begin{bmatrix} u \\ v \\ w \end{bmatrix}$
Rotational dynamics	$\dot{p} = \left(\frac{I_{yy} - I_{zz}}{I_{xx}}\right)qr + \frac{1}{2I_{xx}}\rho V^2 SbC_l$ $\dot{q} = \left(\frac{I_{zz} - I_{xx}}{I_{yy}}\right)pr + \frac{1}{2I_{yy}}\rho V^2 ScC_m$ $\dot{r} = \left(\frac{I_{xx} - I_{yy}}{I_{zz}}\right)pq + \frac{1}{2I_{zz}}\rho V^2 SbC_n$
Rotational kinematics	$\begin{bmatrix} \dot{\phi} \\ \dot{\theta} \\ \dot{\psi} \end{bmatrix} = \begin{bmatrix} 1 & \tan\theta\sin\phi & \tan\theta\cos\phi \\ 0 & \cos\phi & -\sin\phi \\ 0 & \sec\theta\sin\phi & \sec\theta\cos\phi \end{bmatrix} \begin{bmatrix} p \\ q \\ r \end{bmatrix}$

response of aircraft are expected in the presence of wind, which could typically be characterized as having steady, temporally and spatially varying components. A complex wind model will usually have all three components.

8.4.3 Sources of Nonlinearity

The aircraft dynamics equations in Table 8.1 or 8.2 contain several non-linear terms arising from different sources:

- **Kinematic coupling:** These appear as product terms in the translational dynamics equations—terms like rv, qw and so on, or $p\sin\alpha$, $r\cos\alpha$. For instance, when an airplane flying at an angle of attack α rolls, this is the effect that converts the angle of attack into a sideslip angle and vice versa. Kinematic coupling therefore couples the angle-of-attack and sideslip dynamics in the presence of angular velocity.

TABLE 8.2
Aircraft Equations of Motion with Components along Wind Axis

Label	Equations
Translational dynamics	$\dot{V}=\frac{1}{m}\left[T\cos\alpha\cos\beta-\bar{q}S\left(C_D\cos\beta-C_Y\sin\beta\right)-mg\sin\gamma\right]$ $\dot{\alpha}=q-\frac{1}{\cos\beta}\left\{\left(p\cos\alpha+r\sin\alpha\right)\sin\beta-\frac{g}{V}\cos\gamma\cos\mu+\frac{\bar{q}SC_L}{mV}+\frac{T\sin\alpha}{mV}\right\}$ $\dot{\beta}=\left(p\sin\alpha-r\cos\alpha\right)+\frac{1}{mV}\begin{bmatrix}-T\cos\alpha\sin\beta+\bar{q}S\left(C_Y\cos\beta+C_D\sin\beta\right)\\+mg\cos\gamma\sin\mu\end{bmatrix}$
Translational kinematics	$\begin{bmatrix}\dot{x}_E\\\dot{y}_E\\\dot{z}_E\end{bmatrix}=\begin{bmatrix}V\cos\gamma\cos\chi\\V\cos\gamma\sin\chi\\-V\sin\gamma\end{bmatrix}$
Rotational dynamics	$\dot{p}=\left(\frac{I_{yy}-I_{zz}}{I_{xx}}\right)qr+\frac{1}{2I_{xx}}\rho V^2SbC_l$ $\dot{q}=\left(\frac{I_{zz}-I_{xx}}{I_{yy}}\right)pr+\frac{1}{2I_{yy}}\rho V^2ScC_m$ $\dot{r}=\left(\frac{I_{xx}-I_{yy}}{I_{zz}}\right)pq+\frac{1}{2I_{zz}}\rho V^2SbC_n$
Rotational kinematics	$\begin{bmatrix}\dot{\mu}\\\dot{\gamma}\\\dot{\chi}\end{bmatrix}=\begin{bmatrix}1&\tan\gamma\sin\mu&\tan\gamma\cos\mu\\0&\cos\mu&-\sin\mu\\0&\sec\gamma\sin\mu&\sec\gamma\cos\mu\end{bmatrix}\begin{bmatrix}p_w\\q_w\\r_w\end{bmatrix}$

- **Inertia coupling:** These terms arise in the rotational dynamics as a product of angular rates—terms like qr, rp, pq, which come with an inertia coefficient. These are also sometimes called the gyroscopic terms.
- **Gravity:** The gravity terms come with non-linear trigonometric functions of the Euler angles.
- **Nonlinear aerodynamics:** Finally, the aerodynamic forces themselves are quadratic functions of the velocity V, and the aerodynamic force and moment coefficients are usually non-linear functions of the angle of attack and sideslip angle beyond a narrow linear range. Further, especially in the transonic regimes of speed, the aerodynamic force and moment coefficients are also non-linear functions of the Mach number.

Nonlinearities in the aircraft model manifest themselves as resulting non-linear dynamic behaviour of an aircraft in flight. In Figure 8.7, the aircraft critical flight regime (representing non-linear behaviour of aircraft) is demarcated from normal

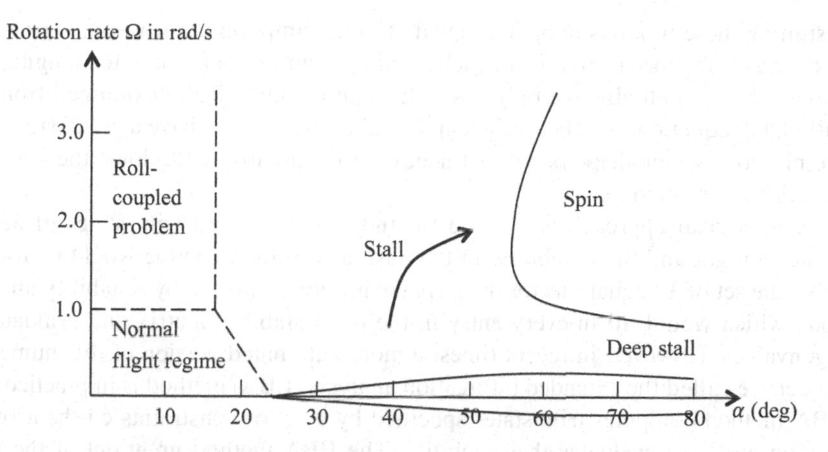

FIGURE 8.7 Flight regimes of aircraft. (Reprinted from Goman, M.G., Zagaynov, G.I. and Khramtsovsky, A.V., Application of bifurcation methods to non-linear flight dynamics problems, *Progress in Aerospace Sciences*, Vol 33, 1997, pp. 539–586. © 1997, With Permission from Elsevier.)

flight regime. The classification of different flight regimes shown in Figure 8.7 is based on the angle of attack and the rotation rate of aircraft. An aircraft enters into critical flight regimes from a normal flight as a result of mild or abrupt loss of stability. Motion of an aircraft in critical flight regimes may be controllable or uncontrollable and unusual response of the aircraft to control inputs is possible. There is a variety of non-linear phenomena associated with aircraft dynamics, each distinct from the other. A number of different types of non-linear behaviour in longitudinal and lateral motions are usually connected with stall, such as 'pitch up', 'pitch bucking', 'tumbling', 'wing drop', 'roll off', 'wing rock', 'nose slice', 'yaw off' and so on. In the post-stall region at moderate-to-high angles of attack, post-stall gyrations, spin and deep stall behaviour are dominant. On the other hand, at low angles of attack, the possibility of loss of stability and controllability at high roll rate (due to roll coupling problem) is well-known. In rapid roll manoeuvres, jump in roll rate, autorotation and roll reversal are commonly observed.

8.5 NUMERICAL ANALYSIS OF AIRCRAFT MOTIONS

The complete set of 12 equations for the airplane dynamics in Table 8.1 or 8.2, or the slightly reduced set of 8 equations used to study airplane stability, are still quite formidable for manual analysis. Except the simplest of manoeuvres such as straight and level flight or a level turn, it is quite difficult to obtain analytical solutions for airplane manoeuvres starting from the full set of 12 equations. For stability analysis, the set of 8 equations is usually linearized about a selected trim state and an 8×8 stability matrix is written down. The eigenvalues of this stability matrix contain information about the frequency/damping and growth/decay rates of the airplane modes. Even this is usually not analytically tractable; hence, the 8×8 matrix is split into two 4×4 matrices, one each for longitudinal and lateral-directional motion, assuming these motions to be decoupled. That assumption, it turns out, is true only in the case of some special trims, such as straight and level flight in the longitudinal plane. Thus, eventually, the only cases that can be analytically examined from the full 6DOF equations are the straight and level flights that we have already studied in the previous seven chapters without needing to explicitly write down the complete equations of motion.

A numerical approach is adopted to study more general trim states of an airplane in flight and their stability. In the past, a trimming routine would be used to solve the set of 12 equations for the airplane motion, followed by a stability analysis code which would fill in every entry in the 8×8 stability matrix and evaluate the eigenvalues. However, in recent times, a more automated version of this numerical procedure called the extended bifurcation analysis (EBA) method is in practice. The EBA method computes trim states specified by a set of constraints on the airplane motion and also evaluates their stability. The EBA method grew out of the standard bifurcation analysis (SBA) method, which was originally proposed to study airplane stability at high angles of attack and other non-linear flight regimes. The SBA method could not specify constraints on the airplane state variables; hence, the trims considered in the SBA method were somewhat unrealistic. Once it became possible to specify particular trim states, the EBA method could be used to analyse airplane motion and stability as desired. The EBA then is no longer a specialized high-angle-of-attack or non-linear analysis tool—it can be used across all segments of the flight spectrum, particularly for airplane performance, trim and stability analysis.

In the following sections, we first define and discuss the notions of trim and stability for general higher-order systems (you may remember that we limited ourselves to first- and second-order systems in Chapter 2). Then, we introduce the standard bifurcation method and demonstrate its working with an example.

8.5.1 Generalized Airplane Trim and Stability Analysis

In general, rigid airplane dynamics can be divided into two kinds of motions, natural or in response to disturbances/inputs from external sources, and intended, that is, in response to pilot inputs. Numerically, however, the problem of analysing airplane motions is taken up in two steps as below.

i. Trim flight and small disturbances from the trim state—one issue here is to find entire families of all possible trim states and classify them according to their stability characteristics. This is called *local dynamic behaviour* of airplane around an equilibrium state. We have already examined one particular trim state, namely straight and level flight, and studied the airplane small-perturbation response about this trim state in terms of its dynamic modes (short period, phugoid, roll, Dutch roll and spiral). Soon, we shall numerically examine other trim states such as level turning flight and their stability. A second issue of interest here is airplane response to small pilot inputs or to disturbances. These are usually studied in terms of transfer functions between the airplane state and the input or between the airplane state and the disturbance.

ii. The transition between different sets of trim states, either as a result of an instability or deliberately by a piloted input. This is called *global dynamic behaviour.* Start and end states of a possible transition may be identified from a bifurcation analysis, but the path followed by the transition in state space, or the time history of the state variables, can only be obtained by numerical integration of the airplane dynamic equations (Table 8.1 or 8.2).

8.5.1.1 Local Dynamic Behaviour: Trim and Stability Analysis

The computation of all possible equilibrium states in the prescribed range of parameters and their stability is central to any numerical investigation of aircraft dynamics and to control design. The equations of airplane dynamics in Table 8.1 or 8.2 appear as a set of first-order differential equations that may be compactly written as:

$$\underline{\dot{x}} = \underline{f}\left(\underline{x}, \underline{U}\right) \quad (8.64)$$

where $\underline{x}$ are the airplane flight dynamic states and $\underline{U}$ are the control inputs.

The computation of equilibrium states amounts to solving the set of simultaneous algebraic equations:

$$\underline{\dot{x}}^* = \underline{f}\left(\underline{x}^*, \underline{U}^*\right) = 0 \quad (8.65)$$

where the starred variables indicate equilibrium values.

If nothing is done, that is, no disturbance hits the aircraft, the aircraft remains in the equilibrium state; therefore, you should not be surprised if you see no changes at all in aircraft states when you integrate all the equations together from an initial condition which is an equilibrium state. A small change in initial condition from the equilibrium state caused by a small disturbance (due to external wind or the pilot), let us say $\Delta\underline{x}$, will change the aircraft state to a new state:

$$\underline{x} = \underline{x}^* + \Delta\underline{x}$$

and considering that small change in inputs from an equilibrium state can also come from the pilot, the perturbed values of control inputs can be written as:

$$\underline{U} = \underline{U}^* + \Delta\underline{U}$$

Introducing these new variables in the equations of motion, we obtain:

$$\underline{\dot{x}}^* + \Delta\underline{\dot{x}} = \underline{f}\left(\underline{x}^* + \Delta\underline{x}, \underline{U}^* + \Delta\underline{U}\right) \quad (8.66)$$

Equation 8.66 can be expanded using the Taylor series expansion about the equilibrium state $(\underline{x}^*, \underline{U}^*)$ as follows:

$$\begin{aligned}\dot{\underline{x}}^* + \Delta\dot{\underline{x}} &= \underline{f}(\underline{x}^* + \Delta\underline{x}, \underline{U}^* + \Delta\underline{U}) \\ &= \underline{f}(\underline{x}^*, \underline{U}^*) + \left.\frac{\partial \underline{f}}{\partial \underline{x}}\right|_{(\underline{x}^*, \underline{U}^*)} \Delta\underline{x} + \left.\frac{\partial \underline{f}}{\partial \underline{U}}\right|_{(\underline{x}^*, \underline{U}^*)} \Delta\underline{U} + \left.\frac{\partial^2 \underline{f}}{\partial \underline{x}^2}\right|_{(\underline{x}^*, \underline{U}^*)} \frac{\Delta\underline{x}^2}{2!} \\ &\quad + \left.\frac{\partial^2 \underline{f}}{\partial \underline{U}^2}\right|_{(\underline{x}^*, \underline{U}^*)} \frac{\Delta\underline{U}^2}{2!} + H.O.T.\end{aligned}$$

H.O.T. means higher-order terms, other quadratic, cubic and those of higher order. One of the assumptions in a stability analysis is that perturbations or disturbances in variables are small, so that higher powers of disturbances are further small and can be dropped as negligible terms. Thus, only terms linear in $\Delta\underline{x}$ and $\Delta\underline{U}$ are retained:

$$\dot{\underline{x}}^* + \Delta\dot{\underline{x}} = \underline{f}(\underline{x}^*, \underline{U}^*) + \left.\frac{\partial \underline{f}}{\partial \underline{x}}\right|_{(\underline{x}^*, \underline{U}^*)} \Delta\underline{x} + \left.\frac{\partial \underline{f}}{\partial \underline{U}}\right|_{(\underline{x}^*, \underline{U}^*)} \Delta\underline{U}$$

As starred quantities are equilibrium states at which $\dot{\underline{x}}^* = 0$, $\underline{f}(\underline{x}^*, \underline{U}^*) = 0$ or $\dot{\underline{x}}^* = \underline{f}(\underline{x}^*, \underline{U}^*)$ is identically true; the above equation is further simplified to:

$$\Delta\dot{\underline{x}} = \left.\frac{\partial \underline{f}}{\partial \underline{x}}\right|_{(\underline{x}^*, \underline{U}^*)} \Delta\underline{x} + \left.\frac{\partial \underline{f}}{\partial \underline{U}}\right|_{(\underline{x}^*, \underline{U}^*)} \Delta\underline{U} = A\Delta\underline{x} + B\Delta\underline{U} \tag{8.67}$$

Equation 8.67, known as linear model, describes the local dynamics of perturbations from an equilibrium state. *A*, called the system matrix, is a matrix of constant terms obtained by evaluating the *Jacobian* matrix $(\partial \underline{f} / \partial \underline{x})$ at a particular equilibrium state. Thus, in general, *A* evaluates differently from one equilibrium state to another. The general form of *A* for the aircraft equations of motion is given in Appendix 8.1. *B* in control theory parlance is referred to as control matrix and like *A* must be evaluated for every equilibrium state.

For uncontrolled disturbances, that is, $\Delta\underline{U} = 0$:

$$\Delta\dot{\underline{x}} = A\Delta\underline{x} \tag{8.68}$$

Linear system theory tells us that the dynamic behaviour of Equation 8.68 depends on its eigenvalues λ obtained by solving:

$$\det(\lambda I - A) = 0 \tag{8.69}$$

Equation 8.69 is known as the characteristic equation. The eigenvalues can either be real or complex conjugate pairs. As long as the real part of all the eigenvalues of A are negative, starting from an initial condition $\Delta\underline{x}(0) \neq 0$, $\Delta\underline{x}(t) \rightarrow 0$ as $t \rightarrow \infty$, which means that perturbations die out in time and the equilibrium state $\left(\underline{x}^*, \underline{U}^*\right)$ is stable, else the state $\left(\underline{x}^*, \underline{U}^*\right)$ is effectively unstable.

Bifurcation methods automate the process of solving for entire families of equilibrium solutions with one or more varying parameters and simultaneously evaluate the eigenvalues of the local A matrix at each equilibrium point, thus establishing its stability. By piecing together all the equilibrium points and their stability, bifurcation methods aim to provide a global picture of the system dynamics.

8.6 STANDARD BIFURCATION ANALYSIS

This methodology is based on using a continuation algorithm[1] to compute the steady states of a system of non-linear ordinary differential equations (ODEs):

$$\dot{\underline{x}} = \underline{f}\left(\underline{x}, \underline{U}\right) \tag{8.70}$$

where x are the states and $\underline{U}$ are the parameters including control inputs.

A continuation algorithm computes steady states of Equation 8.70 by solving a simultaneous set of non-linear algebraic equations:

$$\dot{\underline{x}} = \underline{f}\left(\underline{x}, u, \underline{p}\right) = 0 \tag{8.71}$$

as a function of a varying parameter of the system $u \in \underline{U}$, while the other parameters of the system $\underline{p} \in \underline{U}$ remain fixed in a continuation. In short, starting from a given steady state $\left(\underline{x}^*, u^*\right)$, the continuation technique uses a predictor step, which finds the next point $\left(\underline{x}_1, u_1\right)$ (refer to Figure 8.8a) by solving:

$$\frac{\partial \underline{f}}{\partial \underline{x}} \Delta \underline{x} + \frac{\partial \underline{f}}{\partial u} \Delta u = 0$$

Or equivalently,

$$\underline{x}_1 - \underline{x}^* = -\left(\frac{\partial \underline{f}}{\partial \underline{x}}\right)_*^{-1} \left(\frac{\partial \underline{f}}{\partial u}\right)_* \left(u_1 - u^*\right)$$

In the corrector step, the correct solution $\left(\underline{x}_1^*, u_1\right)$ satisfying the steady-state condition Equation 8.71 is found by applying the Newton–Raphson scheme. As shown in Figure 8.8b, the corrector step starts from the solution $\left(\underline{x}_1, u_1\right)$ obtained in the predictor step and marches towards the correct solution $\left(\underline{x}_1^*, u_1\right)$, which satisfies the steady-state condition $\underline{f}\left(\underline{x}_1^*, u_1\right) = 0$. As part of the continuation process, the Jacobian matrix $J = \left(\partial \underline{f} / \partial \underline{x}\right)$, its eigenvalues and eigenvectors are simultaneously computed at every equilibrium state.

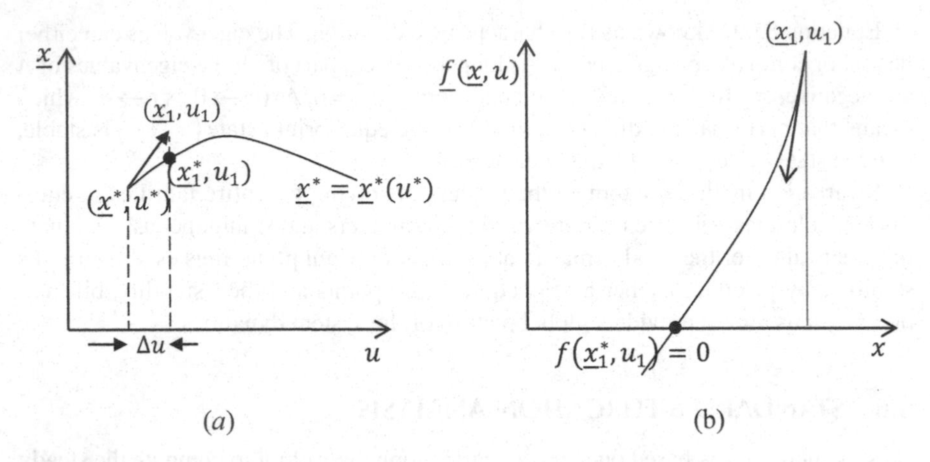

FIGURE 8.8 (a) Predictor step and (b) corrector step in a continuation.

A curve of steady states, $\underline{x}^* = \underline{x}^*(u)$, as shown in Figure 8.8a, along with stability information can thus be traced as a function of the varying parameter u as a result of a continuation run. When u is varied in a continuation, loss of stability on the equilibrium solution branch may occur due to crossing over of eigenvalues from the left of the imaginary axis to the right in the complex plane. The loss of stability at a critical value of $u = u_{cr}$ indicates a bifurcation, which results in the number and/or type of steady states changing when u is increased beyond u_{cr}. Depending on the non-linearities in the model represented by the ODEs, non-linear dynamic behaviour of the system is manifested as bifurcations which may be predicted using bifurcation theory[2] methodology.

The basic bifurcations of equilibrium states as a result of the loss of stability when one of the real eigenvalues of the system crosses the imaginary axis through the origin from the left half complex plane to the right half are shown in Figure 8.9.

In Figure 8.9a, the time response of disturbance or perturbation around the equilibrium state $x^* = 0$ for three different locations of eigenvalues are plotted, which tells us that for $u < u_{cr}$, disturbance decays exponentially, and eventually, the original equilibrium state $x^* = 0$ is recovered by the system. Hence the equilibrium state $x^* = 0$ is stable for $u < u_{cr}$. In the second case, at $u = u_{cr}$, disturbance stays unchanged and nothing can be said about the stability of the equilibrium state $x^* = 0$. For $u > u_{cr}$, disturbance grows exponentially and hence the equilibrium state $x^* = 0$ for $u > u_{cr}$ is unstable. The loss of stability via crossing of a single eigenvalue on the real axis gives rise to three different types of (popularly known as 'static') bifurcations depending upon the non-linearities present in the system. They are transcritical, pitchfork and saddle node bifurcations shown in Figures 8.9b–8.9d.

In the case of a transcritical bifurcation, two solution branches (one stable and the other unstable) intersect at the bifurcation point at $u = u_{cr}$ with exchange of stability as shown in Figure 8.9b. Thus, the equilibrium state $x^* = 0$ with increasing u encounters a loss of stability at critical value of $u = u_{cr}$, while the other equilibrium ($x^* \neq 0$) gains stability at $u = u_{cr}$. Beyond u_{cr}, perturbations take the system away

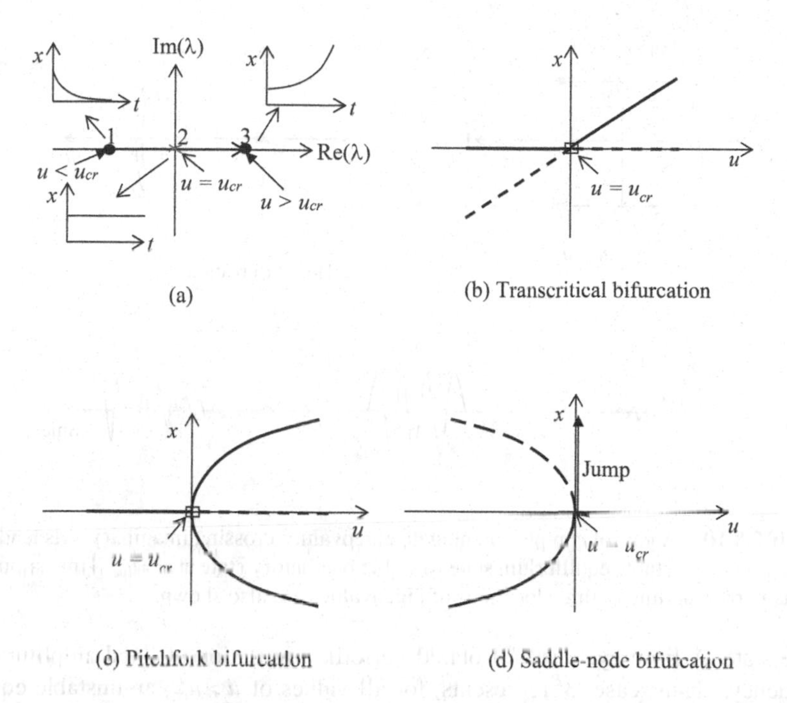

FIGURE 8.9 (a) Perturbed state time history for three different locations of eigenvalues, and branch of stable (solid line) and unstable (dashed line) equilibrium states as functions of varying system parameter u showing (b) transcritical, (c) pitchfork and (d) saddle-node bifurcations.

from $x^* = 0$ to the other stable state indicating a 'departure' from the equilibrium ($x^* = 0$) to a secondary equilibrium state with non-zero values of x^*.

In a pitchfork bifurcation, shown in Figure 8.9c, a single stable branch of equilibria at $x^* = 0$ gives rise to two stable equilibrium branches at the bifurcation point $u = u_{cr}$ while losing its own stability. Beyond this point, a 'departure' takes the system to the upper equilibrium branch (positive values of x^* now) or to the lower equilibrium branch (negative values of x^* now) depending on the sign of the disturbance.

In the case of a saddle node bifurcation, two equilibrium branches (one stable and the other unstable) annihilate each other at the bifurcation point $u = u_{cr}$ as depicted in Figure 8.9d. Beyond u_{cr} no equilibrium state is seen in close proximity; therefore, it is possible that the system may 'jump' to a faraway stable equilibrium for parameter values $u > u_{cr}$.

Moving on now to complex eigenvalues, in Figure 8.10, time responses corresponding to three different locations of a pair of complex–conjugate eigenvalues are shown. Case '1' represents the case of stable equilibrium states for $u < u_{cr}$. Disturbances in this case die down with time as shown in the time history, and thus the original equilibrium condition is recovered. In case '2', the perturbation induced by external disturbances stays unchanged and the response oscillates with constant amplitude and frequency, so the aircraft cannot return to its original equilibrium condition. In case '3', for $u > u_{cr}$, disturbances grow and eventually acquire a limit

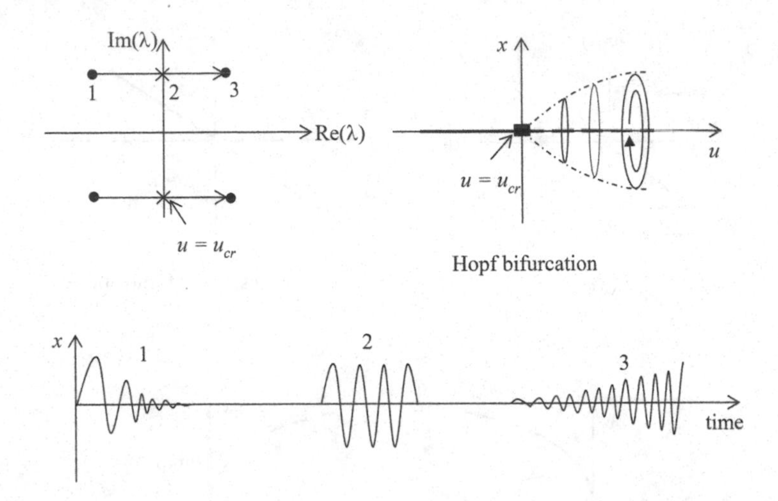

FIGURE 8.10 A pair of complex–conjugate eigenvalues crossing imaginary axis leading to bifurcation from stable equilibrium state to stable oscillatory state at $u = u_{cr}$. Time simulation results corresponding to three locations of eigenvalues are also shown.

cycle state. A limit cycle is an isolated periodic oscillation of fixed amplitude and frequency.[2] Thus, case '3' represents, for all values of $u > u_{cr}$, an unstable equilibrium branch $x^* = 0$ surrounded by stable steady states that are periodic in nature. With increasing u, in this case, the system experiences a departure into a stable oscillatory state. This type of bifurcation resulting from loss of stability due to a pair of complex–conjugate eigenvalues crossing the imaginary axis is known as Hopf bifurcation. Further bifurcations of oscillatory states into still higher-order motions are possible, but that is beyond scope of this book.

8.6.1 Application of SBA to F-18/HARV Dynamics

We now apply SBA to a model of the F-18/HARV aircraft. Eight equations are selected as follows: The translational equations in V, α, β from Table 8.2, and the rotational dynamics and kinematics equations in the variables p, q, r and ϕ, θ from Table 8.1. The velocity variable V is scaled by the speed of sound 'a' and $V/a = Ma$, the Mach number is used instead of V. Thus,

$$\underline{x} = \left[Ma \ \alpha \ \beta \ p \ q \ r \ \phi \ \theta\right]'$$

are the state variables, and

$$\underline{U} = \left[\eta \ \delta e \ \delta a \ \delta r\right]'$$

is the vector of the aircraft control parameters, where η $(=(T/T_{max}))$ is the fraction of maximum available engine thrust. Elevator deflection $u = \delta e$ is used as the varying parameter as it directly affects the angle of attack of the aircraft. Other parameters $\underline{p} = \left[\eta \ \delta a \ \delta r\right]'$ are kept fixed in the continuation run.

8.6.1.1 Stall and Post-Stall Solutions

The result obtained from using a numerical continuation algorithm is usually presented in the form of a *bifurcation diagram* such as the one shown in Figure 8.11 for F-18/HARV.

In Figure 8.11, equilibrium states for the F-18/HARV model (see Appendix 8.2) from a continuation run are plotted as a function of varying elevator deflection. Full lines are stable equilibrium states (trajectories starting 'nearby' converge to these states) and dashed lines are unstable equilibrium states (trajectories starting 'nearby' diverge from these states). The following observations about the aircraft dynamic behaviour can be made from Figure 8.11:

- In the low-to-moderate range of angle of attack up to 0.75 rad (≈43°), the solution branch consists of stable equilibrium states with very short stretches of unstable equilibrium states bounded on both sides by Hopf bifurcation points (filled squares). The roll and yaw rates for these equilibrium states are zero (as also the angles of sideslip and bank, not plotted here), indicating symmetric flight condition in the longitudinal plane.
- A pitchfork bifurcation (empty square) at $\alpha \approx 0$ rad is seen, below which the longitudinal trim states become unstable and a new branch of stable trim states appear, which represent aircraft departure from longitudinal

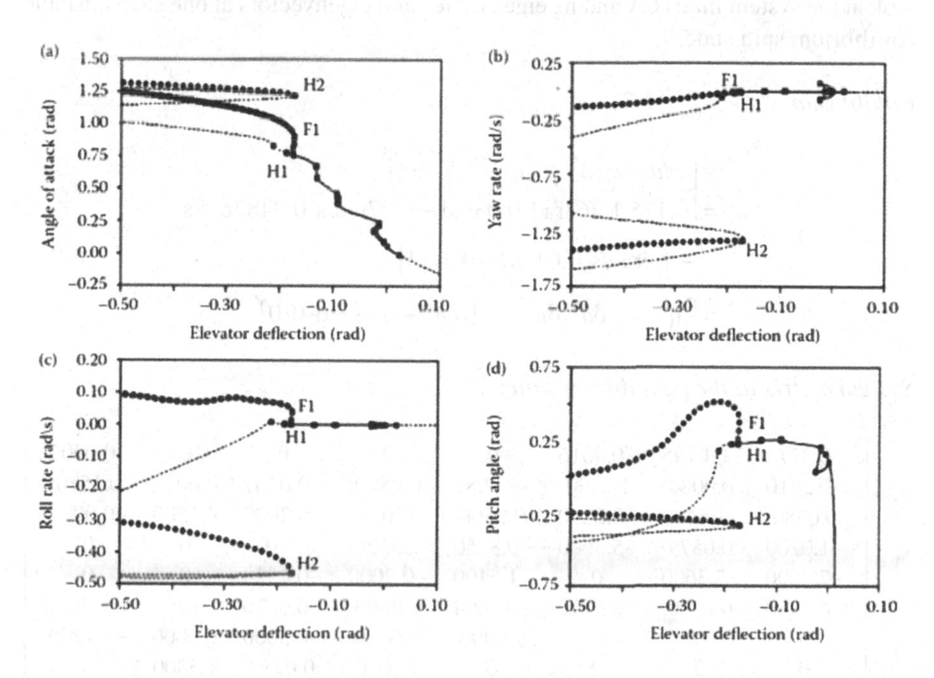

FIGURE 8.11 Bifurcation plots of state variables: (a) angle of attack α, (b) yaw rate r, (c) roll rate p and (d) pitch angle, θ for the F-18/HARV model as function of the elevator deflection (solid line: stable trim, dashed line: unstable trim, solid circle: stable oscillatory state, empty circle: unstable oscillatory state, empty square: pitchfork bifurcation, solid square: Hopf bifurcation).

dynamics (clearly visible in the bifurcation plot for the yaw rate). This bifurcation corresponds to an instability of the spiral mode.

- For a range of elevator deflection, multiple steady states exist, both stable and unstable, representing different types of dynamics. The equilibrium branches are unstable and the circles surrounding them represent the maximum amplitude of oscillatory (periodic) states about the mean unstable trim state. These periodic solutions arise at Hopf bifurcations (labelled H1 and H2) with the filled circles being stable periodic states and the empty ones being unstable.
- In the post-stall regime of angle of attack, phenomena such as 'pitch bucking' and 'spin' appear as a result of the Hopf bifurcations. The uppermost unstable solution branch (dashed line) surrounded by stable limit cycles (filled circles) with large yaw rates of the order of 1.5 rad/s represents oscillatory spin states.

Continuation algorithms evaluate the Jacobian matrix at each equilibrium point along with its eigenvalues and eigenvectors—these can be used to understand the local behaviour of the aircraft around that point. From the bifurcation diagram in Figure 8.11, we observe that spin equilibrium states are unstable, around which, stable oscillatory states have been created through a Hopf bifurcation. Let us have a look at the system matrix A and its eigenvalues and eigenvectors at one such unstable equilibrium spin state.

Equilibrium state:

$$\begin{aligned}\underline{x}^* &= \left[Ma^* \ \alpha^* \ \beta^* \ p^* \ q^* \ r^* \ \phi^* \ \theta^* \right]' \\ &= [0.175 \ 1.263\text{rad} \ 0.03\text{rad} - 0.47\text{rad/s} \ 0.0487\text{rad/s} \\ &\quad - 1.5\text{rad/s} \ 0.03\text{rad} - 0.3\text{rad}]'\end{aligned}$$

$$\underline{U}^* = \left[\eta^* \ \delta e^* \ \delta a^* \ \delta r^* \right]' = [0.38 - 0.39\text{rad} \ 0 \ 0]'$$

System matrix at the equilibrium state:

$$A = \begin{bmatrix} -0.3270 & -0.0295 & -0.0016 & 0 & 0 & 0 & 0.0015 & -0.0002 \\ -0.2410 & 0.0080 & 1.6500 & -0.0082 & 0.9880 & -0.0267 & 0.0012 & 0.1570 \\ 0.0380 & -1.6400 & -0.2290 & 0.9560 & 0 & -0.2930 & 0.1500 & -0.0012 \\ -1.0700 & 0.0579 & -5.3600 & -0.8550 & 1.0300 & 0.1610 & 0 & 0 \\ -7.7300 & -2.3600 & 0 & -1.5100 & -0.2000 & -0.4520 & 0 & 0 \\ -0.3080 & 0.6320 & -0.3190 & -0.0604 & 0.3800 & -0.0070 & 0 & 0 \\ 0 & 0 & 0 & 1.0000 & 0.0084 & -0.3000 & 0.0000 & -1.7200 \\ 0 & 0 & 0 & 0 & 1.0000 & 0.0280 & 1.5800 & 0 \end{bmatrix}$$

Eigenvalues of A:

$$\lambda_s = [-0.487 \mp j2.482 \ 0.044 \mp j2.351 - 0.234 - 0.17 - 0.178 - 0.155 \mp j1.637]$$

Eigenvectors of A:

$E =$					
Ma:	$-0.0001 \pm j0.0003$	$-0.0030 \pm j0.0007$	0.0984	−0.0487	$0.0006 \pm j0.0015$
α:	$-0.0261 \pm j0.0061$	$0.0906 \pm j0.2306$	−0.3091	0.2479	$-0.1210 \pm j0.0289$
β:	$0.0363 \pm j0.2594$	$-0.1552 \pm j0.1810$	0.0503	−0.0425	$0.0136 \pm j0.0228$
p:	−0.7669	0.6339	−0.2719	0.1391	$-0.0782 \pm j0.0153$
q:	$-0.0673 \pm j0.5028$	$-0.2884 \pm j0.4898$	0.0856	−0.0501	$0.0592 \pm j0.2592$
r:	$-0.1042 \pm j0.0231$	$0.1651 \pm j0.0207$	0.8458	−0.9234	$-0.0625 \pm j0.0385$
ϕ:	$0.0429 \pm j0.2665$	$-0.1698 \pm j0.1992$	−0.0233	0.0206	−0.6924
θ:	$-0.0313 \pm j0.0071$	$0.0703 \pm j0.2361$	−0.3084	0.2438	$-0.0968 \pm j0.6423$

For the chosen spin equilibrium state, we have three pairs of complex–conjugate eigenvalues and two real eigenvalues. From the set of eigenvalues, we observe that one pair of complex–conjugate eigenvalues has a positive real part indicating an unstable mode. One cannot directly relate these eigenvalues to standard aircraft dynamic modes because the motion is highly coupled in the longitudinal-lateral dynamics in a spin. For example, if we look at the first pair of complex–conjugate eigenvalues and corresponding (first) column of eigenvectors, it appears that the magnitudes of both roll and pitch rate variables are large and comparable indicating that the roll rate and pitch rate motions are predominant and coupled in spin. The unstable pair of eigenvalues (second entry in λ) and corresponding eigenvectors (second column of E) indicates similar comparable magnitudes in the roll and pitch rates. The real set of eigenvalues and corresponding eigenvectors (columns 3 and 4 in E) are also not distinct and show large participation of the variables α, p, r, θ once more, the coupling between the longitudinal (α, θ) and lateral (p, r) variables is apparent. The last pair of complex–conjugate eigenvalues and the corresponding eigenvectors (column 5 in E) indicate the largest magnitudes in ϕ and θ variables. It is interesting to note that the magnitude of velocity in the 5th column of eigenvector matrix above is much less overall and it can be assumed to be very little perturbed from equilibrium value. This explains why many a times the velocity equation is dropped while studying spin dynamics.

8.6.1.2 Roll Manoeuvres

To carry out the bifurcation analysis of roll manoeuvres, it is required to compute the steady states of the aircraft roll rate p as a function of aileron deflection δa. As pointed out earlier in Chapter 7, a non-zero aileron deflection results in a non-zero roll rate initiating aircraft rolling motion with bank angle ϕ changing continuously. As ϕ changes continuously in this manoeuvre, a steady state in the sense of equilibrium solution of the eight-state model of aircraft is not possible. Therefore, steady states in roll manoeuvres are not really trim states; they are rather called pseudo-trims or pseudo-steady states (PSS).

A fifth-order aircraft model popularly known as the PSS model has been found to be useful to study roll manoeuvres. The PSS model (given in Appendix 8.3) ignores the slower timescale variables V, θ, ϕ and deals only with the fast timescale variables

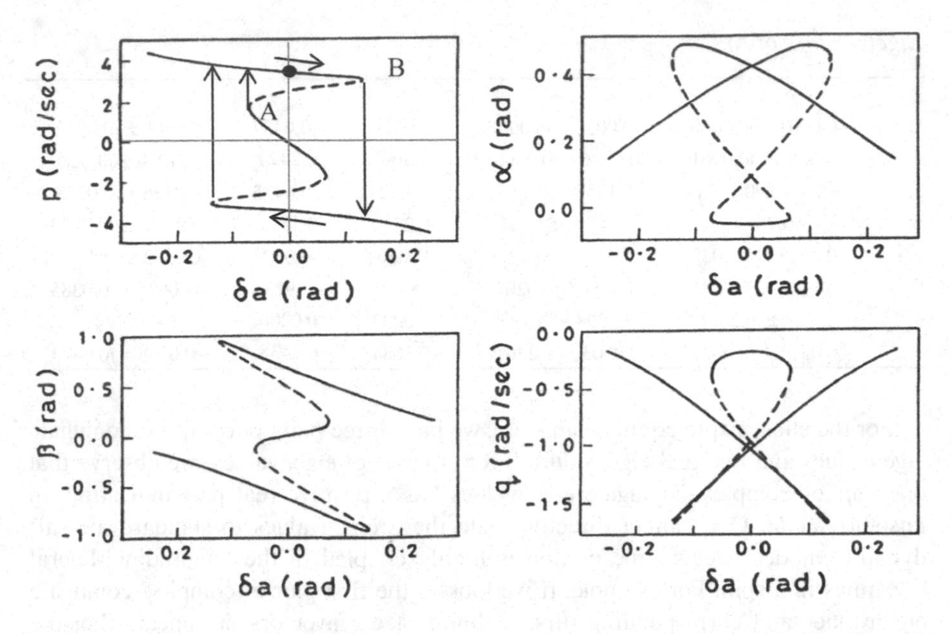

FIGURE 8.12 PSS roll rate and other variables as a function of aileron deflection δa, with $\delta e = -2°$, $\delta r = 0$ (solid lines: stable states; dashed lines: unstable states).

α, β, p, q, r. The eigenvalues of the PSS model thus roughly correspond to the fast dynamic modes—short period, roll and Dutch roll subject to the coupling between the lateral and longitudinal motions. The roll manoeuvre of the aircraft model given in Appendix 8.3 initiated from a level trim corresponding to $\delta e = -2°$ is considered. The bifurcation results obtained by numerical continuation of the PSS model are shown in Figure 8.12.

Nonlinear behaviour of aircraft associated with roll coupling or inertia coupling problems is well-known. The bifurcation results shown in Figure 8.12 indicate three typical non-linear behaviour of the aircraft in roll dynamics: jump in roll rate, autorotation and roll reversal. Arrows in the roll rate p versus aileron deflection δa plot indicate how the roll rate changes with respect to aileron deflection. Reading the plots in Figure 8.12, the following observations can be made:

- When the aileron is deflected in a negative sense, the roll rate increases (in a positive sense) more or less linearly, as expected, up to point 'A'.
- Non-linear effects due to inertia coupling manifest as a saddle-node bifurcation at 'A', beyond which a 'jump' in roll rate is observed. A jump in roll rate is associated with a simultaneous jump in all other states.
- A further increase in the aileron deflection angle moves the roll rate on the upper stable branch to the left. Interestingly, a recovery process from this high roll rate branch by deflecting aileron towards its neutral position keeps the aircraft on the high roll rate solution branch and, at zero aileron deflection, still positive roll rate is observed (represented by the filled circle), which indicates an 'autorotation' condition.

- A further decrease in aileron below the neutral position (increasing positive value now), still results in a positive roll rate, which indicates a 'roll-reversal' condition.
- A jump in the roll rate is again observed at a saddle-node bifurcation point 'B', beyond which the roll rate drops to a large negative value, which initiates a 'hysteresis' loop (marked by the arrows) in roll rate plot.

As you may have noticed from the above examples of bifurcation diagrams, they have been obtained by varying one control parameter while keeping other parameters at their fixed values. An aircraft usually flies in conditions that are constrained such that more than one control is required to be simultaneously employed. For example, a typical mission of a transport aircraft consists of flight phases such as take-off, cruise, loiter and landing. All of these are constrained flight conditions. Cruise is a longitudinal equilibrium flight condition in which the aircraft wings are levelled (zero bank), the flight path is straight and at a constant altitude (zero flight path angle) and the sideslip angle is zero. In cruise, throttle and elevator deflection are coordinated while changing the speed and/or the angle of attack of the aircraft. In the following, we study such practical flight conditions of aircraft using EBA methodology.

8.7 EXTENDED BIFURCATION ANALYSIS (EBA)

In previous chapters, we have studied individual dynamic modes of aircraft, namely short period, phugoid, roll, spiral and Dutch roll, through the analysis of simple models. In what follows, we revisit these modes using the full 6DOF model and set up a numerical procedure to compute the modes through an example simulation. As explained earlier, dynamic modes represent aircraft small-perturbation behaviour around equilibrium points.

We first carry out this computation for a straight and level flight trim. The results ought to match closely with the model approximations we obtained in earlier chapters using a less formal approach. Then we apply the EBA method to a level turning trim flight which is not easily amenable to approximate analysis. The EBA comes to the rescue in that case though it can only give a numerical solution, not an analytical one.

To carry out bifurcation analysis of a constrained trim condition, it is obvious that more than one control has to be moved in a coordinated manner such that constraints are always satisfied. For example, to maintain a straight and level flight, engine throttle has to be regulated (to change speed) as the elevator is used to change the angle of attack in order to maintain the trim condition defined by $L = W; T = D; M_{CG} = 0$. Similarly, to change the angular rates in a level turn trim flight, all controls are to be operated simultaneously in a cooperative manner to satisfy the constrained condition. To study the stability of the aircraft in such constrained trim conditions and to investigate possible loss of stability and departure (bifurcation) from constrained trim flight, it becomes imperative to compute the *bifurcation diagram* including the constrained conditions. A methodology for computation of the bifurcation diagram for aircraft in constrained flight was first proposed in Ref. 3. The methodology involves two steps. In the first step, constrained trims are computed by solving an augmented set of non-linear algebraic equations including the constraint conditions, $\underline{g}(\underline{x})=0$:

$$\underline{f}(\underline{x},\underline{p}_1,\underline{p}_2,u)=0; \quad \underline{g}(\underline{x})=0 \tag{8.72}$$

Naturally, to take care of the added equations representing constraints, that many additional unknowns are required for well-posedness of the problem, otherwise the increased number of equations cannot be solved. Hence some of the control parameters $\underline{p}_1 \in \underline{U}$ are freed for this purpose, while the other parameters $\underline{p}_2 \in \underline{U}$ are kept fixed. A continuation of the augmented set of Equations 8.72 then gives the relation $\underline{p}_1(u)$ between the control parameters such that the constraints of the flight condition are met. In a continuation for straight and level flight trim over a range of angles of attack, thus one gets the relation $\eta=\eta(\delta e)$ for the variation of the throttle with varying elevator deflection in the first step of the methodology. In the second step, to compute the correct stability of the constrained trim condition and also to compute the bifurcation diagram (which may indicate a departure from constrained flight conditions at the point of loss of stability), the relations computed in the first step are included in the model and a second continuation of the equation:

$$\dot{\underline{x}}=\underline{f}(\underline{x},\underline{p}_1(u),\underline{p}_2,u) \tag{8.73}$$

is carried out. Note that only the freed set of parameters $\underline{p}_1(u)$ is changed with varying u, and $\underline{p}_2$ are still fixed. This step results in the computation of the *bifurcation diagram* for the system in constrained conditions.

8.7.1 Straight and Level Flight Trim

To compute the equilibrium solution branches with the stability of the aircraft in level flight trim condition, a two-step continuation procedure as explained before via Equations 8.72 and 8.73 is carried out. In the first step, aircraft equations of motion along with constraints, that is:

$$\dot{\underline{x}}=\underline{f}(\underline{x},\delta e,\eta,\delta a,\delta r)=0; \quad \phi=0; \quad \beta=0; \quad \gamma=0 \tag{8.74}$$

are solved together with elevator deflection δe varying as the continuation parameter. In this step, the other control parameters $\eta,\delta a,\delta r$ are left free as extra variables so that the mathematical problem is well defined. Thus, from this step, one gets all possible equilibrium solutions that satisfy the constraints of straight and level trim condition and also the variation of the freed control parameters $\eta,\delta a,\delta r$ as a function of elevator deflection, that is, $\eta(\delta e),\delta a(\delta e),\delta r(\delta e)$, which are known as control schedules. This tells us, for example, how the throttle should be scheduled with the elevator to change the angle of attack or velocity in a straight and level flight condition to move from one trim to another. The first step however does not give the correct stability information because presently available continuation algorithms do not distinguish between the aircraft equations and the constraint equations. This results in the computation of a bigger Jacobian matrix (11×11 in size in this case) and its 11 eigenvalues, which is more than the order of the dynamic equations. A second continuation is therefore carried out to determine the correct stability of the constrained trims. In the second step, a continuation of the aircraft model with the schedule $\eta(\delta e), \delta a(\delta e), \delta r(\delta e)$ included:

$$\dot{\underline{x}} = \underline{f}\left(\underline{x}, \delta e, \eta(\delta e), \delta a(\delta e), \delta r(\delta e)\right) \tag{8.75}$$

is carried out. The results of this step are equilibrium branches satisfying the straight and level flight trim condition with correct stability information. This step also captures the bifurcated solution branches which arise as a result of one of the dynamic modes becoming unstable.

Bifurcation analysis results for straight and level flight trim states of F-18/HARV obtained using the two-step continuation procedure are shown in Figure 8.13. In Figure 8.13a, control schedules $\eta(\delta e), \delta a(\delta e), \delta r(\delta e)$ are plotted where η is the ratio of thrust T to maximum available thrust. The straight and level flight trims are presented in plots in Figure 8.13b. On the straight and level trim branch ($\gamma = \beta = \phi = 0$), several bifurcations can be noticed at critical values of the varying paramotor δe. Equilibrium branches not satisfying the level flight constraints emerge at some bifurcation points indicating departure from constrained flight. Simultaneously eigenvalues and eigenvectors at each trim state are computed, from which, one can determine the characteristics of the dynamic modes. In the following, we arbitrarily select two straight and level flight trim states from Figure 8.13, one stable and the other unstable, and study the typical modes.

Equilibrium state:

$$\underline{x}^* = \left[Ma^* \ \alpha^* \ \beta^* \ p^* \ q^* \ r^* \ \phi^* \ \theta^*\right]'$$

$$[0.386 \ 0.078\text{rad} \ 0 \ 0 \ 0 \ 0 \ 0 \ 0.078\text{rad}]'$$

$$\underline{U}^* = \left[\eta^* \ \delta e^* \ \delta a^* \ \delta r^*\right]' = [0.25 - 0.038\text{rad} \ 0 \ 0]'$$

System matrix:

$A =$	Ma	α	β	p	q	r	ϕ	θ
	–0.016	–0.010	0	0	0	0	0	–0.0287
	–0.382	–1.080	0	0	0.990	0	0	0
	0	0	–0.199	0.0787	0	–0.995	0.0741	0
	0	0	–12.9	–2.57	0	0.792	0	0
	0	–0.832	0	0	–0.365	0	0	0
	0	0	1.62	–0.049	0	–0.134	0	0
	0	0	0	1.0	0	0.079	0	0
	0	0	0	0	1.0	0	0	0

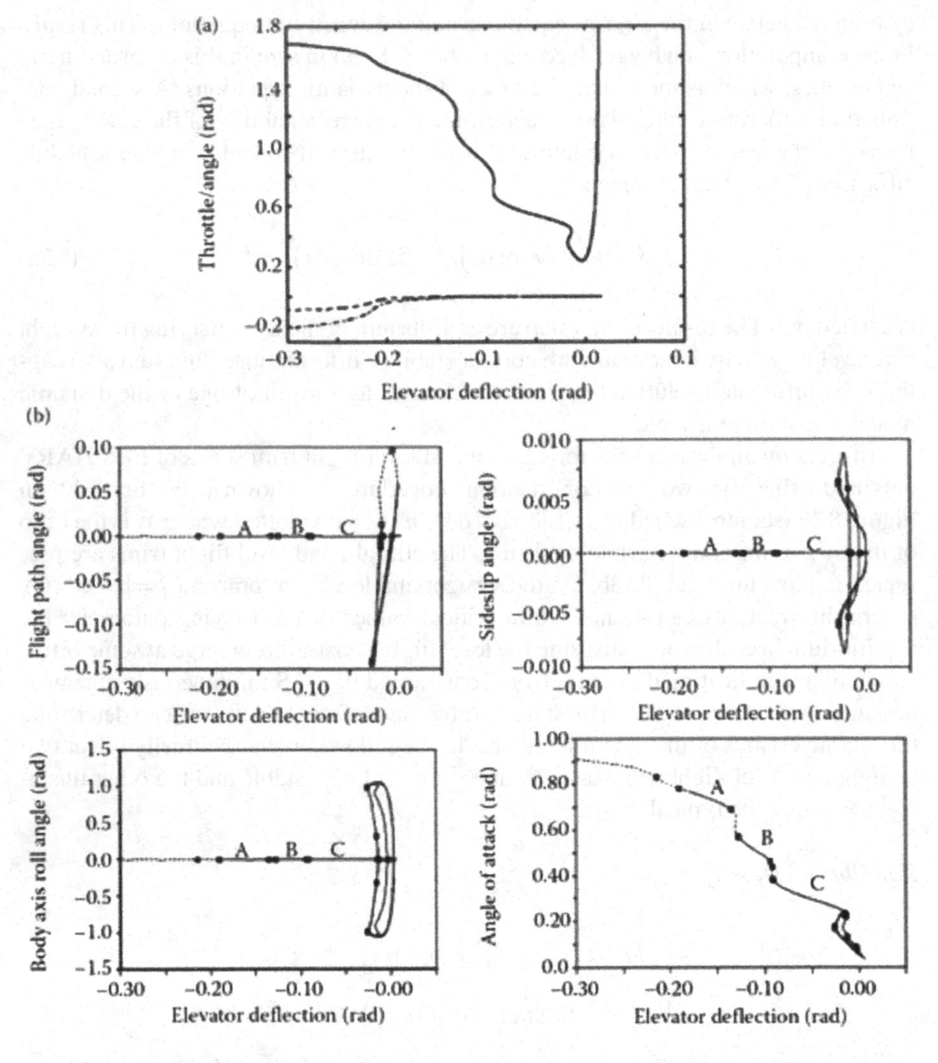

FIGURE 8.13 (a) Thrust fraction (full line), aileron (dash-dot line), rudder (dashed line) versus elevator deflection schedule and (b) Bifurcation plots of Straight and level flight trim states as function of elevator deflection (solid lines: stable trims, dashed lines: unstable trims, empty squares: pitchfork bifurcations, solid squares: Hopf bifurcation).

Eigenvalues of A:

$$\lambda = \left[\underbrace{-0.727 \pm j0.83}_{\text{Short period}} \ \underbrace{-0.003 \pm j0.086}_{\text{Phugoid}} \ \underbrace{-2.43}_{\text{Roll}} \ \underbrace{-0.236 \pm j1.53}_{\text{Dutch roll}} \ \underbrace{-0.0009}_{\text{Spiral}} \right]$$

Eigenvectors of A:

$E=$					
Ma:	$-0.0169 \pm 0.0008i$	$-0.0419 \pm 0.3091i$	0.0	0.0	0.0
α:	0.6287	$0.0096 \pm 0.0353i$	0.0	0.0	0.0
β:	0.0	0.0	−0.0084	$-0.15 \pm 0.08i$	−0.0058
p:	0.0	0.0	0.9241	0.8108	0.0066
q:	$-0.23 \pm 0.5285i$	$-0.0026 \pm 0.0817i$	0.0	0.0	0.0
r:	0.0	0.0	0.0258	$-0.0766 \pm 0.1914i$	−0.0726
ϕ:	0.0	0.0	−0.3811	$-0.0695 \pm 0.5150i$	−0.9973
θ:	$-0.223 \pm 0.4718i$	−0.9459	0.0	0.0	0.0

Clearly, for this level flight equilibrium state, all the eigenvalues have negative real part; hence, this equilibrium state is stable. All the eigenvalues are quite distinctly located in the complex plane, and it should be easy to identify which eigenvalue corresponds to what mode as described below. Note that the system matrix A and the matrix E consisting of eigenvectors are distinctly divided into two 4×4 matrices (non-zero entries) representing decoupled motion in longitudinal and lateral variables. For clarity, the system matrix A in this case is rewritten below with longitudinal and lateral-directional variables clearly separated.

$A=$	Ma	α	q	θ	β	p	r	ϕ
	0.016	−0.010	0	−0.287	0	0	0	0
	−0.382	−1.080	0.99	0	0	0	0	0
	0	−0.832	−0.365	0	0	0	0	0
	0	0	1.0	0	0	0	0	0
	0	0	0	0	−0.199	−0.0787	−0.995	0.074
	0	0	0	0	−12.9	−2.57	0.792	0
	0	0	0	0	1.62	−0.049	−0.134	0
	0	0	0	0	0	1.0	0.079	0

The rearrangement clearly shows the decoupling of longitudinal and lateral-directional dynamics in straight and level flight trim conditions.

Further observations can be made as follows:

- The first pair of complex–conjugate eigenvalues is far to the left with the highest damping and the highest frequency/smallest time period values and must be the short-period mode eigenvalue. The corresponding eigenvectors (first column of E) indicate the largest magnitudes in variables α, q, θ and motion predominantly in pitch with hardly any change in velocity.
- The second pair of eigenvalues (second entry in λ) with the lowest damping and the longest time period/lowest frequency value corresponds to the phugoid mode. This is further confirmed by the corresponding eigenvector in the second column of E which suggests a motion predominant in variables Ma, θ with small changes in variables α, q.
- The third eigenvalue is on the real axis and the farthest among all; our immediate guess should be that this corresponds to the highly damped roll mode.

The corresponding eigenvector, confirming our observation, indicates a motion predominant in variables p, ϕ with minor change in variables β, r.

- The third pair of complex–conjugate eigenvalues belongs to the oscillatory Dutch roll mode. The corresponding eigenvector suggests that the motion is predominant in all lateral variables β, r, p, ϕ.
- The last eigenvalue on the real axis and close to the origin in the complex plane corresponds to the spiral mode.

In the next example, we consider an unstable level flight trim from Figure 8.13.

Equilibrium state:

$$\underline{x}^* = \left[Ma^* \ \alpha^* \ \beta^* \ p^* \ q^* \ r^* \ \phi^* \ \theta^*\right]' = [0.265 \ 0.157rad \ 0 \ 0 \ 0 \ 0 \ 0 \ 0.157rad]'$$

$$\underline{U}^* = \left[\eta^* \ \delta e^* \ \delta a^* \ \delta r^*\right]' = [0.32 - 0.02rad \ 0 \ 0]'$$

System matrix:

A =	Ma	α	β	p	q	r	ϕ	θ
	−0.0317	−0.018	0	0	0	0	0	−0.0287
	−0.796	−0.693	0	0	0.991	0	0	0
	0	0	−0.15	0.157	0	−0.986	0.107	0
	0	0	−7.82	−1.61	0	0.798	0	0
	−0.0002	−0.360	0	0	−0.233	0	0	0
	0	0	0.838	−0.0294	0	−0.0932	0	0
	0	0	0	1.0	0	0.158	0	0
	0	0	0	0	1.0	0	0	0

Eigenvalues of matrix A:

$$\lambda = \left[\underbrace{-0.48 \pm j0.538}_{\text{Short period}} \ \underbrace{0.002 \pm j0.125}_{\text{Phugoid}} \ \underbrace{-1.348}_{\text{Roll}} \ \underbrace{-0.256 \pm j1.319}_{\text{Dutch roll}} \ \underbrace{0.008}_{\text{Spiral}}\right]$$

Eigenvectors of A:

E =					
Ma:	−0.0343 ± 0.007i	−0.046 ± 0.2139i	0.0	0.0	0.0
α:	−0.6936	0.041 ± 0.0798i	0.0	0.0	0.0
β:	0.0	0.0	−0.0233	0.142 ± 0.1206i	−0.0124
p:	0.0	0.0	0.8	−0.7846	0.0083
q:	−0.176 ± 0.383i	−0.0019 ± 0.121i	0.0	0.0	0.0
r:	0.0	0.0	0.0343	−0.0623 ± 0.1155i	−0.1046
ϕ:	0.0	0.0	−0.5978	0.0967 ± 0.5684i	−0.9944
θ:	−0.2329 ± 0.5348i	0.9641	0.0	0.0	0.0

From the set of eigenvalues for this equilibrium state, we observe that one real eigenvalue and a pair of complex–conjugate eigenvalues have positive real parts, indicating that the equilibrium state under consideration is unstable. The unstable eigenvalues correspond to phugoid and spiral dynamic modes. The corresponding eigenvectors indicating contribution from the different variables to each mode are also listed above.

8.7.2 Coordinated (Zero Sideslip) Level Turn Trim

Level turn is a manoeuvre in which the curved flight path is in a horizontal plane parallel to the ground at constant altitude. Level turn demands that the velocity vector V is confined to the constant altitude plane, thus the flight path angle, $\gamma = 0$, needs to be satisfied at all time. In a steady (unaccelerated) level turn, a circular path of constant radius is essentially initiated by banking the aircraft at a constant bank angle, ϕ. The banking results in a component of lift, $Lsin\phi$, directed towards the centre of the circular path (as shown in Figure 1.10c), which balances out the centrifugal force; thus,

$$Lsin\phi = m\frac{V^2}{R} \tag{8.76}$$

is satisfied at all times, and R is the radius of the circular path. The other component of lift, $Lcos\phi$, balances the weight of the aircraft; thus,

$$Lcos\phi = W \Rightarrow cos\phi = \frac{1}{(L/W)} = \frac{1}{n} \tag{8.77}$$

where n is the load factor. In a steady level turn, a fixed bank angle also indicates a fixed load factor. Further, using Equation 8.76 and Equation 8.77, the following relation is obtained:

$$\tan\phi = \frac{V^2}{Rg} \Rightarrow \frac{V^2}{Rg} = \sqrt{n^2 - 1} \Rightarrow \frac{\omega^2 R}{g} = \sqrt{n^2 - 1} \tag{8.78}$$

where $\omega(= V/R)$ is the turn rate.

With a fixed load factor, a fixed velocity in trim indicates a constant radius and a constant turn rate as per Equation 8.78. Remember (from Chapter 6) that as the aircraft banks into the turn a component of weight $Wsin\phi$ acts along the inner wing causing the aircraft to sideslip, which needs to be suppressed for the turn to remain level. For this purpose a constraint on the sideslip angle $\beta = 0$ is exercised with the help of the rudder. Thus, two of the constraint equations in level turn trim flights are:

$$\gamma = 0, \quad \beta = 0$$

TABLE 8.3
Summary of Constraints in Level Turn Manoeuvre (Corresponding to Figure 8.15)

Branch	Constraints	Free Parameters	Fixed Parameters
A	$\gamma = 0,\ \beta = 0$	$\delta a,\ \delta r$	$\eta = 1.0$
B	$\gamma = 0,\ \beta = 0,\ \alpha = 0.63 rad\ (= \alpha_{CLmax})$	$\delta a,\ \delta r,\ \eta$	—
C	$\gamma = 0,\ \beta = 0,\ \mu = 1.38 rad\ (n_{max} = 5.4)$	$\delta a,\ \delta r,\ \eta$	—

The third constraint equation can be defined variously as shown in Table 8.3—either by fixing the thrust or the angle of attack or the load factor (equivalently, the bank angle).

Perturbed motions from a level turn trim are coupled in longitudinal and lateral dynamics—for example, if R decreases by a small amount, everything else remaining the same, then the centripetal force will increase; so in Equation 8.76, either V, ϕ or both must increase—so phugoid, roll, Dutch roll and spiral modes can be excited together.

A level turn trim state corresponding to maximum available thrust $(\eta = 1.0)$ is taken here as an illustrative example.

Equilibrium state:

$$\underline{x}^* = \left[Ma^*\ \alpha^*\ \beta^*\ p^*\ q^*\ r^*\ \phi^*\ \theta^* \right]'$$

$$= \left[0.24\ 0.372\text{rad}\ 0\ 0.033\text{rad/s}\ 0.107\text{rad/s} - 0.084\text{rad/s} - 0.905\text{rad}\ 0.236\text{rad} \right]'$$

$$\underline{U}^* = \left[\eta^*\ \delta e^*\ \delta a^*\ \delta r^* \right]' = [1.0 - 0.103\text{rad}\ 0.044\text{rad}\ 0.013\text{rad}]'$$

This equilibrium state corresponds to the maximum stable sustained turn rate (STR) for the F-18/HARV model as we will see soon in Figure 8.15; it turns out that this equilibrium state is stable as we conclude from the eigenvalue analysis below.

System matrix:

$A =$	Ma	α	β	p	q	r	ϕ	θ
	–0.0987	–0.0341	0.0219	0	0	0	–0.008	–0.0275
	–0.9120	–0.2720	0.0000	0	0.9920	0	–0.0845	0.0260
	–0.3760	–0.0016	–0.1430	0.3640	0	–0.9320	0.0712	–0.0219
	–0.3930	0.4090	–11.40	–0.8810	–0.0552	0.8540	0	0
	0.1220	–0.3820	0	0.0804	–0.2260	–0.0314	0	0
	0.0130	–0.0299	0.2810	–0.0849	0.0264	–0.1040	0	0
	0	0	0	1.0000	0.1900	0.1490	0	0.1440
	0	0	0	0	0.6170	–0.7870	–0.1360	0

Eigenvalues of matrix A:

$$\lambda = \left[\underbrace{-0.303 \pm j2.073}_{\text{Coupled Dutch roll}} \ \underbrace{-0.056 \pm j0.181}_{\text{Coupled Phugoid}} \ \underbrace{-0.498}_{\text{Coupled roll}} \ \underbrace{-0.25 \pm j0.585}_{\text{Coupled Short period}} \ \underbrace{-0.0084}_{\text{Coupled Spiral}} \right]$$

The real parts of all the eigenvalues at this trim condition are negative; hence, we conclude that this level turn trim state is stable.

Eigenvectors of A:

E =					
Ma:	0.001 ± *j*0.0002	−0.0263 ± *j* 0.1446	0.0136	−0.0245 ± *j* 0.0544	−0.152
α:	0.0002 ∓ *j*0.0017	0.0336 ± *j*0.0166	−0.0135	−0.6603	0.0595
β:	−0.0464 ∓ *j*0.1583	0.0086 ∓ *j*0.0128	0.0054	−0.0324 ± *j*0.0118	0.0097
p:	0.8891	−0.0894 ± *j*0.0743	−0.438	−0.0894 ∓ *j*0.0743	0.0441
q:	−0.0016 ∓ *j*0.0346	−0.0072 ± *j*0.118	0.0922	−0.0531 ∓ *j*0.4257	−0.202
r:	−0.0176 ± *j*0.0444	−0.0162 ∓ *j*0.0511	−0.1058	0.0079 ∓ *j*0.0143	−0.0274
ϕ:	−0.0617 ∓ *j*0.4183	0.4074 ∓ *j*0.2788	0.8869	−0.3165 ± *j*0.2169	−0.7187
θ:	−0.0012 ∓ *j*0.0101	0.8381	−0.0392	−0.408 ± *j*0.1676	0.6428

Even though the lateral and longitudinal motions are coupled in this flight condition, the dynamic modes at this equilibrium state are distinctly placed in the complex plane. The eigenvalues may therefore be notionally labelled as 'short-period-like', 'phugoid-like' and so on. However, it is not easy to qualify these eigenvalues as representing the five typical aircraft dynamic modes from an analysis of the eigenvectors.

8.7.3 Performance and Stability Analysis

The EBA method brings together the analysis of airplane point performance and airplane stability. After all, the airplane's performance deals with a variety of airplane trim (steady) states, for example, different straight and level trims with varying velocity, or straight climbing trims with different climb angle, or level turns with different load factors and so on. One may be interested in finding the trim velocity where aerodynamic efficiency is a maximum, or the climb angle for which the climb rate is a maximum. In that case, we usually obtain entire branches of steady states related by a common condition, for example, straight, climbing flight in the longitudinal plane. It would be ideal if the stability of these steady states could be computed simultaneously so that along with trim states with optimal performance parameters, one could also judge their suitability from the stability point of view. This is precisely what the EBA method is capable of doing. It computes an entire family of trims prescribed by a set of constraints and also the stability of each trim state. Thus, the EBA method brings together airplane performance and stability in a single procedure. The EBA also identifies the points on the family of trims where stability is lost and the new family of trims (which do not satisfy the original set of constraints) that are stable instead.

8.7.3.1 Straight and Level Flight Trim

Level flight trims with different throttle settings are usually studied in airplane performance. Owing to the quadratic nature of the drag coefficient C_D with the lift coefficient C_L, usually two sets of trim states are obtained for every throttle setting—one at a higher velocity (or Mach number) and one at a lower velocity. The set of lower-velocity solutions are sometimes referred to as 'back of the power curve' trims and are usually considered to be unstable from a 'speed stability' point of view. That is, when the airplane speed is reduced from a 'back of the power curve' trim state, the change in drag is more than the change in thrust, resulting in a further decrease in velocity, and so on, hence the instability. This ad hoc concept of 'speed stability' has been prevalent in airplane performance texts.

Let us look at the EBA-generated results for straight and level flight trims shown in Figure 8.14. The EBA procedure also computes the stability at the same time, and this information is also marked on the plots in Figure 8.14. In Figure 8.14a; a plot of thrust required versus the Mach number in straight and level flight condition for a low-angle-of-attack model of F-18/HARV (Appendix 8.2) is shown. The curve D–C–B–A–E looks quadratic following the parabolic drag polar law. The minimum thrust required or the minimum drag trim condition at the bottom of this curve is one of the performance parameters in aircraft design. But, note that the 'back of the power curve' segments up to B and then between C and D are actually stable. The stability indicated here means that all the airplane dynamic modes—short period, phugoid, roll, Dutch roll and spiral—are stable. The Hopf bifurcation at point B suggests that a pair of complex–conjugate eigenvalues have crossed over to the right half-plane (see Figure 8.14d). This turns out to be the phugoid eigenvalues; since the velocity is a key ingredient of the phugoid eigenvector, the phugoid instability may even be referred to casually as 'speed instability'. However, as can be seen from Figure 8.14a, not all trims on the 'back of the power curve' are unstable, only some are. Hence, the prevalent notion of 'speed stability' in airplane performance based only on comparing change in thrust and drag is not a sound one. Instead, it is more appropriate and correct to evaluate all the dynamic modes and then to judge the stability of a trim state, as is being done by EBA in Figure 8.14.

Traditionally, trims on the 'front of the power curve' are usually considered to be stable. But, as Figure 8.14 shows, for the present case, most of these trim states (between points A and E) are actually unstable. From Figure 8.14d, we can conclude that the mode that goes unstable at point A is the spiral. Instead, another branch of steady states between A and F is stable, but these are not level trims. From Figure 8.14b, we can see that the flight path angle γ is negative over this branch, which suggests that these are descending flights. From Figure 8.14c, the reasonably large bank angle ϕ with small values of all the angular rates p, q, r indicates that these are turning flights. Thus, the branch A–F represents descending turns that are spiralling down in space, and these are the stable trims.

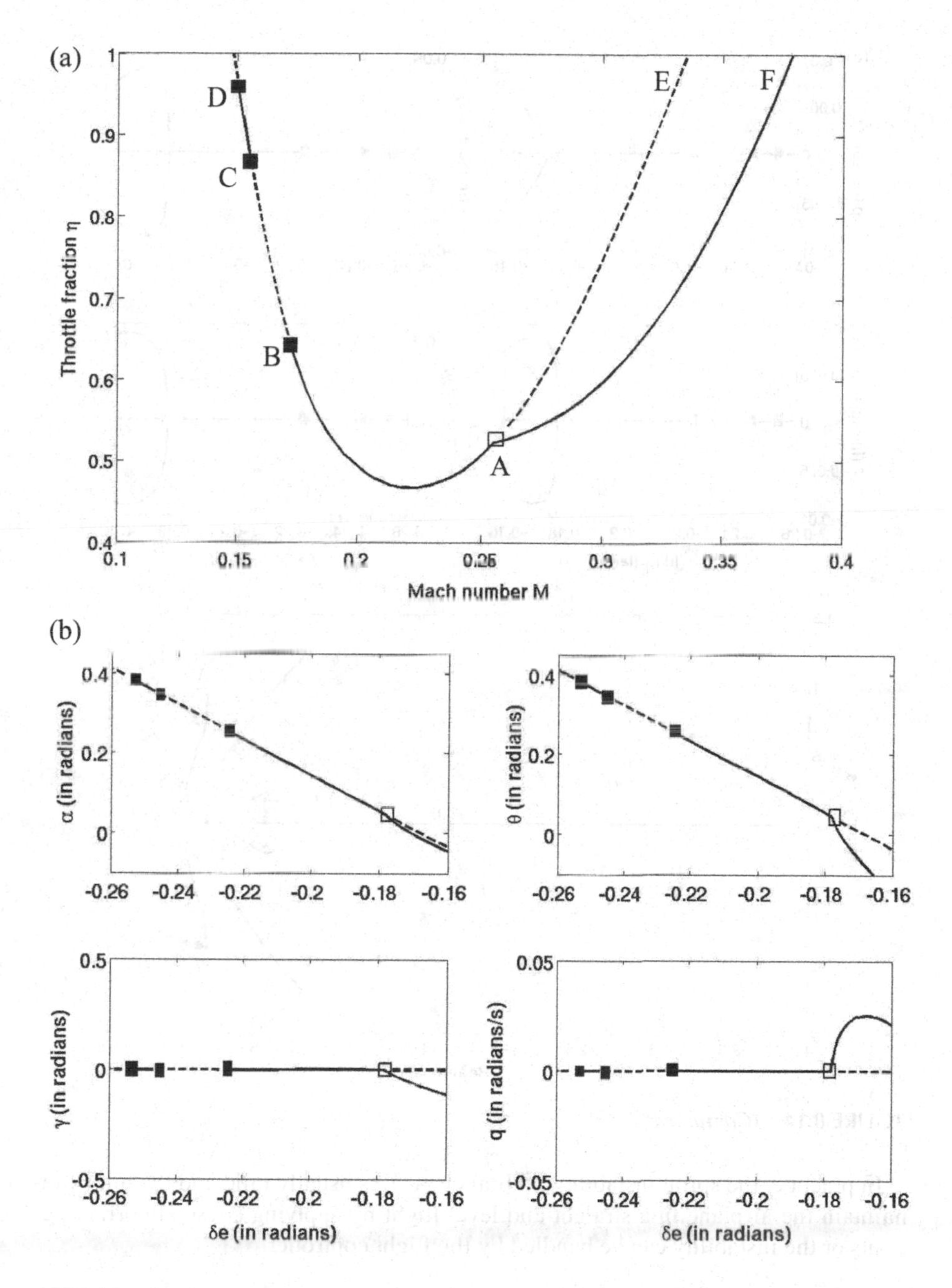

FIGURE 8.14 Bifurcation diagram of: (a) throttle fraction in straight and level flight trim condition as a function of the Mach number, (b) and (c) other variables as functions of elevator deflection (solid lines: stable trims; dashed lines: unstable trims; empty squares: pitchfork bifurcations; solid squares: Hopf bifurcation) and (d) root locus plot with variation in trim angle of attack.

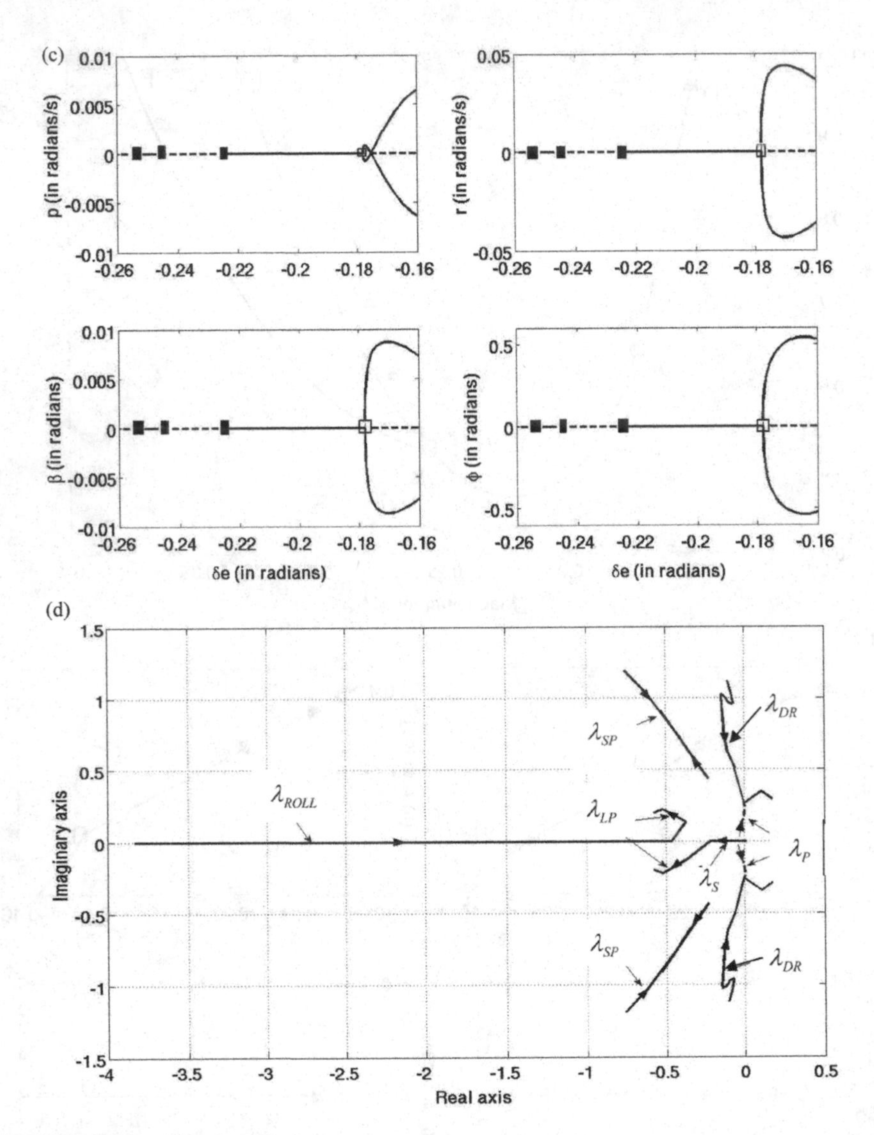

FIGURE 8.14 (*Continued*)

In practice, the spiral instability on branch A–E is usually mild and the pilot can maintain the airplane in a straight and level flight by applying constant correcting inputs or the instability can be handled by the flight controller.

8.7.3.2 Level Turn Manoeuvre

In this manoeuvre performance curves are typically presented as plots between the turn rate and the speed or as turn radius versus speed. Of interest are the maximum possible turn rate (also called fastest turn) and the minimum possible turn radius (also called tightest turn). Two kinds of turn rate are used as performance parameters—the maximum sustained turn rate (STR), where 'sustained' implies that the thrust is equal to the drag; hence, the velocity (and the turn rate) can be maintained with time, and maximum instantaneous turn rate (ITR), where the thrust–drag equality is forsaken, and thrust being usually less than drag, this implies that the velocity is bled off and hence the turn rate cannot be sustained with time.

Thus, three different solutions are usually found for turn performance depending on the constraint applied. In every case, the level flight condition and the zero sideslip condition are imposed: $\gamma = 0$ and $\beta = 0$. The difference arises in the nature of the third constraint, which can either be: (i) thrust limited—fixed level of thrust, usually maximum military thrust; (ii) lift limited—fixed level of C_L, usually maximum C_L corresponding to stall, or marginally less than stall; or (iii) load factor limited—fixed level of load factor, usually maximum load factor. These constraints for the bifurcation analysis of level turn manoeuvre are listed in Table 8.3 along with free and fixed parameters according to the computed level turn trim solution branches as shown in Figure 8.15.[4]

In Figure 8.15a, the turn rates for the F-18/HARV model corresponding to three different constrained conditions in level turn manoeuvres are presented as functions of the Mach number. Branch 'A' consists of level turn trim solutions corresponding to fixed maximum available thrust, branch 'B' consists of level turn trim solutions corresponding to stall angle of attack α_{stall} defining the aerodynamic boundary and branch 'C' consists of level turn trim solutions corresponding to maximum load factor (n_{max}) value defining the structural limit. Plots in Figure 8.15b show the thrust required (in terms of throttle fraction) exceeding the maximum value of 1.0 for the level turn trim solution branches 'B' and 'C', but for the solution branch 'A', throttle is fixed at its maximum value of 1.0. Each point on solution branch 'A' represents a level turn trim corresponding to a particular trim velocity and a fixed value of load factor at maximum available thrust. Branch 'A' thus represents the maximum STR at each velocity that the aircraft can achieve. Region '1' below branch 'A' is the region of possible STR manoeuvre provided all level turn trim solutions in this region are stable. If we ignore stability, point 'T' on the branch 'A' is the level turn trim solution corresponding to maximum STR. The maximum STR considering stability is the one corresponding to the point 'S'. Between points 'P' and 'Q', and 'S' and 'U', on branch 'A', level turn trims are unstable. Region '2' in Figure 8.15a represents the region of instantaneous turn bounded by the maximum turn rate solution branches 'B' (corresponding to stall limit) and 'C' (corresponding to structural limit). For a particular value of the Mach number, the maximum ITR can thus be found on branches 'B' and 'C'. The corner point 'CP' is the level turn trim state corresponding to the maximum ITR and is stable. But note that trim has been artificially enforced on branches 'B' and 'C' by allowing $\eta > 1$.

8.7.3.3 Maximum Roll Rate in a Roll Manoeuvre

We have seen earlier from the bifurcation analysis results of the PSS model (Figure 8.12) that the maximum pre-jump roll rate is limited to a finite value due to inertia coupling instability and the occurrence of a saddle node bifurcation at a critical aileron deflection angle. The increase in the sideslip angle with increasing aileron deflection is known to have mainly contributed to a loss of stability. To limit the growth of the sideslip angle in a roll manoeuvre, many strategies have been suggested;

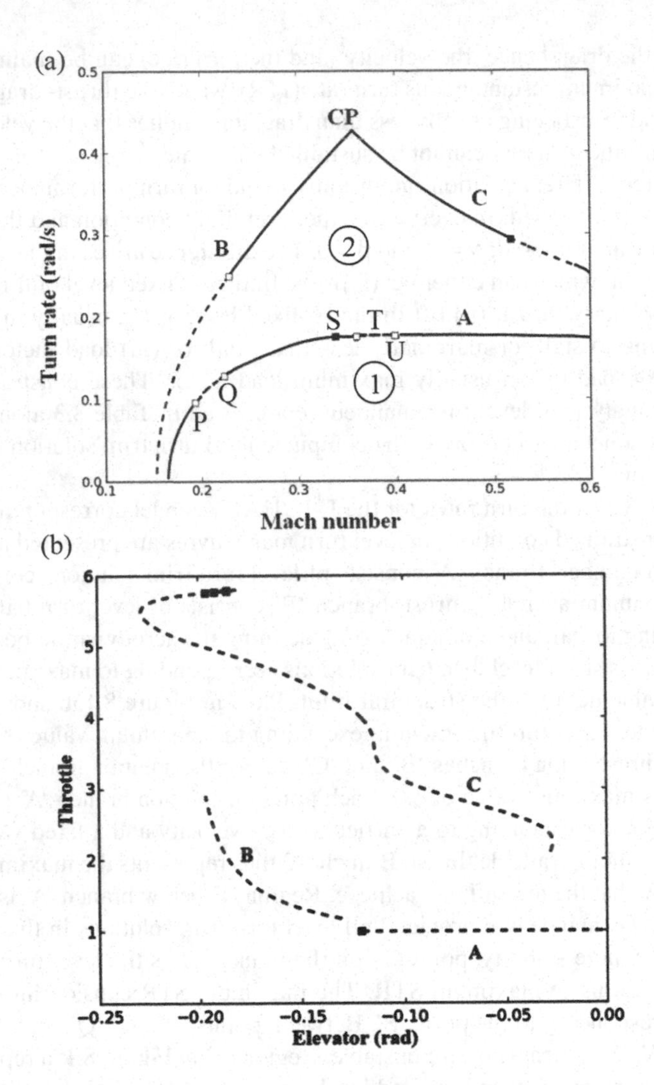

FIGURE 8.15 Bifurcation plots of: (a) turn rate as function of Mach number and (b) throttle as function of elevator deflection in level turn flight (solid lines: stable trims; dashed lines: unstable trims; empty squares: pitchfork bifurcations; solid squares: Hopf bifurcation).

among them, simultaneous deployment of rudder according to laws known as aileron-to-rudder interconnect (ARI) is popular. The development of ARI laws based on various constraints and computation of PSS roll rates with stability information and discussion on roll performance for the aircraft data given in Appendix 8.3 follows.

In Figure 8.16, the ARI, $\delta r(\delta a)$, for zero sideslip roll manoeuvre (with constraint, $\beta = 0$ on PSS model) is shown in the top left plot. The corresponding roll rate solutions are plotted along with other variables in other plots. From the plots, it is noticed that with the use of ARI, the saddle node bifurcation vanishes; instead, a transcritical bifurcation point appears that avoids jump and the associated non-linear behaviour

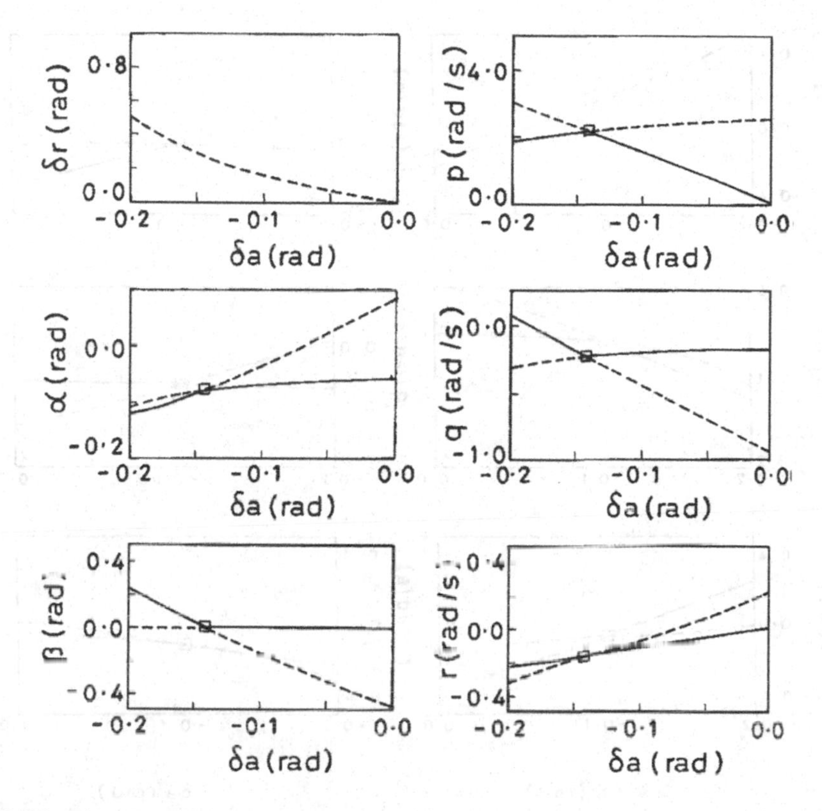

FIGURE 8.16 Zero sideslip aileron–rudder interconnect (ARI) law (top left) and corresponding PSS solutions (solid lines: stable solutions; dashed lines: unstable solutions; empty square: transcritical bifurcation).

in roll manoeuvre. The maximum roll rate is also slightly improved as compared to the no ARI case (Figure 8.12).

Further, the analysis of a velocity vector roll manoeuvre is carried out. The constraint in this case is that the linear and angular velocity vectors must coincide with each other. Mathematically, this constraint can be written as:

$$\frac{\underline{V} \cdot \underline{\omega}}{|\underline{V}||\underline{\omega}|} - 1 = 0 \tag{8.79}$$

which in a simplified form can be written as:

$$\frac{p + q\beta + r\alpha}{\sqrt{p^2 + q^2 + r^2}} - 1 = 0 \tag{8.80}$$

The computation of the velocity-vector roll solutions by carrying out a continuation of the PSS model with the constraint in Equation 8.80 results in ARI as shown in the top left plot in Figure 8.17. The velocity-vector roll does not result in better

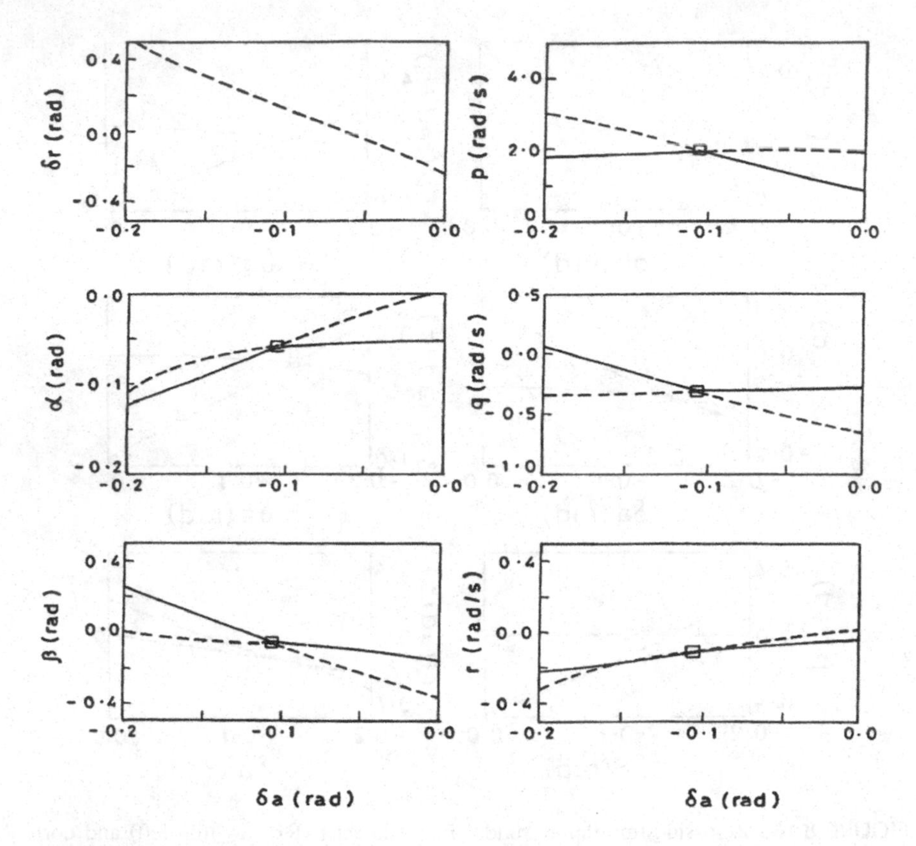

FIGURE 8.17 Velocity-vector roll aileron–rudder interconnect (ARI) law (top left) and corresponding PSS solutions (solid lines: stable solutions; dashed lines: unstable solutions; empty square: transcritical bifurcation).

maximum roll rate, but can be achieved for lower aileron deflection and lower rudder deflection ($\delta a \approx -0.11$ rad, $\delta r \approx 0.12$ rad) as compared to zero-sideslip roll manoeuvre ($\delta a \approx -0.15$ rad, $\delta r \approx 0.25$ rad). Similar maximum value of the roll rate of around 2 rad/s can be observed in both cases.

EXERCISE PROBLEMS

8.1 A thrust-vectored airplane generates a thrust $T = 40\ kN$ along a local vertical during a vertical take-off. Determine the component of the thrust along body-fixed axes, if the attitude of the aircraft is given by Euler angles, $[\phi\ \theta\ \psi] = [5\ 10\ 0]$, where the angles are in degrees.

8.2 Given V (m/s) = $[100\ 0\ 0]'$ in the wind-fixed axes system and orientation of wind-fixed axes system defined by the angles $\alpha = 30°$; $\beta = 30°$, determine the components of the relative wind along the aircraft-fixed body axes.

8.3 Determine a trim condition for the low-angle-of-attack data of F-18 given in Appendix 8.2 in: (1) a straight and level flight condition at $\alpha = 5°$ and (2) a level turn at load factor $n = 1.4$ at $\alpha = 5°$.

8.4 Carry out a stability analysis of the trims calculated in Exercise Problem 8.3 [*Hint:* You can use the linear set of equations in Appendix 8.1 to construct the system matrix A. Find out the eigenvalues of A (using [e,v] = eig(A) command in MATLAB®) and look at their locations in the complex plane.]

8.5 The MATLAB output from Exercise Problem 8.4 also gives the eigenvectors of interest. Identify typical aircraft dynamics modes from the eigenvectors.

8.6 From the eighth-order linear model given in Appendix 8.1, derive the reduced-order models presented in this book for aircraft dynamic modes. Discover the assumptions involved.

8.7 An airplane is performing a zero-sideslip level turn, that is, flying a circular path in the horizontal plane (see Figure 6.6). The bank angle is marked ϕ, the angular velocity vector is ω and the radius of the circle is R. Y^B and Z^B axes are marked in the figure. X^B axis is into the plane of the paper and aligned with the velocity vector V. Write down the equations for force and moment balance along all three axes. Also write down the components of the airplane velocity (u, v, w) and angular velocity (p, q, r) along the three axes.

8.8 Consider an axis system with origin at $(0, -H)$ with axes X^E running left and Z^E running down. An airplane flies a perfect clockwise circular loop of radius R and constant velocity V as sketched in Figure 1.10c and given by the following equations:

$$X^E = R\sin\gamma; \quad Z^E = -H + R\cos\gamma$$

Find the variation of thrust as a function of γ such that the velocity can be maintained constant at all points in the loop.

8.9 The PSS model given in Appendix 8.3 represents the fast dynamics of an aircraft. (a) From the equations, determine a steady-state condition representing the coordinated roll for the data given in Appendix 8.3. (b) Find a linear model of the equations. (c) From the linear model, determine the stability of the steady state determined in (a). (d) Study the eigenvectors computed in (c).

APPENDIX 8.1: SMALL-PERTURBATION EQUATIONS (AT LONGITUDINAL TRIM STATE)

$$\Delta\dot{V} = -\left(\frac{\rho VS}{m}C_D + \frac{\rho V^2 S}{2ma}C_{DMa}\right)^* \Delta V + \left(\frac{T\sin\alpha}{m} - \frac{\rho V^2 S}{2m}C_{D\alpha} + g\cos\gamma\right)^* \Delta\alpha$$

$$-g\cos\gamma^*\Delta\theta + \left(\frac{\cos\alpha}{m}\right)^* \Delta T - \left(\frac{\rho V^2 S}{2m}C_{D\delta e}\right)^* \Delta\delta e$$

$$\Delta\dot{\alpha} = \left(\frac{T\sin\alpha}{mV^2} - \frac{C_L\rho S}{2m} - \frac{\rho SV}{2ma}C_{LMa} - \frac{g}{V^2}\cos\gamma\right)^* \Delta V + \left(\frac{g}{V}sin\gamma - \frac{Tcos\alpha}{mV} - \frac{\rho SV}{2m}C_{L\alpha}\right)^* \Delta\alpha$$

$$+\Delta q - \left(\frac{\rho Sc}{4m}C_{Lq1}\right)^* (\Delta q - \Delta q_w) - \left(\frac{\rho Sc}{4m}C_{Lq2}\right)^* \Delta q_w + \left(-\frac{g\sin\gamma}{V}\right)^* \Delta\theta$$

$$-\left(\frac{\sin\alpha}{mV}\right)^* \Delta T - \left(\frac{\rho SV}{2m}C_{L\delta e}\right)^* \Delta\delta e$$

$$\Delta\dot{\beta} = \left(-\frac{T\cos\alpha}{mV} + \frac{\rho SV}{2m}C_{Y\beta} + \frac{g}{V}\sin\gamma\right)^* \Delta\beta + \left(\frac{\rho Sb}{4m}C_{Yp1}\right)^* (\Delta p - \Delta p_w) + \left(\frac{\rho Sb}{4m}C_{Yp2}\right)^* \Delta p_w$$

$$+\Delta p\sin\alpha^* + \left(\frac{\rho Sb}{4m}C_{Yr1}\right)^* (\Delta r - \Delta r_w) + \left(\frac{\rho Sb}{4m}C_{Yr2}\right)^* \Delta r_w - \Delta r\cos\alpha^*$$

$$+\left(\frac{\rho SV}{2m}C_{Y\delta a}\right)^* \Delta\delta a + \left(\frac{\rho SV}{2m}C_{Y\delta r}\right)^* \Delta\delta r$$

$$\Delta\dot{p} = \left(\frac{\rho SbV^2}{2I_{xx}}C_{l\beta}\right)^* \Delta\beta + \left(\frac{\rho Sb^2V}{4I_{xx}}C_{lp1}\right)^* (\Delta p - \Delta p_w) + \left(\frac{\rho Sb^2V}{4I_{xx}}C_{lp2}\right)^* \Delta p_w$$

$$+\left(\frac{\rho Sb^2V}{4I_{xx}}C_{lr1}\right)^* (\Delta r - \Delta r_w) + \left(\frac{\rho Sb^2V}{4I_{xx}}C_{lr2}\right)^* \Delta r_w + \left(\frac{\rho SbV^2}{2I_{xx}}C_{l\delta a}\right)^* \Delta\delta a + \left(\frac{\rho SbV^2}{2I_{xx}}C_{l\delta r}\right)^* \Delta\delta r$$

$$\Delta\dot{q} = \left(\frac{\rho SV^2c}{2I_{yy}a}C_{mMa}\right)^* \Delta V + \left(\frac{\rho ScV^2}{2I_{yy}}C_{m\alpha}\right)^* \Delta\alpha + \left(\frac{\rho Sc^2V}{4I_{yy}}C_{mq1}\right)^* (\Delta q - \Delta q_w)$$

$$+\left(\frac{\rho Sc^2V}{4I_{yy}}C_{mq2}\right)^* \Delta q_w + \left(\frac{\rho ScV^2}{2I_{yy}}C_{m\delta e}\right)^* \Delta\delta e$$

$$\Delta\dot{r} = \left(\frac{\rho SbV^2}{2I_{zz}}C_{n\beta}\right)^* \Delta\beta + \left(\frac{\rho Sb^2V}{4I_{zz}}C_{np1}\right)^* (\Delta p - \Delta p_w) + \left(\frac{\rho Sb^2V}{4I_{zz}}C_{np2}\right)^* \Delta p_w$$

$$+\left(\frac{\rho Sb^2V}{4I_{zz}}C_{nr1}\right)^* (\Delta r - \Delta r_w) + \left(\frac{\rho Sb^2V}{4I_{zz}}C_{nr2}\right)^* \Delta r_w + \left(\frac{\rho SbV^2}{2I_{zz}}C_{n\delta a}\right)^* \Delta\delta\text{a} + \left(\frac{\rho SbV^2}{2I_{zz}}C_{n\delta r}\right)^* \Delta\delta r$$

$$\Delta\dot{\phi} = \Delta p + (\tan\theta)^* \Delta r$$

$$\Delta\dot{\theta} = \Delta q$$

APPENDIX 8.2: F-18 DATA

Aerodynamic Data

The low-angle-of-attack aerodynamic model of F-18 (all angles and control deflections are in degrees and angular rates in radians per second):

$$C_D = \begin{cases} 0.0013\alpha^2 - 0.00438\alpha + 0.1423 & -5 \le \alpha \le 20 \\ -0.0000348\alpha^2 + 0.0473\alpha - 0.358 & 20 \le \alpha \le 40 \end{cases}$$

$$C_L = \begin{cases} 0.0751\alpha + 0.0144\delta e + 0.732 & -5 \le \alpha \le 10 \\ -0.00148\alpha^2 + 0.106\alpha + 0.0144\delta e + 0.569 & 10 \le \alpha \le 40 \end{cases}$$

$$C_m = -0.00437\alpha - 0.0196\delta e - 0.123(q_b - q_w) - 0.1885$$

$$C_l = C_l(\alpha,\beta) - 0.0315\, p_w + 0.0126(r_b - r_w) + \frac{\delta a}{25}(0.00121\alpha - 0.0628) - \frac{\delta r}{30}(0.000351\alpha - 0.0124)$$

$$C_l(\alpha,\beta) = \begin{cases} (-0.00012\alpha - 0.00092)\beta & -5 \le \alpha \le 15 \\ (0.00022\alpha - 0.006)\beta & 15 \le \alpha \le 40 \end{cases}$$

$$C_Y = -0.0186\beta + \frac{\delta a}{25}(-0.00227\alpha + 0.039) + \frac{\delta r}{30}(-0.00265\alpha + 0.141)$$

$$C_n = C_n(\alpha,\beta) - 0.0142(r_b - r_w) + \frac{\delta a}{25}(0.000213\alpha + 0.00128) + \frac{\delta r}{30}(0.000804\alpha - 0.0474)$$

$$C_n(\alpha,\beta) = \begin{cases} 0.00125\beta & -5 \le \alpha \le 10 \\ (-0.00022\alpha + 0.00342)\beta & 10 \le \alpha \le 25 \\ -0.00201\beta & 25 \le \alpha \le 35 \end{cases}$$

The high-angle-of-attack research vehicle aerodynamic model F-18/HARV (available at http://www.nasa.gov/centers/dryden/history/pastprojects/HARV/Work/NASA2/nasa2.html) consists of tabulated database in the aerodynamic derivatives and coefficients given below in the range of angle of attack $-4° \le \alpha \le 90°$.

Drag	C_{D0}	C_{Dq1}	$C_{D\delta e,r}$	$C_{D\delta e,l}$			
Side force	$C_{Y\beta}$	C_{Yp2}	C_{Yr1}	$C_{Y\delta e,r}$	$C_{Y\delta e,l}$	$C_{Y\delta a}$	$C_{Y\delta r}$
Lift	C_{L0}	C_{Lq1}	$C_{L\delta e,r}$	$C_{L\delta e,l}$			
Roll	$C_{l\beta}$	C_{lp2}	C_{lr1}	$C_{l\delta e,r}$	$C_{l\delta e,l}$	$C_{l\delta a}$	$C_{l\delta r}$
Pitch	C_{m0}	C_{mq1}	$C_{m\delta e,r}$	$C_{m\delta e,l}$			
Yaw	$C_{n\beta}$	C_{np2}	C_{nr1}	$C_{n\delta e,r}$	$C_{n\delta e,l}$	$C_{n\delta a}$	$C_{n\delta r}$

Geometric and Other Data for F-18

$m = 15118.35$ kg; $b = 11.405$ m; $c = 3.511$ m;

$S = 37.16$ m^2; $I_{xx} = 31181.88$ kg-m^2;

$I_{yy} = 205113.07$ kg-m^2; $I_{zz} = 230400.22$ kg-m^2

$T_m = 49817.6$ N; $\rho_{air}(@sea-level) = 1.225$ kg/m^3

$g = 9.81$ m/s^2; a(speed of sound) $= 340.0$ m/s

APPENDIX 8.3: EQUATIONS AND AIRCRAFT DATA USED FOR ROLL MANOEUVRE

$$\dot{\alpha} = q - p\beta - \frac{\rho VS}{2m} C_{L\alpha}\alpha$$

$$\dot{\beta} = p\alpha - r + \frac{\rho VS}{2m} C_{y\beta}\beta$$

$$\dot{p} = \left(\frac{I_{yy} - I_{zz}}{I_{xx}}\right)qr + \left(\frac{\bar{q}Sb}{I_{xx}}\right)C_{l\beta}\beta + \left(\frac{\bar{q}Sb}{I_{xx}}\right)C_{lp2}\,p\left(\frac{b}{2V}\right)$$
$$+\left(\frac{\bar{q}Sb}{I_{xx}}\right)C_{lr1}(r - r_w)\left(\frac{b}{2V}\right) + \left(\frac{\bar{q}Sb}{I_{xx}}\right)C_{l\delta a}\delta a + \left(\frac{\bar{q}Sb}{I_{xx}}\right)C_{l\delta r}\delta r$$

$$\dot{q} = \left(\frac{I_{zz} - I_{xx}}{I_{yy}}\right)pr + \left(\frac{\bar{q}Sc}{I_{yy}}\right)C_{m\alpha}\alpha + \left(\frac{\bar{q}Sc}{I_{yy}}\right)C_{mq1}(q - q_w)\left(\frac{c}{2V}\right) + \left(\frac{\bar{q}Sc}{I_{yy}}\right)C_{m\delta e}\delta e$$

$$\dot{r} = \left(\frac{I_{xx} - I_{yy}}{I_{zz}}\right)pq + \left(\frac{\bar{q}Sb}{I_{zz}}\right)C_{n\beta}\beta + \left(\frac{\bar{q}Sb}{I_{zz}}\right)C_{np2}\,p\left(\frac{b}{2V}\right)$$
$$+\left(\frac{\bar{q}Sb}{I_{zz}}\right)C_{nr1}(r - r_w)\left(\frac{b}{2V}\right) + \left(\frac{\bar{q}Sb}{I_{zz}}\right)C_{n\delta a}\delta a + \left(\frac{\bar{q}Sb}{I_{zz}}\right)C_{n\delta r}\delta r$$

$m = 2718$ kg; $b = 11.0$ m; $c = 1.829$ m;

$S = 20.07$ m^2; $\rho = 1.2256$ kg/m^3; $I_{xx} = 2304.9$ kg-m^2;

$I_{yy} = 16809$ kg-m^2; $I_{zz} = 18436$ kg-m^2

$$C_{L\alpha} = 4.35/\text{rad}; \quad C_{y\beta} = -0.081/\text{rad}; \quad C_{l\beta} = -0.081/rad;$$

$$C_{lp2} = -0.442/\text{rad}; \quad C_{lr1} = 0.0309/\text{rad}; \quad C_{l\delta a} = -0.24/rad;$$

$$C_{l\delta r} = 0.0; \quad C_{m\alpha} = -0.435/\text{rad}; \quad C_{mq1} = -9.73/rad;$$

$$C_{m\delta e} = -1.07/\text{rad}; \quad C_{n\beta} = 0.0218/\text{rad}; \quad C_{np2} = 0.0;$$

$$C_{n\delta a} = 0.0; \quad C_{nr1} = -0.0424/\text{rad}; \quad C_{n\delta r} = -0.01/rad;$$

REFERENCES

1. Cummings, P.A., Continuation methods for qualitative analysis of aircraft dynamics, NASA/CR-2004-213035, NIA Report No. 2004-06, 2004.
2. Strogatz, S.H., *Nonlinear Dynamics and Chaos*, Second Edition, Westview Press, Cambridge, MA, 2014.
3. Ananthkrishnan, N. and Sinha, N.K., Level flight trim and stability analysis using extended bifurcation and continuation procedure, *Journal of Guidance, Control, and Dynamics*, 24(6), 2001, 1225–1228.
4. Paranjape, A. and Ananthkrishnan, N., Airplane level turn performance, including stability constraints, using extended bifurcation and continuation method, AIAA Atmospheric Flight Mechanics Conference, San Francisco, CA, August 2005, AIAA 2005-5898.
5. Goman, M.G., Zagaynov, G.I. and Khramtsovsky, A.V., Application of bifurcation methods to nonlinear flight dynamics problems, *Progress in Aerospace Sciences*, 33, 1997, 539–586.

9 Appendix: Case Studies

As mentioned in Chapter 1, the presentation in this book is by and large applicable to any rigid, fixed-wing airplane. Throughout this book we have used numerical data for the F-18/HARV—a combat aircraft—to numerically illustrate the various concepts discussed. Examples of other aircraft, including missiles, parafoils and airships, have appeared either in the chapter-end Exercise Problems or as Examples within the text. We have also commented in Chapter 7 on the matter of scale and how the developments in this book apply equally well to unmanned aerial vehicles (UAVs). In this Appendix, we provide supplementary material to add to the contents of the first eight chapters; however, it is not necessary to read or be acquainted with any of the contents of this Appendix to appreciate the earlier presentation. We shall take up two case studies in this Appendix: firstly, that of a light General Aviation (GA) airplane, and secondly, an example of a rigid airship.

9.1 EXAMPLE OF GA AIRPLANE

We shall use this case study to present two different aspects of flight dynamics calculations that we could not present in the previous chapters.

1. Estimation of aerodynamic characteristics (the aero coefficients and derivatives) of an airplane—In the examples presented previously, the aerodynamic data (called 'aero data' for short) was presented in the form of tables or figures or curve-fitted formulas. In general, aero data can be obtained by one of three means: by analytical/semi-empirical methods, by the use of numerical methods (CFD), or from wind tunnel tests. Often, during an aircraft design and development process, all three methods are employed at various points of time, and the data obtained from each of the methods may be blended to create a final database. In this section, we shall first take up an example of GA aircraft and demonstrate the use of the analytical/semi-empirical method to estimate the aero data.
2. Modal (longitudinal and lateral-directional) stability analysis using the first-order form of the equations—In Chapters 5 and 6, we have presented the small-perturbation equations for the dynamics of the airplane in longitudinal flight and in lateral-directional flight, respectively, in second-order form. The second-order form is physically meaningful and convenient to derive the literal approximations to the various modes (short period, Dutch roll, etc.), as we have already seen. On the other hand, the first-order form of the linearized equations for the modal dynamics is commonly used to

DOI: 10.1201/9781003096801-9

numerically evaluate the eigenvalues of the 'A' matrix and hence determine the stability of the modes. In the interest of completeness, we present here the small-perturbation equations for the longitudinal and lateral-directional dynamics rewritten in first-order form. Subsequently, we use the aero data estimated for the GA aircraft to populate the 'A' matrices for the longitudinal and lateral-directional dynamics and evaluate the stability of the respective modes.

9.1.1 Aero Data Estimation

The GA aircraft model used here is a six-seater business transport similar to the Piper Cherokee shown in Figure 9.1. The aerodynamic estimation procedure follows along the lines as described in Ref. [1]. The aircraft is a low-wing conventional configuration with a single engine (Textron Lycoming piston type of 360 hp with 3-blade propeller) mounted in a tractor arrangement. A trapezoidal wing planform of taper ratio 0.5 has been used. Wing airfoil section varies along span from NACA23018 at root to NACA23012 at wing tip, however, wing aerodynamic data for the aircraft has been estimated by assuming constant airfoil section NACA23018 throughout the span. NACA0012 airfoil section has been used for both horizontal and vertical tails. Relevant aircraft parameters are provided in Tables 9.1–9.3.

FIGURE 9.1 A six-seater business transport, Piper PA-32-300 Cherokee. (Photo courtesy: Arping Stone; https://upload.wikimedia.org/wikipedia/commons/f/f2/Piper.pa32.cherokee.six.g-bxwp.arp.jpg.)

TABLE 9.1
GA Aircraft Geometry, Mass and Inertia Data

Acceleration due to gravity	$g = 9.81 m/s^2$	Mass of aircraft	$m = 1859.73kg$
Reference area (wing planform area)	$S_w = 16.35m^2$	Moments of inertia (about principal axes)	$I_{xx} = 3355.65kg - m^2$
Wing root chord (NACA 23018)	$c_r = 2.02m$		$I_{yy} = 4180.57kg - m^2$
Wing tip chord (NACA 23012)	$c_t = 1.01m$		$I_{zz} = 6140.66kg - m^2$
Wing span	$b_w = 10.75m$	Wing incidence setting angle	$i_w = 0deg$
Wing dihedral angle	$\Gamma = 5deg$	Wing zero-lift angle of attack	$\alpha_{L=0} = -1.1deg$
Wing quarter-chord sweep angle	$\Lambda_{c/4} = 5.38deg$	CG location from aircraft nose	$X_{CG} = 2.39m$
Fuselage length	$l_f = 7.89m$	Horizontal tail planform area	$S_{HT} = 3.15m^2$
Fuselage side projected area	$S_f = 8.12m^2$	Horizontal tail aspect ratio	$AR_{HT} = 4.0$
Maximum fuselage width	$D_f = 1.37m$	Horizontal tail incidence setting angle	$i_t = 0deg$
Vertical tail planform area	$S_{VT} = 2.16m^2$	HT aerodynamic centre location (from aircraft nose)	$X_{AC}^{HT} = 7.62m$
Vertical tail aspect ratio	$AR_{VT} = 1.5$	VT aerodynamic centre location (from aircraft nose)	$X_{AC}^{VT} = 7.28m$
Aileron surface area	$S_{\delta a} = 0.82m^2$	Elevator surface area	$S_{\delta e} = 1.38m^2$
Aileron span	$l_{\delta a} = 2.15m$	Rudder surface area	$S_{\delta r} = 0.43m^2$

TABLE 9.2
GA Aircraft Flight Data

Density of air at flight altitude	$\rho_{air} = 1.225kg/m^3$	Speed of sound at flight altitude	$a = 340m/s$
Dynamic pressure ratio at the tail	$\eta = 0.95$		

TABLE 9.3
GA Aircraft Engine Data

Maximum engine power available	$P_{max} = 268452N - m/s$	Number of propeller blades	$N = 3$
Propeller diameter	$D = 1.99m$	Propeller disk distance from aircraft CG	$l_{prop} = -2.39m$

Calculated values:

Wing taper ratio: $\lambda = c_t / c_r = 0.5$;

Wing mean aerodynamic chord:

$$\bar{c} = \frac{2}{3} c_r \left(\frac{1+\lambda+\lambda^2}{1+\lambda} \right) = \frac{2}{3} \times 2.02 \times \left(\frac{1+0.5+0.5^2}{1+0.5} \right) = 1.57m;$$

Wing aspect ratio: $AR_w = b_w^2 / S_w = \dfrac{10.75^2}{16.35} = 7.07$;

Horizontal tail span: $b_{HT} = \sqrt{S_{HT} \times AR_{HT}} = \sqrt{4.0 \times 3.45} = 3.71m$;

Vertical tail span: $b_{VT} = \sqrt{S_{VT} \times AR_{VT}} = \sqrt{1.5 \times 2.16} = 1.8m$;

Vertical tail AC height above FRL: $z_{VT} = \dfrac{4}{9} \times b_{VT} = \dfrac{4}{9} \times 1.8 = 0.8m$;

Propeller disc area: $S_{prop} = \dfrac{\pi D^2}{4} = \dfrac{\pi \times 1.99^2}{4} = 3.11m^2$;

Tail arms:

HT moment arm length: $l_{HT} = X_{AC}^{HT} - X_{CG} = 7.62 - 2.39 = 5.23m$;

VT moment arm length: $l_{VT} = X_{AC}^{VT} - X_{CG} = 7.28 - 2.39 = 4.89m$;

HT volume ratio: $V_{HT} = \dfrac{S_{HT} \times l_{HT}}{S_w \times \bar{c}} = \dfrac{3.45 \times 5.23}{16.35 \times 1.57} = 0.7$;

VT volume ratio: $V_{VT} = \dfrac{S_{VT} \times l_{VT}}{S_w \times b_w} = \dfrac{2.16 \times 4.89}{16.35 \times 10.75} = 0.06$;

Lift-curve slope:

Wing airfoil lift-curve slope, $c_{l\alpha,w} = 0.1/deg = 5.73/rad$ (From Ref. [1] for NACA23018);

HT airfoil lift-curve slope, $c_{l\alpha,HT} = 0.109/deg = 6.24/rad$;

VT airfoil lift-curve slope, $c_{l\alpha,VT} = 0.109/deg = 6.24/rad$;

For subsonic speeds, the lift-curve slope of finite aspect ratio straight tapered wings can be determined using the expression:

Wing lift-curve slope: $C_{L\alpha,w} = \dfrac{2\pi AR_w}{2 + \sqrt{\dfrac{AR_w^2 \beta^2}{k^2} \left(1 + \dfrac{tan^2 \Lambda_{c/2}}{\beta^2} \right) + 4}}$; $\beta = \sqrt{1 - Ma^2}$,

$k = C_{l\alpha,w} / 2\pi$, $\Lambda_{c/2}$ is the sweep angle of mid-chord of the wing. At low subsonic speed ($\beta \approx 1$) and $\left(\dfrac{tan^2 \Lambda_{c/2}}{\beta^2} \right) \ll 1$, the above expression simplifies to: $C_{L\alpha,w} = \dfrac{c_{l\alpha,w}}{1 + \left(c_{l\alpha,w} / \pi AR_w \right)}$.

Wing lift-curve slope:

$$C_{L\alpha,w} = \frac{c_{l\alpha,w}}{1+\left(c_{l\alpha,w}/\pi AR_w\right)} = \frac{5.73}{1+(5.73/3.14\times 7.07)} = 4.55/rad;$$

HT lift-curve slope:

$$C_{L\alpha,HT} = \frac{c_{l\alpha,HT}}{1+\left(c_{l\alpha,HT}/\pi AR_{HT}\right)} = \frac{6.24}{1+(6.24/3.14\times 4.0)} = 4.17/rad;$$

VT lift-curve slope: $C_{L\alpha,VT} = \frac{c_{l\alpha,VT}}{1+\left(c_{l\alpha,VT}/\pi AR_{VT}\right)} = \frac{6.24}{1+(6.24/3.14\times 1.5)} = 2.68/rad;$

Wing lift coefficient at zero angle-of-attack:

$$C_{L0,w} = C_{L\alpha,w}\times\alpha_{L=0}\times\pi/180 = 4.55\times 1.1\times\pi/180 = 0.087;$$

SM, AC calculations:

Non-dimensionalized CG location from airplane nose: $h_{CG} = X_{CG}/\bar{c} = \frac{2.39}{1.57} = 1.52$

Assuming static margin: $SM = 15\%$

$$h_{NP} - h_{CG} = 0.15;$$

$$h_{NP} = 1.52 + 0.15 = 1.67; \quad X_{NP} = h_{NP}\times\bar{c} = 1.67\times 1.57 = 2.62m$$

$$h_{AC}^{wb} = h_{NP} - V_{HT}\left\{\frac{C_{L\alpha,HT}}{C_{L\alpha,w}}\right\}\left(1-\frac{d\varepsilon}{d\alpha}\right)$$

Downwash angle: $\varepsilon = \frac{2C_{L,w}}{\pi AR_w} = \underbrace{\frac{2C_{L0,w}}{\pi AR_w}}_{\varepsilon_0} + \underbrace{\frac{2C_{L\alpha,w}}{\pi AR_w}}_{\varepsilon_\alpha}(\alpha + i_w)$

$$= \underbrace{\frac{2\times 0.087}{\pi\times 7.07}}_{0.008} + \underbrace{\frac{2\times 4.55}{\pi\times 7.07}}_{0.409}\times\alpha;$$

$$h_{AC}^{wb} = h_{NP} - V_{HT}\left\{\frac{C_{L\alpha,HT}}{C_{L\alpha,w}}\right\}\left(1-\frac{d\varepsilon}{d\alpha}\right) = 1.67 - 0.7\times\frac{4.17}{4.55}\times(1-0.409) = 1.29;$$

Wing aerodynamic centre location from aircraft nose: $X_{AC}^{wb} = h_{AC}^{wb}\times\bar{c} = 1.29\times 1.57 = 2.02m;$

Oswald efficiency factor: $e = 0.9; K = \frac{1}{\pi AR_w e} = \frac{1}{\pi\times 7.07\times 0.9} = 0.05;$

Residual pitching moment coefficient at aerodynamic centre: $C_{m,ac} = -0.005;$

Pitching moment coefficient and trim calculation:

*Contributions to pitching moment coefficient from fuselage and propeller can be significant and must be modelled. Both these effects are neglected here and only contributions from wing and horizontal tail are considered.

$$C_{m\alpha,w} = \left(\frac{X_{CG} - X_{AC}^{wb}}{\bar{c}}\right) \times C_{L\alpha,w} = \left(\frac{2.39 - 2.02}{1.57}\right) \times 4.55 = 1.07/rad;$$

$$C_{m\alpha,HT} = -\eta' V_{HT} C_{L\alpha,HT}(1 - \varepsilon_\alpha) = -0.95 \times 0.7 \times 4.17 \times (1 - 0.409) = -1.64/rad;$$

$$C_{m\alpha} = C_{m\alpha,w} + C_{m\alpha,HT} = 1.07 - 1.64 = -0.57/rad;$$

$$C_{m0} = C_{mAC,w} + C_{m0,HT}$$

$$= -0.005 + \eta' V_{HT} C_{L\alpha,HT}(-i_t + \varepsilon_0)$$

$$= -0.005 + 0.95 \times 0.7 \times 4.17 \times (0 + 0.008) = 0.017$$

Trim angle of attack: $\alpha^* = -\frac{C_{m0}}{C_{m\alpha}} = -\frac{0.017}{-0.57} = 0.03 rad = 1.7 deg$ (For zero elevator deflection)

Airplane trim lift coefficient:

$$C_L^* = C_{L\alpha,w} \times (\alpha^* - \alpha_{L=0}) + \eta' \frac{S_{HT}}{S_w} C_{L\alpha,HT}\left(\alpha^* - \frac{d\varepsilon}{d\alpha}\alpha^* - \varepsilon_0 + i_t\right)$$

$$= 4.55 \times (1.7 - (-1.1)) \times \frac{\pi}{180} + 0.95 \times \frac{3.45}{16.35} \times 4.17$$

$$\times \left\{1.7 \times \frac{\pi}{180} \times (1 - 0.409) - 0.008 - 0\right\} = 0.22 + 0.008 = 0.23;$$

Trim speed:

$$V^* = \sqrt{\frac{2}{\rho}\frac{W}{S_w}\frac{1}{C_L^*}} = \sqrt{\frac{2}{1.225} \times \frac{1859.73 \times 9.81}{16.35} \times \frac{1}{0.23}} = 89.18 m/s;$$

$$C_{L0} = C_{L\alpha,w}(0 - \alpha_{L=0}) + \eta' \frac{S_{HT}}{S_w} C_{L\alpha,HT}(i_t - \varepsilon_0)$$

$$= 4.55 \times 1.1 \times \frac{\pi}{180} + 0.95 \times \frac{3.45}{16.35} \times 4.17 \times (-0 - 0.008) = 0.08$$

$$C_{L\alpha} = C_{L\alpha,w} + \eta' \frac{S_{HT}}{S_w} C_{L\alpha,HT}\left(1 - \frac{d\varepsilon}{d\alpha}\right)$$

$$= 4.55 + 0.95 \times \frac{3.45}{16.35} \times 4.17 \times (1 - 0.409) = 5.04/rad;$$

$$C_{mq1} = -2\eta' V_{HT} C_{L\alpha,HT} \times (l_{HT}/\bar{c}) = -2 \times 0.95 \times 0.7 \times 4.17 \times \frac{5.23}{1.57} = -18.47/rad$$

$$C_{m\dot{\alpha}} = 0.0/rad; \; C_{L\dot{\alpha}} = 0.0/rad;$$

$$C_{Lq1} = 2\eta' V_{HT} C_{L\alpha,HT} = 2 \times 0.95 \times 0.7 \times 4.17 = 5.56/rad;$$

$$C_{Dq1} = 2KC_L^* C_{Lq1} = 2 \times \frac{1}{\pi \times 0.9 \times 7.07} \times 0.23 \times 5.56 = 0.13/rad;$$

$$C_D^* = C_{D0} + KC_L^{*2} = 0.017 + \frac{0.23^2}{\pi \times 0.9 \times 7.07} = 0.02;$$

Thrust required: $T_R = \frac{1}{2}\rho V^{*2} S C_D^* = 0.5 \times 1.22 \times 89.18^2 \times 16.35 \times 0.02 = 1586.4N;$

Power required: $P_R = T_R V^* = 1586.4 \times 89.18 = 141475.09$ N-m/s;

$$C_{D\alpha}^* = 2KC_L^* C_{L\alpha} = \frac{2 \times 0.23 \times 5.04}{\pi \times 0.9 \times 7.07} = 0.12/rad$$

$$\bar{q}^* = \frac{1}{2}\rho V^{*2} = \frac{1}{2} \times 1.225 \times 89.18^2 = 4851.37 N/m^2$$

$$\left(\frac{\bar{q}^* Sc}{I_{yy}}\right) = \frac{4851.37 \times 16.35 \times 1.57}{4180.57} = 29.78/s^2$$

$$\left(\frac{g}{V^*}\right) = \frac{9.81}{89.18} = 0.11/s; \quad \frac{\bar{q}^* S}{W} = \frac{4851.37 \times 16.35}{1859.73 \times 9.81} = 4.35; \quad Ma^* = \frac{V^*}{a} = \frac{89.18}{340.0} = 0.26$$

$$\left(\frac{g}{V^*}\right)\left(\frac{\bar{q}^* S}{W}\right) = 0.11 \times 4.35 = 0.48/s; \quad \frac{\bar{c}}{2V^*} = \frac{1.57}{2 \times 89.18} = 0.009s;$$

$$C_{LMa} = C_L^* \frac{Ma^*}{1 - Ma^{*2}} = 0.23 \times \frac{0.26}{1 - 0.26^2} = 0.064; \quad C_{DMa} = 0.0$$

Longitudinal mode calculations:
Short-period mode:
From Equations 2.33 and 2.34:

$$\left(\omega_n^2\right)_{SP} = -\left(\frac{\bar{q}^* Sc}{I_{yy}}\right) C_{m\alpha}$$

$$= -29.78 \times -0.57 = 16.97$$

$$\omega_{nSP} = 4.12 rad/s; \quad T_{SP} = \frac{2\pi}{\omega_{SP}} = \frac{2\times 3.14}{4.12} = 1.52s$$

$$(2\zeta\omega_n)_{SP} = -\left(\frac{\overline{q}^* Sc}{I_{yy}}\right)\left(\frac{c}{2V^*}\right)(C_{mq1} + C_{m\dot{\alpha}})$$

$$= -29.78 \times \frac{1.57}{2\times 89.18} \times (-18.47 + 0) = 4.84;$$

$$\zeta_{SP} = \frac{4.84}{2\omega_{nSP}} = \frac{4.84}{2\times 4.12} = 0.59$$

Phugoid mode:
From Equations 5.35 and 5.36:

$$(\omega_n^2)_p = \left(\frac{g}{V^*}\right)^2 \left[\frac{\overline{q}^* S}{W}(Ma^* C_{LMa} + 2C_L^*)\right]$$

$$= 0.11^2 \times [4.35 \times (0.26 \times 0.064 + 2\times 0.23)] = 0.025$$

$$\omega_{nPh} = \sqrt{0.025} = 0.16 rad/s; \quad T_{Ph} = \frac{2\pi}{\omega_{nPh}} = \frac{2\times 3.14}{0.16} = 39.66s$$

$$(2\zeta\omega_n)_{Ph} = \left(\frac{g}{V^*}\right)\left[\frac{\overline{q}^* S}{W}(Ma^* C_{DMa} + 2C_D^*)\right] \approx \left(\frac{g}{V^*}\right)\left[\frac{\overline{q}^* S}{W} 2C_D^*\right]$$

$$= 0.11 \times 4.35 \times 2 \times 0.02 = 0.02;$$

$$\zeta_{Ph} = \frac{0.02}{2\times 0.16} = 0.06.$$

9.1.2 First-Order Form of the Small-Perturbation Longitudinal Dynamics Equations

Beginning with the longitudinal small-perturbation dynamics equations in level flight presented in Chapter 5, we now recast them in first-order form. The state variables of interest are: ΔV, $\Delta\gamma$, $\Delta\alpha$ and $\Delta\dot{\alpha}$.

The velocity dynamics at the slow phugoid timescale comes from Equation 5.42:

$$\frac{\Delta\dot{V}}{V^*} = \left(\frac{g}{V^*}\right)\left\{-\frac{\overline{q}^* S}{W}\left(\Delta C_D + 2C_D^* \frac{\Delta V}{V^*}\right) - \Delta\gamma\right\}$$

$$= \left(\frac{g}{V^*}\right)\left\{-\frac{\overline{q}^* S}{W}\left(Ma^* C_{DMa}\frac{\Delta V}{V^*} + C_{D\alpha}\Delta\alpha + C_{Dq1}\Delta\dot{\alpha}\left(\frac{c}{2V^*}\right) + 2C_D^*\frac{\Delta V}{V^*}\right) - \Delta\gamma\right\}$$

$$=\left(\frac{g}{V^*}\right)\left\{-\frac{\bar{q}^*S}{W}\left(\left\{Ma^*C_{DMa}+2C_D^*\right\}\frac{\Delta V}{V^*}+C_{D\alpha}\Delta\alpha+C_{Dq1}\Delta\dot{\alpha}\left(\frac{c}{2V^*}\right)\right)-\Delta\gamma\right\}$$

$$=-\left(\frac{g}{V^*}\right)\left(\frac{\bar{q}^*S}{W}\right)\left\{Ma^*C_{DMa}+2C_D^*\right\}\frac{\Delta V}{V^*}-\left(\frac{g}{V^*}\right)\left(\frac{\bar{q}^*S}{W}\right)C_{D\alpha}\Delta\alpha$$

$$-\left(\frac{g}{V^*}\right)\left(\frac{\bar{q}^*S}{W}\right)C_{Dq1}\left(\frac{c}{2V^*}\right)\Delta\dot{\alpha}-\left(\frac{g}{V^*}\right)\Delta\gamma \tag{9.1}$$

From Equation 5.43:

$$\Delta\dot{\gamma}=\left(\frac{g}{V^*}\right)\left\{\frac{\bar{q}^*S}{W}\left(\Delta C_L+2C_L^*\frac{\Delta V}{V^*}\right)\right\}$$

$$-\left(\frac{g}{V^*}\right)\left(\frac{\bar{q}^*S}{W}\right)\left\{\left(\left\{Ma^*C_{LMa}+2C_L^*\right\}\frac{\Delta V}{V^*}+C_{L\alpha}\Delta\alpha+C_{Lq1}\Delta\dot{\alpha}\left(\frac{c}{2V^*}\right)\right)\right\}$$

$$=\left(\frac{g}{V^*}\right)\left(\frac{\bar{q}^*S}{W}\right)\left\{Ma^*C_{LMa}+2C_L^*\right\}\frac{\Delta V}{V^*}+\left(\frac{g}{V^*}\right)\left(\frac{\bar{q}^*S}{W}\right)C_{L\alpha}\Delta\alpha$$

$$+\left(\frac{g}{V^*}\right)\left(\frac{\bar{q}^*S}{W}\right)C_{Lq1}\left(\frac{c}{2V^*}\right)\Delta\dot{\alpha} \tag{9.2}$$

Differentiating Eq 9.1 term-wise:

$$\frac{\Delta\ddot{V}}{V^*}=-\left(\frac{g}{V^*}\right)\left(\frac{\bar{q}^*S}{W}\right)\left\{Ma^*C_{DMa}+2C_D^*\right\}\frac{\Delta\dot{V}}{V^*}-\left(\frac{g}{V^*}\right)\left(\frac{\bar{q}^*S}{W}\right)C_{D\alpha}\Delta\dot{\alpha}$$

$$-\left(\frac{g}{V^*}\right)\left(\frac{\bar{q}^*S}{W}\right)C_{Dq1}\left(\frac{c}{2V^*}\right)\Delta\ddot{\alpha}-\left(\frac{g}{V^*}\right)\Delta\dot{\gamma}$$

$$=-\left(\frac{g}{V^*}\right)\left(\frac{\bar{q}^*S}{W}\right)\left\{Ma^*C_{DMa}+2C_D^*\right\}\frac{\Delta\dot{V}}{V^*}-\left(\frac{g}{V^*}\right)\left(\frac{\bar{q}^*S}{W}\right)C_{D\alpha}\Delta\dot{\alpha}$$

$$-\left(\frac{g}{V^*}\right)\left(\frac{\bar{q}^*S}{W}\right)C_{Dq1}\left(\frac{c}{2V^*}\right)\Delta\ddot{\alpha}$$

$$-\left(\frac{g}{V^*}\right)\left\{\left(\frac{g}{V^*}\right)\left(\frac{\bar{q}^*S}{W}\right)\left\{Ma^*C_{LMa}+2C_L^*\right\}\frac{\Delta V}{V^*}+\left(\frac{g}{V^*}\right)\left(\frac{\bar{q}^*S}{W}\right)C_{L\alpha}\Delta\alpha\right.$$

$$\left.+\left(\frac{g}{V^*}\right)\left(\frac{\bar{q}^*S}{W}\right)C_{Lq1}\left(\frac{c}{2V^*}\right)\Delta\dot{\alpha}\right\} \tag{9.3}$$

The pitch dynamics at the fast short period time scale is obtained from Equation 5.44:

$$\Delta\ddot{\theta} = \Delta\ddot{\alpha} + \Delta\ddot{\gamma} = \left(\frac{\bar{q}^* Sc}{I_{yy}}\right)\Delta C_m \tag{9.4}$$

$$\Delta\ddot{\alpha} = \left(\frac{\bar{q}^* Sc}{I_{yy}}\right)\Delta C_m - \Delta\ddot{\gamma}$$

$$= \left(\frac{\bar{q}^* Sc}{I_{yy}}\right)\left\{Ma^* C_{mMa}\frac{\Delta V}{V^*} + C_{m\alpha}\Delta\alpha + C_{mq1}\Delta\dot{\alpha}\left(\frac{c}{2V^*}\right)\right\}$$

$$-\left\{\left(\frac{g}{V^*}\right)\left(\frac{\bar{q}^* S}{W}\right)\left(\left\{Ma^* C_{LMa} + 2C_L^*\right\}\frac{\Delta\dot{V}}{V^*} + C_{L\alpha}\Delta\dot{\alpha} + C_{Lq1}\Delta\ddot{\alpha}\left(\frac{c}{2V^*}\right)\right)\right\}$$

$$\Delta\ddot{\alpha}\left\{1 + \left(\frac{g}{V^*}\right)\left(\frac{\bar{q}^* S}{W}\right)C_{Lq1}\left(\frac{c}{2V^*}\right)\right\} = -\left(\frac{g}{V^*}\right)\left(\frac{\bar{q}^* S}{W}\right)\left\{Ma^* C_{LMa} + 2C_L^*\right\}\frac{\Delta\dot{V}}{V^*}$$

$$+\left(\frac{\bar{q}^* Sc}{I_{yy}}\right)Ma^* C_{mMa}\frac{\Delta V}{V^*} + \left(\frac{\bar{q}^* Sc}{I_{yy}}\right)C_{m\alpha}\Delta\alpha$$

$$+\left\{\left(\frac{\bar{q}^* Sc}{I_{yy}}\right)\left(\frac{c}{2V^*}\right)C_{mq1} - \left(\frac{g}{V^*}\right)\left(\frac{\bar{q}^* S}{W}\right)C_{L\alpha}\right\}\Delta\dot{\alpha}$$

$$\Delta\ddot{\alpha}\left\{1 + \left(\frac{g}{V^*}\right)\left(\frac{\bar{q}^* S}{W}\right)C_{Lq1}\left(\frac{c}{2V^*}\right)\right\}$$

$$= \left(\frac{g}{V^*}\right)^2\left(\frac{\bar{q}^* S}{W}\right)^2\left\{Ma^* C_{LMa} + 2C_L^*\right\}\begin{bmatrix} \left\{Ma^* C_{DMa} + 2C_D^*\right\}\frac{\Delta V}{V^*} \\ +C_{D\alpha}\Delta\alpha \\ +C_{Dq1}\left(\frac{c}{2V^*}\right)\Delta\dot{\alpha} + \left\{\left(\frac{g}{V^*}\right)/\left(\frac{\bar{q}^* S}{W}\right)\right\}\Delta\gamma \end{bmatrix}$$

$$+\left(\frac{\bar{q}^* Sc}{I_{yy}}\right)Ma^* C_{mMa}\frac{\Delta V}{V^*} + \left(\frac{\bar{q}^* Sc}{I_{yy}}\right)C_{m\alpha}\Delta\alpha$$

$$+\left\{\left(\frac{\bar{q}^* Sc}{I_{yy}}\right)\left(\frac{c}{2V^*}\right)C_{mq1} - \left(\frac{g}{V^*}\right)\left(\frac{\bar{q}^* S}{W}\right)C_{L\alpha}\right\}\Delta\dot{\alpha}$$

$$= \left(\frac{g}{V^*}\right)^2\left(\frac{\bar{q}^* S}{W}\right)\left\{Ma^* C_{LMa} + 2C_L^*\right\}\Delta\gamma +$$

$$+\left\{\left(\frac{g}{V^*}\right)^2\left(\frac{\bar{q}^*S}{W}\right)^2\left\{Ma^*C_{LMa}+2C_L^*\right\}C_{D\alpha}+\left(\frac{\bar{q}^*Sc}{I_{yy}}\right)C_{m\alpha}\right\}\Delta\alpha$$

$$+\left\{\left(\frac{g}{V^*}\right)^2\left(\frac{\bar{q}^*S}{W}\right)^2\left\{Ma^*C_{LMa}+2C_L^*\right\}\left\{Ma^*C_{DMa}+2C_D^*\right\}+\left(\frac{\bar{q}^*Sc}{I_{yy}}\right)Ma^*C_{mMa}\right\}\frac{\Delta V}{V^*}$$

$$+\left(\left\{\left(\frac{\bar{q}^*Sc}{I_{yy}}\right)\left(\frac{c}{2V^*}\right)C_{mq1}-\left(\frac{g}{V^*}\right)\left(\frac{\bar{q}^*S}{W}\right)C_{L\alpha}\right\}\right.$$

$$\left.+\left(\frac{g}{V^*}\right)^2\left(\frac{\bar{q}^*S}{W}\right)^2\cdot C_{Dq1}\left(\frac{c}{2V^*}\right)\left\{Ma^*C_{LMa}+2C_L^*\right\}\right)\Delta\dot{\alpha}$$

Let us define:

$$k_{\ddot{\alpha}}=\left\{1+\left(\frac{g}{V^*}\right)\left(\frac{\bar{q}^*S}{W}\right)C_{Lq1}\left(\frac{c}{2V^*}\right)\right\}$$

$$k_\gamma=\left(\frac{g}{V^*}\right)^2\left(\frac{\bar{q}^*S}{W}\right)\left\{Ma^*C_{LMa}+2C_L^*\right\}$$

$$k_\alpha=\left\{\left(\frac{g}{V^*}\right)^2\left(\frac{\bar{q}^*S}{W}\right)^2\left\{Ma^*C_{LMa}+2C_L^*\right\}C_{D\alpha}+\left(\frac{\bar{q}^*Sc}{I_{yy}}\right)C_{m\alpha}\right\}$$

$$k_{\dot{\alpha}}=\left\{\left(\frac{\bar{q}^*Sc}{I_{yy}}\right)\left(\frac{c}{2V^*}\right)C_{mq1}-\left(\frac{g}{V^*}\right)\left(\frac{\bar{q}^*S}{W}\right)C_{L\alpha}\right\}$$

$$+\left(\frac{g}{V^*}\right)^2\left(\frac{\bar{q}^*S}{W}\right)^2C_{Dq1}\left(\frac{c}{2V^*}\right)\left\{Ma^*C_{LMa}+2C_L^*\right\}$$

$$k_V=\left\{\left(\frac{g}{V^*}\right)^2\left(\frac{\bar{q}^*S}{W}\right)^2\left\{Ma^*C_{LMa}+2C_L^*\right\}\left\{Ma^*C_{DMa}+2C_D^*\right\}+\left(\frac{\bar{q}^*Sc}{I_{yy}}\right)Ma^*C_{mMa}\right\}$$

The above equations can be recast as:

$$k_{\ddot{\alpha}}\Delta\ddot{\alpha}=k_V\left(\Delta V/V^*\right)+k_\gamma\Delta\gamma+k_\alpha\Delta\alpha+k_{\dot{\alpha}}\Delta\dot{\alpha} \tag{9.5}$$

Introducing the equality relation:

$$\Delta\dot{\alpha}=\Delta\dot{\alpha} \tag{9.6}$$

and collecting term by term from Equations 9.1–9.6, matrix form of the longitudinal perturbation dynamics can be written as:

$$\begin{bmatrix} \Delta\dot{V}/V^* \\ \Delta\dot{\gamma} \\ \Delta\dot{\alpha} \\ \Delta\ddot{\alpha} \end{bmatrix} = \underbrace{\begin{bmatrix} a_{11} & a_{12} & a_{13} & a_{14} \\ a_{21} & a_{22} & a_{23} & a_{24} \\ a_{31} & a_{32} & a_{33} & a_{34} \\ a_{41} & a_{42} & a_{43} & a_{44} \end{bmatrix}}_{A_{long}} \begin{bmatrix} \Delta V/V^* \\ \Delta\gamma \\ \Delta\alpha \\ \Delta\dot{\alpha} \end{bmatrix} \tag{9.7}$$

Evaluating the elements of matrix A_{long} at the trim condition:

$$a_{11} = -\left(\frac{g}{V^*}\right)\left(\frac{\bar{q}^*S}{W}\right)\left\{Ma^*C_{DMa} + 2C_D^*\right\} = -0.11\times4.35\times(0.26\times0+2\times0.02) = -0.019;$$

$$a_{12} = -\left(\frac{g}{V^*}\right) = -0.11; \quad a_{13} = -\left(\frac{g}{V^*}\right)\left(\frac{\bar{q}^*S}{W}\right)C_{D\alpha} = -0.11\times4.35\times0.12 = -0.058;$$

$$a_{14} = -\left(\frac{g}{V^*}\right)\left(\frac{\bar{q}^*S}{W}\right)C_{Dq1}\left(\frac{c}{2V^*}\right) = -0.11\times4.35\times0.13\times0.009 = 0.0006;$$

$$a_{21} = \left(\frac{g}{V^*}\right)\left(\frac{\bar{q}^*S}{W}\right)\left\{Ma^*C_{LMa} + 2C_L^*\right\} = 0.11\times4.35\times(0.26\times0.064+2\times0.23) = 0.23;$$

$$a_{22} = 0.0; \quad a_{23} = \left(\frac{g}{V^*}\right)\left(\frac{\bar{q}^*S}{W}\right)C_{L\alpha} = 0.11\times4.35\times5.04 = 2.41;$$

$$a_{24} = \left(\frac{g}{V^*}\right)\left(\frac{\bar{q}^*S}{W}\right)C_{Lq1}\left(\frac{c}{2V^*}\right) = 0.11\times4.35\times5.56\times0.009 = 0.024;$$

$$a_{31} = 0; \quad a_{32} = 0; \quad a_{33} = 0; \quad a_{34} = 1.0;$$

$$k_V = \left(\left(\frac{g}{V^*}\right)^2\left(\frac{\bar{q}^*S}{W}\right)^2\left\{Ma^*C_{LMa} + 2C_L^*\right\}\left\{Ma^*C_{DMa} + 2C_D^*\right\} + \left(\frac{\bar{q}^*Sc}{I_{yy}}\right)Ma^*C_{mMa}\right)$$

$$= 0.11^2\times4.35^2\times(0.26\times0.064+2\times0.23)\times(0+2\times0.02) + 29.78\times0.26\times0 = 0.004;$$

$$k_{\ddot{\alpha}} = \left\{1 + \left(\frac{g}{V^*}\right)\left(\frac{\bar{q}^*S}{W}\right)C_{Lq1}\left(\frac{c}{2V^*}\right)\right\} = 1 + 0.11\times4.35\times5.56\times0.009 = 1.024;$$

$$k_\gamma = \left(\frac{g}{V^*}\right)^2 \left(\frac{\bar{q}^* S}{W}\right)\left\{Ma^* C_{LMa} + 2C_L^*\right\}$$

$$= 0.11^2 \times 4.35 \times (0.26 \times 0.064 + 2 \times 0.23) = 0.025;$$

$$k_\alpha = \left(\left(\frac{g}{V^*}\right)^2 \left(\frac{\bar{q}^* S}{W}\right)^2 \left\{Ma^* C_{LMa} + 2C_L^*\right\} C_{D\alpha} + \left(\frac{\bar{q}^* Sc}{I_{yy}}\right) C_{m\alpha}\right)$$

$$= 0.11^2 \times 4.35^2 \times (0.26 \times 0.064 + 2 \times 0.23) \times 0.12 + 29.78 \times -0.57 = -16.96;$$

$$k_{\dot{\alpha}} = \left\{\left(\frac{\bar{q}^* Sc}{I_{yy}}\right)\left(\frac{c}{2V^*}\right) C_{mq1} - \left(\frac{g}{V^*}\right)\left(\frac{\bar{q}^* S}{W}\right) C_{L\alpha}\right\}$$

$$+ \left(\frac{g}{V^*}\right)^2 \left(\frac{\bar{q}^* S}{W}\right)^2 C_{Dq1}\left(\frac{c}{2V^*}\right)\left\{Ma^* C_{LMa} + 2C_L^*\right\}$$

$$= (29.78 \times 0.009 \times -18.47) - (0.11 \times 4.35 \times 5.04) + \left(0.11^2 \times 4.35^2 \times 0.13 \times 0.009\right)$$

$$\times (0.26 \times 0.064 + 2 \times 0.23) = -4.95 - 2.41 + 0.0001 = -7.36$$

$$a_{41} = k_V / k_{\ddot{\alpha}} = \frac{0.005}{1.024} = 0.0049; \quad a_{42} = k_\gamma / k_{\ddot{\alpha}} = \frac{0.025}{1.024} = 0.024;$$

$$a_{43} = k_\alpha / k_{\ddot{\alpha}} = -\frac{16.96}{1.024} = -16.56; \quad a_{44} = k_{\dot{\alpha}} / k_{\ddot{\alpha}} = -\frac{7.36}{1.024} = -7.19;$$

$$A_{long} = \begin{bmatrix} -0.019 & -0.11 & -0.058 & 0.0 \\ 0.23 & 0.0 & 2.41 & 0.024 \\ 0.0 & 0.0 & 0.0 & 1.0 \\ 0.005 & 0.024 & -16.96 & -7.19 \end{bmatrix}$$

Computed eigenvalues of A_{long} are:
Short-period: $\lambda_{SP} = -3.59 \pm j2.01$ which gives $\omega_{nSP} = 4.12\ rad/s$ (4.12); $\zeta_{SP} =$ 0.87 (0.59)
Phugoid: $\lambda_{Ph} = -0.0078 \pm j0.1589$ which gives $\omega_{nPh} = 0.16\ rad/s$ (0.16); $\zeta_{Ph} =$ 0.05 (0.06)

Note: In brackets are the values calculated earlier from the approximations presented in this book. While other parameters match well, significant difference in short-period damping can be observed. This difference is typical which is primarily due to the additional term containing $C_{L\alpha}$ appearing in A_{long}, which acts as a coupling parameter between the two modes.

Homework Exercise: Starting with the form of the perturbed pitch dynamics Eq. 9.5, follow the two-timescale approach presented in Chapter 5 and derive improved approximations to the short period modal parameters. In particular, show that the short period damping parameter, $2\zeta_{SP}\omega_{nSP}$ may be approximated as $-k_{\dot{\alpha}}/k_{\ddot{\alpha}}$. Using this approximation, evaluate the Short Period damping ζ_{SP} for the GA airplane data presented here and show that $\zeta_{SP} = 0.87$ — same as that obtained numerically.

9.1.3 Lateral-Directional Aerodynamics Parameters

(**Note**: Contributions from fuselage, propeller and interference effects are not modelled in the following. On many occasions, their contribution may be significant and may need to be considered.)

Side-force coefficient:

$$\text{Dihedral effect: } C_{Y\beta,w} = -0.0001\Gamma = -0.0001 \times 5 \times \frac{\pi}{180} \approx 0;$$

Vertical location of wing root quarter-chord above CG line, $z_w = 0.305m$;
Sidewash, σ:

$$\left(1 + \frac{d\sigma}{d\beta}\right)\eta'_{VT} = 0.724 + 3.06\frac{S_{VT}/S_w}{1 + cos\Lambda_{\frac{c}{4}}} + \frac{0.4z_w}{D_{f,max}} + 0.009AR_w$$

$$= 0.724 + 3.06 \times \frac{2.16/16.35}{2} + \frac{0.4 \times 0.305}{1.3716} + 0.009 \times 7.07 = 1.08;$$

Vertical tail contribution:

$$C_{Y\beta,VT} = -0.95 \times C_{L\alpha,VT} \times \frac{S_{VT}}{S_w} \times \left(1 + \frac{d\sigma}{d\beta}\right)\eta'_{VT} = -0.95 \times 2.68 \times \frac{2.16}{16.35} \times 1.08 = -0.36;$$

Total: $C_{Y\beta} = C_{Y\beta,w} + C_{Y\beta,VT} = -0.36/rad$;

Yawing moment coefficient:

$$\text{Dihedral effect: } C_{n\beta,w} = -0.075 \times \Gamma \times C_L = -0.075 \times 5 \times \frac{\pi}{180} \times 0.23 = -0.001/rad;$$

$$\text{Vertical tail contribution: } C_{n\beta,VT} = -C_{Y\beta,VT} \times \frac{l_{VT}}{b_w} = -(-0.29) \times \frac{4.88}{10.75} = 0.13/rad;$$

Total: $C_{n\beta} = C_{n\beta,w} + C_{n\beta,VT} = -0.001 + 0.13 \approx 0.13/rad$;

Rolling moment coefficient:

$$z = z_{VT}cos\alpha - l_{VT}sin\alpha = 0.8 \times \cos(1.7) - 4.873 \times \sin(1.7) = 0.66m;$$

Wing contribution:

$$C_{l\beta,w} = -\frac{\Gamma C_{L\alpha,w} c_r b_w}{6S_w} = -5 \times \frac{\pi}{180} \times 4.55 \times 2.02 \times \frac{10.75}{6 \times 16.35} = -0.09/rad;$$

Vertical tail contribution: $C_{l\beta,VT} = (z/b_w) C_{Y\beta,VT} = \frac{0.66}{10.75} \times -0.36 = -0.02/rad;$

Total: $C_{l\beta} = C_{l\beta,w} + C_{l\beta,VT} = -0.09 - 0.02 = -0.11/rad;$

Roll rate derivatives:

$$C_{Yp2,VT} = \frac{2}{b_w} \times (z - z_{VT}) \times C_{Y\beta,VT} = \frac{2}{10.75} \times (0.66 - 0.8) \times -0.36 = 0.01/rad;$$

$$C_{Yp2,w} = 0; \quad C_{Yp2,\Gamma} = 0;$$

$$C_{Yp2} = C_{Yp2,VT} + C_{Yp2,w} + C_{Yp2,\Gamma} = 0.01/rad;$$

$$C_{lp2,VT} = 2 \times \left(\frac{z}{b_w}\right) \times \left(\frac{z - z_{VT}}{b_w}\right) \times C_{Y\beta,VT} = 2 \times \frac{0.66}{10.75} \times \frac{0.66 - 0.8}{10.75} \times -0.36 = -0.0006/rad;$$

$$C_{lp2,w} = -\frac{1}{6}(C_{L\alpha} + C_D) = -\frac{1}{6} \times (5.04 + 0.02) = -0.84/rad;$$

$$C_{lp2} = C_{lp2,VT} + C_{lp2,w} = -0.84/rad;$$

$$C_{np2,w} = -\frac{1}{6}(C_L - C_{D\alpha}) = -\frac{1}{6} \times (0.23 - 0.12) = -0.02/rad;$$

$$C_{np2,VT} = -\frac{2}{b_w} \times (l_{VT} cos\alpha + z_{VT} sin\alpha) \times \left(\frac{z - z_{VT}}{b_w}\right) \times C_{Y\beta,VT}$$

$$= -\frac{2}{10.75} \times \left(4.873 \times cos(1.7) + 0.8 \times sin(1.7)\right) \times \frac{0.66 - 0.8}{10.75} \times -0.36 = -0.004;$$

$$C_{np2} = C_{np2,w} + C_{np2,VT} = -0.02/rad;$$

Yaw rate derivatives:

$$C_{Yr1,VT} = -\frac{2}{b_w} \times (l_{VT} cos\alpha + z_{VT} sin\alpha) \times C_{Y\beta,VT}$$

$$= -\frac{2}{10.75} \times \left(4.873 \times cos(1.7) + 0.8 \times sin(1.7)\right) \times -0.36 = 0.33/rad;$$

$$C_{lr1,w} = C_L/3 = 0.23/3 = 0.08/rad;$$

$$C_{lr1,VT} = -\frac{2}{b_w^2}\times(l_{VT}cos\alpha + z_{VT}sin\alpha)\times(z_{VT}cos\alpha - l_{VT}sin\alpha)\times C_{Y\beta,VT}$$

$$= -\frac{2}{10.75^2}\times\left(4.873cos(1.7)+0.8sin(1.7)\right)\times\left(0.8cos(1.7)-4.873sin(1.7)\right)\times -0.36$$

$$= 0.02/rad;$$

$$C_{lr1} = C_{lr1,w} + C_{lr1,VT} = 0.1/rad;$$

$$C_{nr1,VT} = \frac{2}{b^2}\times(l_{VT}cos\alpha + z_{VT}sin\alpha)^2\times C_{Y\beta,VT}$$

$$= \frac{2}{10.75^2}\times\left(4.873\times cos(1.7)+0.8\times sin(1.7)\right)^2\times -0.36 = -0.15/rad;$$

$$C_{nr1,w} = 0;$$

$$C_{nr1} = C_{nr1,w} + C_{nr1,VT} = -0.15/rad;$$

Lateral-directional mode calculations:

$$\bar{q}^* = \frac{1}{2}\rho V^{*2} = \frac{1}{2}\times 1.22\times 89.18^2 = 4851.37N/m^2$$

$$\left(\frac{\bar{q}^* Sb}{I_{xx}}\right) = \frac{4851.37\times 16.35\times 10.75}{3355.65} = 254.11/s^2$$

$$\left(\frac{\bar{q}^* Sb}{I_{zz}}\right) = \frac{4851.37\times 16.35\times 10.75}{6140.66} = 138.86/s^2$$

$$\left(\frac{\bar{q}^* S}{m}\right) = \frac{4851.37\times 16.35}{1859.73} = 42.65m/s^2$$

$$\frac{b}{2V^*} = \frac{10.75}{2\times 89.18} = 0.06s;\quad \frac{g}{V^*} = \frac{9.81}{89.18} = 0.11s$$

$$C_{lp2} = -0.84/rad;\quad C_{n\beta} = 0.13/rad;$$

$$C_{Y\beta} = -0.36/rad;\quad C_{nr1} = -0.15/rad;\quad C_{l\beta} = -0.11/rad;\quad C_{lr1} = 0.1/rad$$

$$C_{nr2} = -\frac{C_D}{3} = -\frac{0.02}{3} = -0.01/rad$$

$$C_{lr2} = \frac{C_L}{3} = \frac{0.23}{3} = 0.08/rad$$

$N_\beta = \left(\frac{\bar{q}^* S b}{I_{zz}}\right) C_{n\beta}$ $= 138.86 \times 0.13 = 18.05/s^2$	$N_{r2} = \left(\frac{\bar{q}^* S b}{I_{zz}}\right) C_{nr2} \left(\frac{b}{2V^*}\right)$ $= 138.86 \times -0.01 \times 0.06 = -0.08/s$
$Y_\beta = \left(\frac{\bar{q}^* S}{m}\right) C_{Y\beta}$ $= 42.65 \times -0.36 = -15.35 m/s^2$	$\mathcal{L}_\beta = \left(\frac{\bar{q}^* S b}{I_{xx}}\right) C_{l\beta}$ $= 254.11 \times -0.11 = -27.95/s^2$
$\mathcal{L}_{p2} = \left(\frac{\bar{q}^* S b}{I_{xx}}\right) C_{lp2} \left(\frac{b}{2V^*}\right)$ $= 254.11 \times -0.84 \times 0.06 = -12.81/s$	$N_{r1} = \left(\frac{\bar{q}^* S b}{I_{zz}}\right) C_{nr1} \left(\frac{b}{2V^*}\right)$ $= 138.86 \times -0.15 \times 0.06 = -1.25/s$
$\mathcal{L}_{r1} = \left(\frac{\bar{q}^* S b}{I_{xx}}\right) C_{lr1} \left(\frac{b}{2V^*}\right)$ $= 254.11 \times 0.1 \times 0.06 = 1.52/s$	$\mathcal{L}_{r2} = \left(\frac{\bar{q}^* S b}{I_{xx}}\right) C_{lr2} \left(\frac{b}{2V^*}\right)$ $= 254.11 \times 0.08 \times 0.06 = 1.22/s$

Roll mode eigenvalue: $\lambda_r = \mathcal{L}_{p2} = -12.81/s$

Dutch roll mode parameters:

$$\omega_{nDR}^2 = N_\beta + \left(\frac{g}{V^*}\right)\left[Y_\beta N_{r2} + \left(\frac{\mathcal{L}_\beta}{\mathcal{L}_{p2}}\right)\right]$$

$$= 18.05 + \left(\frac{9.81}{89.18}\right)\left[-15.35 \times -0.08 + \left(\frac{-27.95}{-12.81}\right)\right] = 18.42;$$

$$\omega_{nDR} = 4.29 rad/s;$$

$$2\zeta_{DR}\omega_{nDR} = -N_{r1} - \left(\frac{g}{V^*}\right)\left[Y_\beta + \left(\frac{\mathcal{L}_{r1}}{\mathcal{L}_{p2}}\right)\right]$$

$$= 1.25 - \left(\frac{9.81}{89.18}\right)\left[-15.35 + \left(\frac{1.52}{-12.81}\right)\right] = 2.95;$$

$$\zeta_{DR} = \frac{2.95}{2 \times 4.29} = 0.34;$$

Spiral mode eigenvalue:

$$\lambda_S = \left(\frac{g}{V^*}\right)\frac{\left[\mathcal{L}_\beta N_{r2} - N_\beta \mathcal{L}_{r2}\right]}{\lambda_r \omega_{nDR}^2}$$

$$= \left(\frac{9.81}{89.18}\right)\frac{((-27.95\times -0.08)-(18.05\times 1.22))}{-12.81\times 18.42} = 0.009/s;$$

9.1.4 Lateral-Directional Perturbation Dynamics Model

From Table 6.1,

$$\Delta\ddot{\mu} = -\mathcal{L}_{r1}\Delta\dot{\beta} + \left[\mathcal{L}_\beta + \left(\frac{g}{V^*}\right)Y_\beta \mathcal{L}_{r2}\right]\Delta\beta + \mathcal{L}_{p2}\Delta\dot{\mu} + \left(\frac{g}{V^*}\right)\mathcal{L}_{r2}\Delta\mu \tag{9.8}$$

$$\Delta\ddot{\beta} = -\left[-N_{r1} - \left(\frac{g}{V^*}\right)Y_\beta\right]\Delta\dot{\beta} - \left[N_\beta + \left(\frac{g}{V^*}\right)Y_\beta N_{r2}\right]\Delta\beta + \left(\frac{g}{V^*}\right)\Delta\dot{\mu} - \left(\frac{g}{V^*}\right)N_{r2}\Delta\mu \tag{9.9}$$

By introducing equality relations,

$$\Delta\dot{\mu} = \Delta\dot{\mu}, \quad \Delta\dot{\beta} = \Delta\dot{\beta} \tag{9.10}$$

One may arrive at the matrix form of the above equations put together in variables $\left[\Delta\mu\ \Delta\dot{\mu}\ \Delta\beta\ \Delta\dot{\beta}\right]$ as:

$$\begin{bmatrix}\Delta\dot{\mu}\\ \Delta\ddot{\mu}\\ \Delta\dot{\beta}\\ \Delta\ddot{\beta}\end{bmatrix} = \underbrace{\begin{bmatrix} 0 & 1 & 0 & 0 \\ (g/V^*)\mathcal{L}_{r2} & \mathcal{L}_{p2} & \{\mathcal{L}_\beta + (g/V^*)Y_\beta \mathcal{L}_{r2}\} & -\mathcal{L}_{r1} \\ 0 & 0 & 0 & 1 \\ -(g/V^*)N_{r2} & (g/V^*) & \{-N_\beta - (g/V^*)Y_\beta N_{r2}\} & \{N_{r1} + (g/V^*)Y_\beta\} \end{bmatrix}}_{A_{lat-dir}} \begin{bmatrix}\Delta\mu\\ \Delta\dot{\mu}\\ \Delta\beta\\ \Delta\dot{\beta}\end{bmatrix} \tag{9.11}$$

At the chosen trim condition:

$$A_{lat-dir} = \begin{bmatrix} 0 & 1 & 0 & 0 \\ 0.13 & -12.81 & -30.0 & -1.52 \\ 0 & 0 & 0 & 1 \\ 0.008 & 0.11 & -18.185 & -2.94 \end{bmatrix}$$

Computed eigenvalues of $A_{lat-dir}$ are:

$$\lambda_S = 0.009; \quad \lambda_r = -12.83;$$

$\lambda_{DR} = -1.471 \pm j4.033$, which gives $\omega_{nDR} = 4.29\ rad/s$, $\zeta_{DR} = 0.34$

The above values match almost exactly with the ones calculated earlier from modal approximations.

Control effectiveness parameter (from chart in Figure 7.9):

Elevator control effectiveness parameter, $\tau_{\delta e} = \dfrac{\Delta\alpha_t}{\Delta\delta e} = 0.5;$

Rudder control effectiveness parameter, $\tau_{\delta r} = \dfrac{\Delta\alpha_{VT}}{\Delta\delta r} = 0.5;$

Aileron control effectiveness parameter, $\tau_{\delta a} = \dfrac{\Delta\alpha}{\Delta\delta a} = 0.22;$

Control derivatives (calculated at trim $\alpha = 1.7deg$):

$$C_{L,\delta e} = \eta' \times \frac{S_{HT}}{S_w} \times C_{L\alpha,HT} \times \tau_{\delta e} = 0.95 \times \frac{3.45}{16.35} \times 4.17 \times 0.5 = 0.42/rad;$$

$$C_{m,\delta e} = -\eta' V_{HT} C_{L\alpha,HT} \tau_{\delta e} = -0.95 \times 0.7 \times 4.17 \times 0.5 = -1.38/rad;$$

$$C_{l\delta a} = -2\left(\frac{C_{l\alpha,w}\tau_{\delta a}}{Sb}\right)\int_{0.5b}^{0.9b} c(y)\,y\,dy \text{ (for aileron running between 50–90\% of wing span)}$$

$$= -2\left(\frac{C_{l\alpha,w}\tau_{\delta a}}{Sb}\right)c_r \int_{0.5b}^{0.9b}\left[1 + \frac{2y}{b}(\lambda - 1)\right] y\,dy$$

$$= -2\left(\frac{C_{l\alpha,w}\tau_{\delta a}}{Sb}\right)c_r \left[\frac{y^2}{2} + \frac{2y^3}{3b}(\lambda - 1)\right]_{0.5b}^{0.9b};$$

$$= -2 \times \left(\frac{5.73 \times 0.22}{16.35 \times 10.75}\right) \times 2.02 \times \left[\frac{0.9^2 - 0.5^2}{2} - \frac{2(0.9^3 - 0.5^3)}{3} \times (0.5 - 1)\right] \times 10.75^2$$

$$= -0.26/rad;$$

$$C_{n\delta a} \approx -\left(\frac{C_{D\alpha}}{C_{l\alpha,w}}\right) C_{l\delta a} = -\frac{0.12}{5.73} \times 0.26 = -0.01/rad;$$

$$C_{Y\delta a} = 0;$$

$$C_{n\delta r} = -V_{VT} \times \tau_{\delta r} \times C_{L\alpha VT} = -0.06 \times 0.5 \times 2.68 = -0.08/rad;$$

$$C_{Y\delta r} = (S_{VT}/S_w) \times \tau_{\delta r} \times C_{L\alpha VT} = (2.16/16.35) \times 0.5 \times 2.68 = 0.18/rad;$$

$$C_{l\delta r} = (S_{VT}/S_w) \times \tau_{\delta r} \times C_{L\alpha VT} \times (z/b_w) = (2.16/16.35) \times 0.5 \times 2.68 \times \frac{0.66}{10.75} = 0.011/rad$$

9.2 AIRSHIP DYNAMICS

Airships, also known as blimps or dirigibles, are nothing but Helium (inert) or Hydrogen (flammable) filled balloons powered by engine/s producing thrust. Flexible (inflated by pressure equalization on outer and internal sides), semi-rigid (with some stiffening provided by inside structure) and rigid (with rigid structure inside) airships are buoyancy-lifted aerial vehicles floating at altitudes at which weight of airship is neutralized by the buoyancy force. Usual shape of an airship is a doubly symmetric body of revolution as shown in Figure 9.2. Three points of interest on an airship are: centre of volume (CV), which is the geometric centre of the hull, centre of buoyancy (CB) where the buoyancy force acts and the center of mass (CM) which is also the center of gravity (CG). Both CV and CB are, in general, different from the CG. Important features of an airship are the tail control surfaces similar to a rigid airplane and the engine, usually located below the gas-filled hull. The control surfaces, rudder and elevons, provide the control in yaw and pitch/roll, respectively, and engine propellers may be thrust vectored for additional control. Further, ballonets (air-filled bags) fitted inside airship may also be used as a control device via discharge of air. Gondola is the structure which is designed to carry the crew and useful payload.

9.2.1 Airship Equations of Motion

Six-degrees-of-freedom motion of a rigid airship with fixed mass and assuming no aeroelastic effects can be described by three angular rotations and three translational motions similar to airplanes. Airship equations of motion are slightly different from that of airplanes mainly due to different choice of origin of body-fixed axis system, which is conveniently taken to be the centre of volume (CV) of the airship and not the centre of gravity, which for reasons explained later tends to lie closer to the gondola. As pictured in Figure 9.3, the CG is located at distance $\bar{a} = \left[a_x \ a_y \ a_z\right]^T$ from the CV. The location of CB from the CV is $\bar{b} = \left[b_x \ b_y \ b_z\right]^T$. Angular velocity vector $\bar{\omega} = \left[p \ q \ r\right]^T$ and translation velocity vector $\bar{V} = [u \ v \ w]^T$ are defined in the body-fixed axis system. Following the material presented in Chapter 8 (and a good part left as exercise!), now we write down the equations of airship motion.

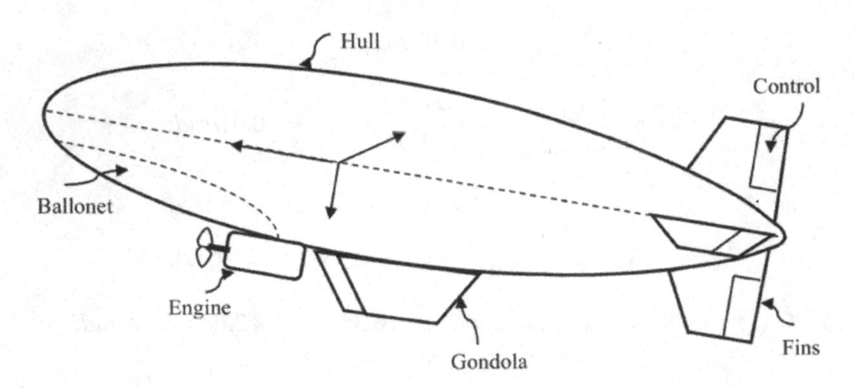

FIGURE 9.2 Airship nomenclature.

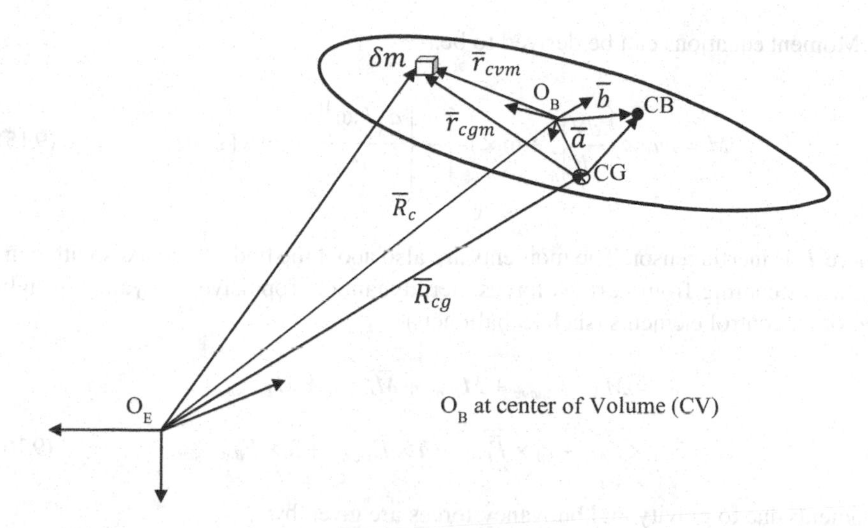

FIGURE 9.3 Body-fixed axis system at the centre of volume.

Force equations (please refer to the original equations in Chapter 8 for comparison):

$$\overline{F} = m\left.\frac{d\overline{V}}{dt}\right|_B + m\left(\overline{\omega}\times\overline{V}\right) + m\left.\frac{d\left(\overline{\omega}\times\overline{a}\right)}{dt}\right|_B + m\left\{\overline{\omega}\times\left(\overline{\omega}\times\overline{a}\right)\right\} \tag{9.12}$$

The additional last two terms in Equation 9.12 appear due to the choice of origin of the body-fixed axis which is not at CG but at the CV of the hull. Accounting for the additional force term due to buoyancy, the force vector is now given as:

$$\overline{F} = \overline{F}_{Gravity} + \overline{F}_{Buoyancy} + \overline{F}_{Aerodynamic} + \overline{F}_{Propulsive}$$

$$= \begin{bmatrix} 0 \\ 0 \\ mg \end{bmatrix}_E + \begin{bmatrix} 0 \\ 0 \\ -B \end{bmatrix}_E + \begin{bmatrix} X_{A/d} \\ Y_{A/d} \\ Z_{A/d} \end{bmatrix}_B + \begin{bmatrix} T \\ 0 \\ 0 \end{bmatrix}_B \tag{9.13}$$

Resultant components of gravity and buoyancy forces (to be used in stability analysis later) along body-fixed axes are:

$$\overline{F}^B_{G+B} = \underbrace{\begin{bmatrix} -mgsin\theta \\ mgsin\phi cos\theta \\ mgcos\phi cos\theta \end{bmatrix}}_{\overline{F}^B_{Gravity}} + \underbrace{\begin{bmatrix} Bsin\theta \\ -Bsin\phi cos\theta \\ -Bcos\phi cos\theta \end{bmatrix}}_{\overline{F}^B_{buoyancy}} = \begin{bmatrix} -(mg-B)sin\theta \\ (mg-B)sin\phi cos\theta \\ (mg-B)cos\phi cos\theta \end{bmatrix} \tag{9.14}$$

Moment equations can be derived to be:

$$\bar{M} = m\bar{a} \times \left\{ \frac{d\bar{V}}{dt}\bigg|_B + \bar{\omega} \times \bar{V} \right\} + \left\{ \frac{d\left(\bar{\bar{I}}.\bar{\omega}\right)}{dt}\bigg|_B + \bar{\omega} \times \left(\bar{\bar{I}}.\bar{\omega}\right) \right\} \tag{9.15}$$

where $\bar{\bar{I}}$ is inertia tensor. The moments are also about the body-fixed axes with contributions coming from various forces: aerodynamic, propulsive, buoyancy, weight and other control elements (such as ballonets).

$$\bar{M} = \bar{M}_{Aerodynamic} + \bar{M}_{Prop} + \bar{M}_{Gravity} + \bar{M}_{Buoyancy}$$

$$= \bar{r}_{A/d} \times \bar{F}_{A/d} + \bar{r}_p \times \bar{F}_{Prop} + \bar{a} \times \bar{F}_{Gravity} + \bar{b} \times \bar{F}_{Buoyancy} \tag{9.16}$$

Moments due to gravity and buoyancy forces are given by:

$$\bar{M}_{G+B} = \begin{bmatrix} \mathcal{L} \\ M \\ N \end{bmatrix}_{G+B} = \bar{a} \times \bar{F}^B_{Gravity} + \bar{b} \times \bar{F}^B_{Buoyancy}$$

$$= \begin{bmatrix} cos\phi cos\theta\left(mga_y - Bb_y\right) + sin\phi cos\theta\left(Bb_z - mga_z\right) \\ sin\theta\left(-mga_z + Bb_z\right) + cos\phi cos\theta\left(Bb_x - mga_x\right) \\ sin\phi cos\theta\left(-Bb_x + mga_x\right) + sin\theta\left(-mga_y + Bb_y\right) \end{bmatrix} \tag{9.17}$$

One may arrive at various configurations of airship with respect to number, location, and orientation of aerodynamic controls and engine plants and their contributions to overall forces and moments can be determined using Equations 9.13 and 9.16. XZ-plane is the plane of symmetry for the airship, thus, $I_{xy} = I_{yx} = I_{zy} = I_{yz} = 0$, and CG lies in the plane of symmetry, that is, $a_y = 0$.

Components of $\bar{F}_{Aerodynamic}$ are defined as:

$$X_{A/d} = \frac{1}{2}\rho V^2 C_X Vol^{2/3}; \quad Y_{A/d} = \frac{1}{2}\rho V^2 C_Y Vol^{2/3}; \quad Z_{A/d} = \frac{1}{2}\rho V^2 C_Z Vol^{2/3}$$

Components of $\bar{M}_{Aerodynamic}$ are:

$$\mathcal{L}_{A/d} = \frac{1}{2}\rho V^2 C_l Vol; M_{A/d} = \frac{1}{2}\rho V^2 C_m Vol; N_{A/d} = \frac{1}{2}\rho V^2 C_n Vol \; \bar{M}_{Aerodynamic}$$

Note that, the volume of the airship is used as reference unlike wing reference area and mean aerodynamic chord (as in the case of airplanes) for arriving at non-dimensionalized aerodynamic force and moment coefficients.

A dominant feature that must be included in airship dynamic models are added/virtual mass and inertia effects. These effects exist for any body submerged in a medium, for example, our own body displaces air around us and we too do feel buoyancy (significantly higher in water) in general. But, as compared to our weight in air, this force is much smaller, actually negligible, and hence can be ignored. For airplanes, which actually displace huge amount of air around them due to their considerable outer volume, the force of buoyancy is still negligible as compared to the much larger weight and lift produced due to relative air speed.

Airships are slow moving vehicles which may not produce large aerodynamic forces, but the unsteady aerodynamic effects due to their large size and being lighter-than-air are predominant. Apparent mass and inertia resulting from the unsteady aerodynamic effects are usually accounted for while modelling dynamics of an airship. These terms having dimension of mass/inertia are added to actual mass/inertia of the airship resulting in virtual/added mass/inertia.

The apparent mass terms subtract from the actual mass giving different mass properties along the three axes as follows:

$$m_x = m - \frac{\partial X}{\partial \dot{u}}; \quad m_y = m - \frac{\partial Y}{\partial \dot{v}}; \quad m_z = m - \frac{\partial Z}{\partial \dot{w}} \tag{9.18}$$

Notice that the additional terms are written in terms of the derivatives of aerodynamic forces X, Y, Z with respect to body acceleration terms $\dot{u}$, $\dot{v}$, $\dot{w}$ and hence have the unit of mass.

Similarly, moment of inertia terms are modified as:

$$J_{xx} = I_{xx} - \frac{\partial \mathcal{L}}{\partial \dot{p}}; \; J_{yy} = I_{yy} - \frac{\partial M}{\partial \dot{q}}; \; J_{zz} = I_{zz} - \frac{\partial N}{\partial \dot{r}}; \; J_{xz} = I_{xz} + \frac{\partial \mathcal{L}}{\partial \dot{r}} \tag{9.19}$$

In order to clearly see where added mass and inertia terms appear in the equations of motion, consider one example each from the force as well as moment equations. The X-axis force equation is (following expansion of Equation (9.12)):

$$\begin{bmatrix} X \\ Y \\ Z \end{bmatrix} = m \left\{ \frac{d}{dt} \begin{bmatrix} u \\ v \\ w \end{bmatrix} + \begin{bmatrix} 0 & -r & q \\ r & 0 & -p \\ -q & p & 0 \end{bmatrix} \begin{bmatrix} u \\ v \\ w \end{bmatrix} + \frac{d}{dt} \left(\begin{bmatrix} 0 & -r & q \\ r & 0 & -p \\ -q & p & 0 \end{bmatrix} \begin{bmatrix} a_x \\ a_y \\ a_z \end{bmatrix} \right) \right.$$

$$\left. + \begin{bmatrix} 0 & -r & q \\ r & 0 & -p \\ -q & p & 0 \end{bmatrix} \begin{bmatrix} 0 & -r & q \\ r & 0 & -p \\ -q & p & 0 \end{bmatrix} \begin{bmatrix} a_x \\ a_y \\ a_z \end{bmatrix} \right\}$$

$$= m \left\{ \begin{bmatrix} \dot{u} \\ \dot{v} \\ \dot{w} \end{bmatrix} + \begin{bmatrix} 0 & -r & q \\ r & 0 & -p \\ -q & p & 0 \end{bmatrix} \begin{bmatrix} u \\ v \\ w \end{bmatrix} + \begin{bmatrix} 0 & -\dot{r} & \dot{q} \\ \dot{r} & 0 & -\dot{p} \\ -\dot{q} & \dot{p} & 0 \end{bmatrix} \begin{bmatrix} a_x \\ a_y \\ a_z \end{bmatrix} \right.$$

$$\left. + \begin{bmatrix} -(r^2+q^2) & pq & pr \\ pq & -(r^2+p^2) & qr \\ pr & qr & -(p+q^2) \end{bmatrix} \begin{bmatrix} a_x \\ a_y \\ a_z \end{bmatrix} \right\}$$

From above, virtual mass terms, m_x for example, will come out when suitable term from the aerodynamic model for X are transferred to the other side, as shown below.

$$m(\dot{u}+qw-rv)+ma_z\dot{q}-ma_y\dot{r}-ma_x\left(r^2+q^2\right)+ma_zpr+ma_ypq=X \quad (9.20)$$

$$X=X_{A/d}-(mg-B)\sin\theta+T_R+T_L \quad (9.21)$$

In Eq 9.21, T_R and T_L are thrust produced by the right and left engines, respectively. $X_{A/d}$ is the resultant aerodynamic force along body x-axis, which is a function of the states and their rate of change:

$$X_{A/d}=X_{A/d}\left(u,v,w,p,q,r,\dot{u},\dot{v},\dot{w},\dot{p},\dot{q},\dot{r},\delta\right) \quad (9.22)$$

Where δ is aerodynamic control surface deflection. X can be separated as sum of the term 'T', which consists of the rate derivative terms to be taken to the LHS and another term 'X_R' consisting of the remaining terms of the expansion of X, as follows:

$$X=\underbrace{X_{\dot{u}}\dot{u}+X_{\dot{q}}\dot{q}+X_{\dot{r}}\dot{r}}_{T}+X_R \quad (9.23)$$

Where $X_{(.)}=\dfrac{\partial X}{\partial(.)}$.

Term 'T' when taken to the left-hand side of Equation 9.20 and setting $a_y=0$, results in:

$$(m-X_{\dot{u}})\dot{u}+mqw-mrv+\left(ma_z-X_{\dot{q}}\right)\dot{q}-X_{\dot{r}}\dot{r}-ma_x\left(r^2+q^2\right)+ma_zpr=X_R$$

Which is:

$$m_x\dot{u}+mqw-mrv+\left(ma_z-X_{\dot{q}}\right)\dot{q}-X_{\dot{r}}\dot{r}-ma_x\left(r^2+q^2\right)+ma_zpr=X_R \quad (9.24)$$

Similarly, Force equations along body fixed Y and Z axes may be written as:

$$m_y\dot{v}+mru-mpw+(ma_x-Y_{\dot{r}})\dot{r}-\left(ma_z+Y_{\dot{p}}\right)\dot{p}+ma_xpq+ma_zqr=Y_R \quad (9.25)$$

$$m_z\dot{w}+mpv-mqu-Z_{\dot{p}}\dot{p}-\left(ma_x+Z_{\dot{q}}\right)\dot{q}-ma_z\left(p^2+q^2\right)+ma_xpr=Z_R \quad (9.26)$$

Likewise, from expansion of Equation 9.15:

$$\bar{M}=m\bar{a}\times\left\{\left.\frac{d\bar{V}}{dt}\right|_B+\bar{\omega}\times\bar{V}\right\}+\left\{\left.\frac{d\left(\bar{\bar{I}}.\bar{\omega}\right)}{dt}\right|_B+\bar{\omega}\times\left(\bar{\bar{I}}.\bar{\omega}\right)\right\}$$

The moment equations about the body x-axis, for instance, can be written as (after dropping the cross inertia terms $I_{xy} = I_{yx} = I_{zy} = I_{yz} = 0$ and $a_y = 0$):

$$-ma_z\left\{\dot{v} + (ur - pw)\right\} + I_{xx}\dot{p} - I_{xz}\dot{r} - I_{xz}pq + I_{zz}rq - rqI_{yy} = \mathcal{L}_R + \mathcal{L}_{(\dot{.})}(\dot{.})$$

Collecting the terms with common derivatives, the above equation can be rewritten as:

$$-\dot{v}(ma_z + \mathcal{L}_{\dot{v}}) - ma_z(ur - pw) + \underbrace{(I_{xx} - \mathcal{L}_{\dot{p}})}_{J_{xx}}\dot{p} - \underbrace{(I_{xz} + \mathcal{L}_{\dot{r}})}_{J_{xz}}\dot{r} - I_{xz}pq + I_{zz}rq - rqI_{yy} = \mathcal{L}_R \tag{9.27}$$

Similarly, the other moment equations including added inertia effect can be written as:

$$(ma_z + M_{\dot{u}})\dot{u} + ma_z(qw - rv) - (ma_x + M_{\dot{w}})\dot{w} - ma_x(pv - uq) + \underbrace{(I_{yy} - M_{\dot{q}})}_{J_{yy}}\dot{q} + I_{xx}pr - I_{xz}(p^2 - r^2) - I_{zz}pr = M_R \tag{9.28}$$

$$(ma_x - N_{\dot{v}})\dot{v} + ma_x(ur - pw) - \underbrace{(I_{zx} + N_{\dot{p}})}_{J_{xz}}\dot{p} + \underbrace{(I_{zz} - N_{\dot{r}})}_{J_{zz}}\dot{r} - I_{xx}pq + I_{xz}rq + pqI_{yy} = N_R \tag{9.29}$$

Kinematic equations are (From Equation 8.3):

$$\dot{\phi} = p + tan\theta(qsin\phi + rcos\phi)$$

$$\dot{\theta} = qcos\phi - rsin\phi$$

Equations 9.24 through 9.29 along with these kinematic equations make a set of eight equations which is used for the analysis of rigid airship flight dynamics.

Trim condition: Our equilibrium condition of interest is a cruise trim state, thus $v^* = w^* = p^* = q^* = r^* = \phi^* = 0; u^* = const \neq 0$. Also assume, $\theta^* = 0$.
Forces in longitudinal flight conditions are (**Homework Exercise!**):

$$X = Lsin\alpha - Dcos\alpha + T - (mg - B)sin\theta;\; Y = 0;\; Z = (mg - B)cos\theta - Lcos\alpha - Dsin\alpha$$

Equilibrium condition (with $\alpha^* = 0$) requires that

$$X = -(mg - B)sin\theta + T - D = 0;\; T = T_R + T_L$$

Which is, $T = D$ in either case, whether $\theta^* = 0$ or $(mg - B) = 0$.

$$Y = 0$$

$$Z = (mg - B)cos\theta - L = 0$$

Which is, $mg - B = L$ at $\theta^* = 0$. Trim angle of attack is zero and, therefore, aerodynamic lift produced on the airship is taken as zero at this trim condition; the Z-force balance then satisfies $mg - B = 0$. Further, moment balance requires that:

$$\mathcal{L} = M = N = 0.$$

Introducing perturbation in the state variables and control inputs, i.e.,

$$u = u^* + \Delta u;\quad v = v^* + \Delta v;\quad w = w^* + \Delta w;\quad p = p^* + \Delta p;\quad q = q^* + \Delta q;\quad r = r^* + \Delta r;$$

$$\phi = \phi^* + \Delta\varphi;\quad \theta = \theta^* + \Delta\theta;\quad \delta(.) = \delta(.)^* + \Delta\delta(.)$$

in Equations 9.24–9.26 and Equations 9.27–9.29 and substituting values of equilibrium conditions, one can arrive at the small-perturbation dynamics equations.

9.2.2 Longitudinal Small-Perturbation Equations

As an example of the perturbation procedure, let us linearise the equation for forward motion (Equation 9.24) at the chosen longitudinal trim condition. Substituting perturbed variables in Equation 9.24, one may arrive at (dropping higher order terms):

$$m_x \Delta\dot{u} + \left(ma_z - X_{\dot{q}}\right)\Delta\dot{q} = \Delta X_R$$

Which is,

$$m_x \Delta\dot{u} + \left(ma_z - X_{\dot{q}}\right)\Delta\dot{q} = \frac{\partial X}{\partial u}\Delta u + \frac{\partial X}{\partial w}\Delta w + q + \frac{\partial X}{\partial \theta}\Delta\theta + \frac{\partial X}{\partial \delta}\Delta\delta \tag{9.30}$$

already assumed $\theta^* = 0$ where, $\frac{\partial X}{\partial \theta} = -(mg - B)cos\theta^*$. Similarly, the other longitudinal perturbation dynamics equations can be derived to be:

$$m_z \Delta\dot{w} - mu^*\Delta q - \left(ma_x + Z_{\dot{q}}\right)\Delta\dot{q} = \Delta Z_R$$

$$m_z \Delta\dot{w} - mu^*\Delta q - \left(ma_x + Z_{\dot{q}}\right)\Delta\dot{q} = \frac{\partial Z}{\partial u}\Delta u + \frac{\partial Z}{\partial w}\Delta w + \frac{\partial Z}{\partial q}\Delta q + \frac{\partial Z}{\partial \theta}\Delta\theta + \frac{\partial Z}{\partial \delta}\Delta\delta \tag{9.31}$$

Where $\frac{\partial Z}{\partial \theta} = -(mg - B)sin\theta^*$.

$$\left(ma_z + M_{\dot{u}}\right)\Delta\dot{u} - \left(ma_x + M_{\dot{w}}\right)\Delta\dot{w} + ma_x u^*\Delta q + \underbrace{\left(I_{yy} - M_{\dot{q}}\right)}_{J_{yy}}\Delta\dot{q} = \Delta M_R$$

$$\left(ma_z + M_{\dot{u}}\right)\Delta\dot{u} - \left(ma_x + M_{\dot{w}}\right)\Delta\dot{w} + ma_x u^*\Delta q + \underbrace{\left(I_{yy} - M_{\dot{q}}\right)}_{J_{yy}}\Delta\dot{q}$$

$$= \frac{\partial M}{\partial u}\Delta u + \frac{\partial M}{\partial w}\Delta w + \frac{\partial M}{\partial q}\Delta q + \frac{\partial M}{\partial \theta}\Delta\theta + \frac{\partial M}{\partial \delta}\Delta\delta \tag{9.32}$$

Where $\frac{\partial M}{\partial \theta} = (-mga_z + Bb_z)cos\theta^* + (mga_x - Bb_x)sin\theta^*$ (From Equation 9.17 with $\phi^* = 0$).

$$\Delta\dot{\theta} = \Delta q \tag{9.33}$$

Rearranging and collecting the four linear equations 9.30 through 9.33 for longitudinal dynamics, one arrives at the matrix form:

$$[M]\Delta\dot{x} = A\Delta x + C\Delta\delta \tag{9.34}$$

All Δ quantities are perturbation from equilibrium flight condition under study. $[M]$ is the mass matrix, which is:

$$M = \begin{bmatrix} m_x & 0 & (ma_z - X_{\dot{q}}) & 0 \\ 0 & m_z & -(ma_x + Z_{\dot{q}}) & 0 \\ (ma_z + M_{\dot{u}}) & -(ma_x + M_{\dot{w}}) & J_{yy} & 0 \\ 0 & 0 & 0 & 1 \end{bmatrix} \tag{9.35}$$

$$A = \begin{bmatrix} X_u & X_w & X_q & -(mg - B)cos\theta^* \\ Z_u & Z_w & (Z_q + mu^*) & -(mg - B)sin\theta^* \\ M_u & M_w & (M_q - ma_x u^*) & (-mga_z + Bb_z)cos\theta^* + (mga_x - Bb_x)sin\theta^* \\ 0 & 0 & 1 & 0 \end{bmatrix}$$

(9.36)

And matrix C, is:

$$C = \begin{bmatrix} X_\delta \\ Z_\delta \\ M_\delta \\ 0 \end{bmatrix} \tag{9.37}$$

As before, for stability analysis, it suffices to determine eigenvalues by solving the characteristic equation

$$|M\lambda - AI| = 0 \tag{9.38}$$

This is the numerical procedure to determine stability of equilibrium/trim states. Seems repetition assuming a level trim condition at which $w^* = 0$ (which also means angle of attack, $\alpha^* = 0$) and $\theta^* = 0$ (thus $\gamma^* = 0$), and also cruise condition at which $(mg - B) = 0$, one can arrive at further simplified form of the linear longitudinal dynamics equations.

Longitudinal dynamic modes: These are the modes characterized by dynamics of longitudinal perturbation variables $[\Delta u\ \Delta w\ \Delta q\ \Delta\theta]^T$ from the cruise condition as described earlier. Let us first look at decoupled form of equations for each mode, assuming the motions to be constrained. Let us assume that the airship is constrained to have only pure forward motion or only heave or only pitch motion (of the three degrees of freedom in longitudinal dynamics).

1. **Surge mode**: This is a first-order mode representing change primarily in forward velocity Δu. Following the perturbed equation 9.30,

$$m_x \Delta\dot{u} - \frac{\partial X}{\partial u}\Delta u = 0 \tag{9.39}$$

$$\lambda_{surge} = \frac{X_u}{m_x} = \frac{X_u}{m - X_{\dot{u}}} \tag{9.40}$$

With the denominator positive always as $X_{\dot{u}}$ is significantly smaller than the mass m, this mode is stable when the derivative X_u is negative. Expression for X_u, at any longitudinal equilibrium state, can be obtained by taking derivative of the resultant axial force X given by

$$X = L sin\alpha - D cos\alpha + T - (mg - B) sin\theta \tag{9.41}$$

$$\frac{\partial X}{\partial u} = \frac{\partial L}{\partial u} sin\alpha - \frac{\partial D}{\partial u} cos\alpha + \frac{\partial T}{\partial u} \tag{9.42}$$

Equilibrium condition, $(mg - B) = 0$, has been used in the above and further using $w^* = 0$ which is $\alpha^* = 0$, Eq 9.42 simplifies to:

$$\frac{\partial X}{\partial u} = -\frac{\partial D}{\partial u} + \frac{\partial T}{\partial u} = -u^* \rho_{air} S C_D^* + \frac{1}{2}\rho_{air} u^{*2} S\left(C_{T_u} - C_{D_u}\right) \tag{9.43}$$

Where $S = Vol^{2/3}$. The second term is insignificantly small at speeds at which airships operate. The first term is clearly negative indicating that the surge mode is stable. This is the slowest of the three longitudinal modes. At the cruise condition, due to the drag dependency, variation of axial force with speed negative, hence stabilizing.

2. **Heave mode**: This is a first-order mode representing change primarily in vertical velocity Δw governed by (Eq. 9.31):

$$m_z \Delta\dot{w} - \frac{\partial Z}{\partial w}\Delta w = 0 \tag{9.44}$$

Thus, approximate expression for heave mode eigenvalue is:

$$\lambda_{heave} = \frac{Z_w}{m_z} = \frac{Z_w}{m - Z_{\dot{w}}} \tag{9.45}$$

$$Z = (mg - B)cos\theta - Lcos\alpha - Dsin\alpha \tag{9.46}$$

From which expression for Z_w can be derived as:

$$Z_w = \frac{\partial Z}{\partial w} = \frac{1}{u^*}\frac{\partial Z}{\partial \alpha} = \frac{1}{u^*}(Lsin\alpha - L_\alpha cos\alpha - Dcos\alpha - D_\alpha sin\alpha) \tag{9.47}$$

At equilibrium state $\alpha^* = 0$, the above expression reduces to:

$$Z_w = \frac{1}{u^*}(-L_\alpha - D) \tag{9.48}$$

Since D and L_α are both usually positive, they contribute to the stability of the heave mode. With increasing trim speed, however, these above approximate expressions in Eqs. 9.40 and 9.45 can be modified by including more coupling effects.

3. **Pitch oscillatory mode**: This is a second-order mode representing oscillatory change in pitch variable, $\Delta\theta$ or $\Delta\alpha$. Usually, heave and the pitch oscillatory modes are strongly coupled.

In order to gain a realistic feel for this mode, we consider pure pitching motion of airship governed by (Eq. 9.32):

$$(ma_z + M_{\dot{u}})\Delta\dot{u} - (ma_x + M_{\dot{w}})\Delta\dot{w} + ma_x u^* \Delta q + \underbrace{(I_{yy} - M_{\dot{q}})}_{J_{yy}}\Delta\dot{q}$$

$$= M_u \Delta_u + M_w \Delta w + M_{q1}\Delta q + M_\theta \Delta\theta \tag{9.49}$$

Since $\Delta\alpha = \Delta\theta$ and $\Delta\dot{\alpha} = \Delta\dot{\theta} = \Delta q$, and dropping u terms, Eq. 9.49 is reduced to

$$J_{yy}\Delta\dot{q} = M_w \Delta w + M_{\dot{w}}\Delta\dot{w} + M_{q1}\Delta\dot{\alpha} - ma_x u^* \Delta\dot{\alpha} + M_\theta \Delta\alpha \tag{9.50}$$

Which (with $\Delta\alpha \approx \frac{\Delta w}{u^*}$; $\Delta\dot{\alpha} = \frac{\Delta\dot{w}}{u^*}$) can be rewritten as:

$$J_{yy}\Delta\dot{q} = M_\alpha \Delta\alpha + M_{\dot{\alpha}}\Delta\dot{\alpha} + (M_{q1} - ma_x u^*)\Delta\dot{\alpha} + M_\theta \Delta\alpha \tag{9.51}$$

Which is,

$$J_{yy}\Delta\ddot{\alpha} = M_{\alpha}\Delta\alpha + \left(M_{\dot{\alpha}} + M_{q1} - ma_x u^*\right)\Delta\dot{\alpha} + M_{\theta}\Delta\alpha \tag{9.52}$$

Where

$$M_{\theta} = \left.\frac{\partial M}{\partial \theta}\right|_{*} = \left(-mga_z + Bb_z\right)$$

Rearranging Equation 9.52, we get:

$$J_{yy}\Delta\ddot{\alpha} - \left(-ma_x u^* + M_{q1} + M_{\dot{\alpha}}\right)\Delta\dot{\alpha} - \left(M_{\alpha} - mga_z + Bb_z\right)\Delta\alpha = 0 \tag{9.53}$$

A quick inspection of Eq. 9.53 tells us that it is similar to Eq. 2.32 except for a few additional terms. The condition for stability in pitch of airship at hover ($u^* = 0$) is given by:

$$\left(M_{q1} + M_{\dot{\alpha}}\right) < 0 \text{ and } \left(M_{\alpha} - mga_z + Bb_z\right) < 0 \tag{9.54}$$

The aerodynamic pitching moment for bodies of revolution such as the hull shape of an airship (also applicable to airplane fuselages) may be modelled using Munk's model as follows:

$$M = \frac{1}{2}\rho V^2 (Vol)(k_2 - k_1) sin2\alpha \tag{9.55}$$

from which the derivative M_{α} may be obtained as below by differentiation:

$$M_{\alpha} = \rho V^2 (Vol)(k_2 - k_1) cos2\alpha \tag{9.56}$$

The proportionality factor, $(k_2 - k_1)$, related to Lamb's inertia ratios [2], is always positive, hence M_{α} is positive, which implies that the airship hull contribution to pitch stability is destabilizing. The CG is always aft of the center of pressure for airship with hull of elliptic cylindrical shape. From the Eq. 9.56, expression for lift produced (known to be significantly small) and lift-curve slope can be estimated. Similar to airplanes, aft tails (fins) are designed to contribute to stability of airship. Other sources of positive stiffness and damping result from purely geometric considerations:

- If CG is placed slightly aft of the reference Z-axis in the longitudinal plane, it may result in positive damping, refer corresponding term in Eq. 9.53 and its sign (but the contributions from the tail fin depending on the moment arm length may reduce in that case), and
- $a_z > b_z$ for positive stiffness in pitch (Refer to the stiffness term in Eq. 9.53, in which, the buoyancy contribution is $mg(a_z - b_z)$ owing to $mg - B = 0$ at our equilibrium condition). What it means is, CG must lie below the point at which force of buoyancy acts. This is in general true as most of the payload is concentrated near bottom of the hull.

From Equation 9.53, one may obtain approximate expressions for pitch oscillatory mode frequency and damping ratio from:

$$\omega_{np}^2 = -\frac{(M_\alpha - mga_z + Bb_z)}{J_{yy}}; \quad 2\zeta_p\omega_{np} = -\frac{(ma_x u^* + M_{q1} + M_{\dot{\alpha}})}{J_{yy}} \tag{9.57}$$

9.2.3 Small-Perturbation Equations for Lateral-Directional Modes

Equations 9.25, 9.27, 9.29 and $\dot{\phi}$ equation representing the lateral dynamics are now linearized at the previous trim condition for stability analysis of off-longitudinal modes.

$$m_y\Delta\dot{v} + (ma_x - Y_{\dot{r}})\Delta\dot{r} - (ma_z + Y_{\dot{p}})\Delta\dot{p} = \frac{\partial Y}{\partial v}\Delta v + \frac{\partial Y}{\partial p}\Delta p + \left(\frac{\partial Y}{\partial r} - mu^*\right)\Delta r + \frac{\partial Y}{\partial \phi}\Delta\phi \tag{9.58}$$

$$(ma_x - N_{\dot{v}})\Delta\dot{v} - J_{xz}\Delta\dot{p} + J_{zz}\Delta\dot{r} = \frac{\partial N}{\partial v}\Delta v + \frac{\partial N}{\partial p}\Delta p + \left(\frac{\partial N}{\partial r} - ma_x u^*\right)\Delta r + \frac{\partial N}{\partial \phi}\Delta\phi \tag{9.59}$$

$$-\Delta\dot{v}(ma_z + \mathcal{L}_{\dot{v}}) + J_{xx}\Delta\dot{p} - J_{xz}\Delta\dot{r} = \frac{\partial \mathcal{L}}{\partial v}\Delta v + \frac{\partial \mathcal{L}}{\partial p}\Delta p + \left(\frac{\partial \mathcal{L}}{\partial r} + ma_z u^*\right)\Delta r + \frac{\partial \mathcal{L}}{\partial \phi}\Delta\phi \tag{9.60}$$

$$\Delta\dot{\phi} = \Delta p \tag{9.61}$$

These equations can be put in matrix form as $M\Delta\dot{x} = A\Delta x$ (with $\Delta x = [\Delta v \ \Delta p \ \Delta r \ \Delta\phi]^T$):

$$M = \begin{bmatrix} m_y & -(ma_z - Y_{\dot{p}}) & (ma_x - Y_{\dot{r}}) & 0 \\ -(ma_z + L_{\dot{v}}) & J_{xx} & -J_{xz} & 0 \\ (ma_x - N_{\dot{v}}) & -J_{xz} & J_{zz} & 0 \\ 0 & 0 & 0 & 1 \end{bmatrix} \tag{9.62}$$

$$A = \begin{bmatrix} Y_v & Y_p & (Y_r - mu^*) & (mg - B)cos\theta^* \\ L_v & L_p & (L_r + ma_z u^*) & (-mga_z + Bb_z)cos\theta^* \\ N_v & N_p & (N_r - ma_x u^*) & (mga_x - Bb_x)cos\theta^* \\ 0 & 1 & 0 & 0 \end{bmatrix} \tag{9.63}$$

Lateral-directional modes: These are the modes characterized by dynamics in off-longitudinal perturbation variables $[\Delta v \ \Delta p \ \Delta r \ \Delta\varphi]^T$. It is not straightforward to derive modal approximations in this case, however, some conclusions may be drawn from simplified analysis.

1. **Roll pendulum mode**: This mode is characterized by lightly damped oscillation in $\Delta\phi$. From pure rolling motion governed by:

$$J_{xx}\Delta\dot{p} = \frac{\partial\mathcal{L}}{\partial p}\Delta p + \frac{\partial\mathcal{L}}{\partial\phi}\Delta\phi$$

$$\Delta\dot{\phi} = \Delta p$$

$$J_{xx}\Delta\ddot{\phi} - \frac{\partial\mathcal{L}}{\partial p}\Delta\dot{\phi} - \frac{\partial\mathcal{L}}{\partial\phi}\Delta\phi = 0 \tag{9.64}$$

   One can see that the stiffness in this mode is provided by the derivative $\frac{\partial\mathcal{L}}{\partial\phi}$, which at the trim condition $\phi^* = \theta^* = 0$ is $\frac{\partial\mathcal{L}}{\partial\phi} = (Bb_z - mga_z)$. Further, at hover condition $mg = B$, hence $\frac{\partial\mathcal{L}}{\partial\phi} = mg(b_z - a_z)$. For positive stiffness in roll $\frac{\partial\mathcal{L}}{\partial\phi}$ must be negative, therefore, $b_z < a_z$, a condition which was also required for positive stiffness in pitch.
2. **Yaw subsidence mode**: Characterized by first-order fast dynamics in Δr, and

3. **Sideslip subsidence mode**: Characterized by first-order dynamics in Δv or $\Delta\beta$. Yaw subsidence mode is the fastest among the three, while roll oscillation and sideslip subsidence modes have a different slower timescale of dynamics.

9.2.4 Useful Empirical Relations

Some estimates for the virtual mass and inertia terms can be obtained from the relations:

$$X_{\dot{u}} = -k_1 \frac{B}{g};\ \ Y_{\dot{v}} = -k_2 \frac{B}{g};\ \ Z_{\dot{w}} = Y_{\dot{v}};\ \ M_{\dot{q}} = -\frac{k'\left(\frac{B}{g}\right)\left(l^2 + d^2\right)}{20};\ \ N_{\dot{r}} = M_{\dot{q}}$$

where l is the length and d is the maximum diameter of the hull. The factors k_1, k_2, k' known as Lamb's inertia ratios are function of $\left(\frac{l}{d}\right)$ or the fineness ratio.

9.2.5 Numerical Example

Data for a Gertler[2] shape airship operating at stratospheric altitude of 20 km ($\rho_{air} = 0.0767\ kg/m^3$) at $\alpha^* = 0$, $u^* = 15\ m/s$ are given as:

$$m = 25093.172\ kg;\ \ m_x = 27602.492\ kg;\ \ m_y = 43044.16\ kg;\ \ m_z = 43044.16\ kg;$$

$$J_{xx} = 17400000.0\ kgm^2;\ \ J_{yy} = 245264282.0\ kgm^2;$$

$$J_{zz} = J_{yy};\ \ J_{xz} = 1920000.0\ kgm^2$$

$$Vol = 327160\ m^3;\ \ l = 226\ m;\ \ d = 56.5\ m;\ \ l/d = 4;$$

$$b_z = 16.43\ m;\ \ a_z = 18.0\ m;\ \ a_x = 0;\ \ b_x = 0;$$

$$k_1 = 0.1;\ \ k_2 = 0.9;\ \ k' = 0.6$$

$$B = Vol\rho_{air} g;\ \ B/g = 327160 \times 0.0767 = 25093.172\ kg$$

$$X_{\dot{u}} = -k_1 \frac{B}{g} = -2509.32;\ \ Y_{\dot{v}} = -k_2 \frac{B}{g} = -22583.85;\ \ Z_{\dot{w}} = Y_{\dot{v}}$$

$$M_{\dot{q}} = -k' \frac{B}{g} \times \frac{\left(l^2 + d^2\right)}{20} = -0.6 \times 25093.17 \times \frac{\left(226^2 + 56.5^2\right)}{20} = -40852875.94$$

$$N_{\dot{r}} = M_{\dot{q}};$$

$$M_\alpha = \frac{1}{2}\rho_{air}u^{*2}Vol(k_2 - k_1)2 = 0.0767 \times 15^2 \times 327160 \times (0.9 - 0.1) = 4516770.96;$$

$$M_w = \frac{M_\alpha}{u^*} = 301118.0;$$

$$M_q = \frac{1}{2}\rho u^{*2}VolC_{mq1} = 0.5 \times 0.0767 \times 15^2 \times 327160 \times -0.065 = -183493.82$$

$$Z_q = -\frac{1}{2}\rho u^{*2}Vol^{\frac{2}{3}}C_{Lq1} = -0.5 \times 0.0767 \times 15^2 \times 327160^{\frac{2}{3}} \times 0.021 = -860.35$$

$$X_u = -\rho u^* Vol^{\frac{2}{3}}C_D^* = -0.0767 \times 15 \times 327160^{\frac{2}{3}} \times 0.03 = -163.88$$

$$Z_w = -\frac{1}{2}\rho u^* Vol^{\frac{2}{3}}C_D^* = \frac{1}{2}X_u = -81.94$$

$$M_\theta = (-mga_z + Bb_z) = mg(b_z - a_z) = -386477.51$$

$$Y_\beta = \frac{1}{2}\rho u^{*2}Vol^{2/3}C_{y\beta} = 0.5 \times 0.0767 \times 15^2 \times 327160^{2/3} \times -0.0109 = -446.56$$

$$Y_v = \frac{Y_\beta}{u^*} = -\frac{446.561}{15} = -29.77; \quad Y_p = 0$$

$$Y_r = \frac{1}{2}\rho u^{*2}Vol^{2/3}C_{yr} = 0.5 \times 0.0767 \times 15^2 \times 327160^{2/3} \times 0.021 = 860.35; \quad Y_\varphi = 0$$

$$\mathcal{L}_\beta = \frac{1}{2}\rho u^{*2}VolC_{l\beta} = \frac{1}{2}\rho u^{*2}Vol \times 0.0 = 0.0$$

$$\mathcal{L}_p = \frac{1}{2}\rho u^{*2}VolC_{lp} = 0.5 \times 0.0767 \times 15^2 \times 327160 \times -0.002 = -5645.96$$

$$\mathcal{L}_r = \frac{1}{2}\rho u^{*2}VolC_{lr} = 0.5 \times 0.0767 \times 15^2 \times 327160 \times 0.0 = 0.0$$

$$\mathcal{L}_\varphi = mg(b_z - a_z) = -386477.51$$

$$N_\beta = \frac{1}{2}\rho u^{*2}VolC_{n\beta} = 0.5 \times 0.0767 \times 15^2 \times 327160 \times -0.00251 = -7085.68$$

$$N_v = \frac{N_\beta}{u^*} = -472.38$$

$$N_\varphi = mg(a_x - b_x) = 0$$

Longitudinal modes:

$$M_{long} = \begin{bmatrix} 25032.64 & 0 & 416610.0 & 0 \\ 0 & 43044.16 & 0 & 0 \\ 416610.0 & 0 & 245264282.0 & 0 \\ 0 & 0 & 0 & 1.0 \end{bmatrix}$$

$$A_{long} = \begin{bmatrix} -163.875 & 0 & 0 & 0 \\ 0 & -81.937 & 860.347 & 0 \\ 0 & 301118.0 & -183493.82 & -386477.51 \\ 0 & 0 & 1.0 & 0 \end{bmatrix}$$

Computed eigenvalues of matrix $M_{long}^{-1} A_{long}$:

$$\underbrace{-0.0005 + j0.04}_{pitch\ pendulum},\ \underbrace{-0.0066}_{surge},\ \underbrace{-0.0019}_{heave}$$

Lateral-directional modes:

$$M_{lat} = \begin{bmatrix} 43044.16 & -416610 & 0 & 0 \\ -416610 & 17400000 & -1920000 & 0 \\ 0 & -1920000 & 245264282 & 0 \\ 0 & 0 & 0 & 1 \end{bmatrix}$$

$$A_{lat-lat} = \begin{bmatrix} -29.77 & 0 & 860.35 & 0 \\ 0 & -5645.96 & 0 & -386477.51 \\ -472.38 & 0 & 0 & 0 \\ 0 & 1 & 0 & 0 \end{bmatrix}$$

Computed eigenvalues of $M_{lat-lat}^{-1} A_{lat}$:

$$\underbrace{-0.0003 \pm j0.17}_{Roll-pendulum},\ \underbrace{-0.0006}_{yaw\ subsidence},\ \underbrace{-0.0001}_{sideslip\ subsidence}$$

REFERENCES

1. Khatri, A.K., Singh, J. and Sinha, N.K., Aircraft design using constrained bifurcation and continuation method, *Journal of Aircraft*, 51(5), 2014, 1647–1652.
2. Rana, V., Development of flight dynamic simulation model of stratospheric airship, *Masters Dissertation*, Department of Aerospace Engineering, IIT Madras, 2014.

Index

Note: Bold page numbers refer to tables and Italic page numbers refer to figures.

A

E

F

J

K

L

M

N

O

P

T

Printed in the United States
by Baker & Taylor Publisher Services

Printed in the United States
by Baker & Taylor Publisher Services